Forest Soils: Functions, Threats, Management

Forest Soils: Functions, Threats, Management

Editor

Klaus von Wilpert

MDPI • Basel • Beijing • Wuhan • Barcelona • Belgrade • Manchester • Tokyo • Cluj • Tianjin

Editor
Klaus von Wilpert
Forest Research Institute of Baden-Württemberg
Germany

Editorial Office
MDPI
St. Alban-Anlage 66
4052 Basel, Switzerland

This is a reprint of articles from the Special Issue published online in the open access journal *Soil Systems* (ISSN 2571-8789) (available at: https://www.mdpi.com/journal/soilsystems/special_issues/ForestSoils_FunctionsThreatsManagement).

For citation purposes, cite each article independently as indicated on the article page online and as indicated below:

LastName, A.A.; LastName, B.B.; LastName, C.C. Article Title. *Journal Name* **Year**, *Volume Number*, Page Range.

ISBN 978-3-0365-5063-3 (Hbk)
ISBN 978-3-0365-5064-0 (PDF)

Cover image courtesy of Klaus von Wilpert.

Contents

About the Editor

Klaus von Wilpert

Klaus von Wilpert worked as a soil scientist and Head of Department at the Forest Research Institute Freiburg, Germany (Department of Soil and Environment) and at Albert-Ludwigs University Freiburg for thirty years, until his retirement. He was involved in running and developing environmental monitoring systems (soil-, tree nutrition-, deposition-, and element flux survey). Monitoring gas exchange between soil and atmosphere under different soil and forest stand conditions was an important item of his scientific work, as was statistically based regionalization approaches of point-related monitoring data as a precondition for their integrated evaluation at landscape level.

soil systems

MDPI

Editorial

Forest Soils: Functions, Threats, Management

Klaus von Wilpert

Department of Soil and Environment, Forest Research Institute of Baden-Wuerttemberg (FVA-BW), D-79100 Freiburg, Germany; klaus.von-wilpert@online.de

The specific characteristic of forest soils is their long-term development under a more-or-less continuous vegetation cover. With deeper-reaching rooting zones and high activity of microbes, soil fauna and plant roots have high humus contents and above-average high porosity and continuity of the soil pore system compared with other land use types [1,2]. Contamination with pesticides is comparably low in forest soils, since forests are nature-near ecosystems [3], and the high demand from trees and soil biota for essential nutrients such as phosphorous and nitrogen leads to low leaching rates of those elements [4]. Thus, forests and forest soils are sources of predominantly pure drinking water [5].

However, some of the functions of forest soils are endangered under the influence of environmental and climate changes, and because of inadequate forest-management measures. The high crown surface of forests combs out acids and nitrogen from air pollution, which leads to severe soil acidification in parts of Central Europe and other industrialized regions [6]. Additionally, the use of heavy forest machinery can cause soil compaction, leading to deficits in soil aeration which can restrict the rooting space for forest trees in the uppermost soil layers [7–10]. Moreover, C-sequestration and greenhouse gas balances in forest soils are highly relevant topics which are also represented in the present Special Issue.

This Special Issue addresses the specific functions of forest soils, the processes which endanger the integrity of these functions, and potential management approaches to counteract the processes which threaten soil functions. This Special Issue comprises one review article [11] and nine research articles [12–20]. The order of the research articles starts with process-oriented studies on the specific functions and threats of forest soils [12–17] and ends with management-oriented studies deriving from and evaluating management approaches which can maintain or recover specific forest soil functions [18–20].

The review article was conceptualized as the leading article of the Special Issue, giving a comprehensive overview on the structures and processes differentiating forest soils from soils of other land use types and defining their specific value. From 208 relevant articles, this literature review revealed that forest soils provide a predominantly differentiated soil structure—the basis for their high ecological functionality. The review also revealed that active management measures must be set in motion to preserve the vulnerable functional structures of forest soils under fast-changing environmental conditions [11].

Schäffer examined 11 forest sites in the federal state of Baden-Wuerttemberg (Germany) with time intervals of 6–37 years between soil compaction from heavy forest machines and observation [12]. Fine root distribution, macroporosity, and apparent gas diffusion coefficients were used to characterize the status of recovery from former soil compaction. He observed a high persistence of damages below 10 cm soil depth, stating that "time spans up to almost four decades are not sufficient for the restoration of soil functionality in compacted silt loam soils". This contribution is an example of management-related threats for forest soil functions.

Takahashi et al. contributed a meta-analysis on the storage and stoichiometry of nutrients—nitrogen, phosphorus, potassium, calcium, and magnesium—in the humus layer of coniferous forests in Japan with different regional climate and soil characteristics [13]. In "cedar and cypress plantations on fertile sites, the forest floor stored low N and P with

Citation: von Wilpert, K. Forest Soils: Functions, Threats, Management. *Soil Syst.* **2022**, *6*, 60. https://doi.org/10.3390/soilsystems6030060

Received: 30 June 2022
Accepted: 7 July 2022
Published: 11 July 2022

high C:N and C:P ratios, suggesting that the forest floor plays there only a minor role as a nutrient reservoir. Subalpine coniferous forests and fir plantations in cool climates had large N and P storage with low C:N and C:P ratios in the forest floor". Thus, the relevance of the storage of essential nutrients in the humus layer for a consistent stand nutrition, even under the impact of climate change, was evaluated.

Rwibasira et al. contributed a study from Rwanda on the long-term effects of different tree species on soil functions characterized by the proxy entities soil acidity, soil organic matter, and exchangeable base cations [14]. They found that Eucalyptus species caused soil acidification, whereas soil-exchangeable cations and pH were higher under native species. In "selecting forest trees, priority should be given to the species which do not negatively alter chemical soil quality".

The study of Zhuang et al. deals with the "home-field advantage of litter decomposition", indicating that litter decomposition is enhanced through specifically adapted decomposer communities in areas where tree species are established—the "home-field". The context of the study is the silvicultural strategy, which shifts tree species composition from mono-cultural spruce stands to nature-near beech stands in the Eifel national park (West Germany). In a "litter transplant experiment" in stands of Norway spruce and European beech and adjacent clear-cuts, the authors found that litter of "spruce decomposed faster in spruce forest while beech-litter decomposed faster in clear-cut, indicating the occurrence of a home-field decomposition advantage at forest" stands for spruce "and clear-cut" for beech. They concluded that "clearcutting modifies the litter-field affinity and helps promote the establishment or regeneration of European beech in this and similar forest mountain upland areas" [15].

Melnichuk et al. studied the introduction of invasive earthworms, finding that it initiates physical and chemical alterations in previously earthworm-free forest soils, which "triggers an ecological cascade", leading to an "apparent shift in the herbaceous ground vegetation" [16]. An interesting result was that Arisaema triphyllum resisted earthworm invasion—because this species was able "to produce insoluble oxalate as an herbivory deterrent, in the presence of earthworms".

The contribution of Jandl et al. aimed to quantify the effect of long-term environmental change on soil acidification, nitrogen enrichment, and the loss of soil organic carbon due to climate change. They evaluated data from two soil survey campaigns comprising a time period of 20 years [17]. They found changes in the stocks of soil organic carbon, soil nitrogen, and soil pH. However, the changes were inconsistent. The authors conclude that "changes in the evaluated soil chemical properties are mainly driven by forest management activities", and that climate change effects have not changed the soil organic carbon stock until now in an unambiguous way.

Since recent studies investigating the plant-mediated alleviation of soil compaction with black alder showed promising results, Warlo et al. used the "characteristics of soil structure and greenhouse gas fluxes to measure soil recovery and GHG fluxes on machine tracks with and without black alders in North-East Switzerland". Unexpectedly, they found that, ten years after machine impact, "alder had no beneficial impact on soil physical parameters". Moreover, the symbiotic nitrogen fixation by alder led to elevated cumulative N_2O emission, and thus has "the potential to deteriorate the GHG balance of the investigated forest stand" [18].

Ahrends et al. studied the possible recovery from acid deposition as depending on tree species, evaluating data from limed and unlimed plots of 21 long-term and extensively instrumented ecosystem studies in Lower Saxony, Germany [19]. The data allowed for trend analyses of the acid-base status over a period of 30–50 years. The recovery, if indicated by an increase in soil pH and base saturation, of soils from limed plots and "plots with deciduous trees appears to have occurred faster than in coniferous forest stands". As the recovery from soil acidification is slow and the acid-base status still shows considerable soil acidification, "mitigation measures such as forest liming still appear to be necessary for accelerating the regeneration process".

The study on "merits and limitations of element balances as a forest planning tool for harvest intensities and sustainable nutrient management" by Ahrends et al. "provides valuable information for practitioners and environmental policy makers to enable spatiotemporal adaptive ecosystem management on the reliable and quality-assured basis of monitoring data" [20]. "The effects of conventional stem harvesting, stem harvesting without bark, and whole-tree harvesting on Ca, Mg and K balances were studied. The nutrient balances were calculated using regular forest monitoring data supplemented by additional data from scientific projects. Effective mitigation management strategies and options are discussed and calculations for the compensation of the potential depletion of nutrients in the soil are presented".

The author would like to thank all contributing authors in this Special Issue on "Forest soils: functions, threats, management" and all reviewers for their constructive criticisms and engagement, which improved the quality of the science and the presentation during the review process.

Funding: One contribution [20] to this Special Issue was supported by the "Fachagentur Nachwachsende Rohstoffe, Germany" who carried the publication costs.

Informed Consent Statement: Not applicable.

Conflicts of Interest: The author declares no conflict of interest.

References

1. Sokołowska, J.; Józefowska, A.; Woznica, K.; Zaleski, T. Succession from meadow to mature forest: Impacts on soil biological, chemical and physical properties—Evidence from the Pieniny Mountains, Poland. *CATENA* **2020**, *189*, 104503. [CrossRef]
2. Ma, S.; De Frenne, P.; Boon, N.; Brunet, J.; Cousins, S.A.O.; Decocq, G.; Kolb, A.; Lemke, I.; Lemke, J.; Naaf, T.; et al. Plant species identity and soil characteristics determine rhizosphere soil bacteria community composition in European temperate forests. *FEMS Microbiol. Ecol.* **2019**, *95*, fiz063. [CrossRef] [PubMed]
3. Orlinskiy, P.; Münze, R.; Beketov, M.; Gunold, R.; Paschke, A.; Knillmann, S.; Liess, M. Forested headwaters mitigate pesticide effects on macroinvertebrate communities in streams: Mechanisms and quantification. *Sci. Total Environ.* **2015**, *524*, 115–123. [CrossRef] [PubMed]
4. Makowski, V.; Julich, S.; Feger, K.-H.; Julich, D. Soil phosphorus translocation via preferential flow pathways: A comparison of two sites with different phosphorus stocks. *Front. For. Glob. Chang.* **2020**, *3*, 48. [CrossRef]
5. Fiquepron, J.; Garcia, S.; Stenger, A. Land use impact on water quality: Valuing forest services in terms of the water supply sector. *J. Environ. Manag.* **2013**, *126*, 113–121. [CrossRef] [PubMed]
6. Waldner, P.; Thimonier, A.; Graf Pannatier, E.; Etzold, S.; Schmitt, M.; Marchetto, A.; Rautio, P.; Derome, K.; Nieminen, T.M.; Nevalainen, S.; et al. Exceedance of critical loads and of critical limits impacts tree nutrition across Europe. *Ann. For. Sci.* **2015**, *72*, 929–939. [CrossRef]
7. Hildebrand, E.E.; Schack-Kirchner, H. The influence of compaction on soil structure and functions in forest sites. In *Modern Trends in Applied Terrestrial Ecology*; Ambasht, N.K., Ambasht, R.S., Eds.; Springer-Science+Business Media: New York, NY, USA, 2002; pp. 1–11.
8. Gaertig, T.; Schack-Kirchner, H.; Hildebrand, E.E.; von Wilpert, K. The impact of soil aeration on oak decline in south-western Germany. *For. Ecol. Manag.* **2002**, *159*, 15–25. [CrossRef]
9. Ampoorter, E. Soil Compaction Due to Mechanized Forest Harvesting: Quantification of Ecosystem Effects and Exploration of Recovery Potential. Ph.D. Thesis, Ghent University, Ghent, Belgium, 2011; p. 182.
10. Schäffer, J.; von Wilpert, K.; Kublin, E. Analysis of fine rooting below skid trails using linear and generalized additive models. *Can. J. For. Res.* **2009**, *39*, 2047–2058. [CrossRef]
11. Wilpert, K.V. Forest Soils—What's their peculiarity? *Soil Syst.* **2022**, *6*, 5. [CrossRef]
12. Schäffer, J. Recovery of soil structure and fine root distribution in compacted forest soils. *Soil Syst.* **2022**, *6*, 49. [CrossRef]
13. Takahashi, M. Nutrient storage and stoichiometry of the forest floor organic matter in Japanese forests. *Soil Syst.* **2021**, *5*, 51. [CrossRef]
14. Rwibasira, P.; Naramabuye, F.X.; Nsabimana, D.; Carnol, M. Long-term effects of forest plantation species on chemical soil properties in Southern Rwanda. *Soil Syst.* **2021**, *5*, 59. [CrossRef]
15. Zhuang, L.; Schnepf, A.; Unger, K.; Liang, Z.; Bol, R. Home-field advantage of litter decomposition faded 8 years after spruce forest clearcutting in western Germany. *Soil Syst.* **2022**, *6*, 26. [CrossRef]
16. Melnichuk, R.D.S.; Tecimen, H.B.; Görres, J.H. Do the invasive earthworms *Amynthas agrestis* (Oligochaeta: Megascolecidae) and *Lumbricus rubellus* (Oligochaeta: Lumbricidae) stimulate oxalate-based browser defenses in jack-in-the-pulpit (*Arisaema triphyllum*) by their presence or their soil biogeochemical activity? *Soil Syst.* **2022**, *6*, 11. [CrossRef]

17. Jandl, R.; Leitgeb, E.; Englisch, M. Decadal changes of organic carbon, nitrogen, and acidity of Austrian forest soils. *Soil Syst.* **2022**, *6*, 28. [CrossRef]
18. Warlo, H.; Zimmermann, S.; Lang, F.; Schack-Kirchner, H. Characteristics of soil structure and greenhouse gas fluxes on ten-year old skid trails with and without black alders (*Alnus glutinosa* (L.) *Gaertn.*). *Soil Syst.* **2022**, *6*, 43. [CrossRef]
19. Ahrends, B.; Fortmann, H.; Meesenburg, H. The influence of tree species on the recovery of forest soils from acidification in lower saxony, Germany. *Soil Syst.* **2022**, *6*, 40. [CrossRef]
20. Ahrends, B.; von Wilpert, K.; Weis, W.; Vonderach, C.; Kändler, G.; Zirlewagen, D.; Sucker, C.; Puhlmann, H. Merits and limitations of element balances as a forest planning tool for harvest intensities and sustainable nutrient management—A case study from Germany. *Soil Syst.* **2022**, *6*, 41. [CrossRef]

soil systems

MDPI

Review

Forest Soils—What's Their Peculiarity?

Klaus von Wilpert

Department Soil and Environment, Forest Research Institute, D-79100 Freiburg, Germany; klaus.von-wilpert@online.de

Abstract: Mankind expects from forests and forest soils benefits like pure drinking water, space for recreation, habitats for nature-near biocenoses and the production of timber as unrivaled climate-friendly raw material. An overview over 208 recent articles revealed that ecosystem services are actually the main focus in the perception of forest soil functions. Studies on structures and processes that are the basis of forest soil functions and ecosystem services are widely lacking. Therefore, additional literature was included dealing with the distinct soil structure and high porosity and pore continuity of forest soils, as well as with their high biological activity and chemical soil reaction. Thus, the highly differentiated, hierarchical soil structure in combination with the ion exchange capacity and the acid buffering capacity could be described as the main characteristics of forest soils confounding the desired ecosystem services. However, some of these functions of forest soils are endangered under the influence of environmental change or even because of forest management, like mono-cultures or soil compaction through forest machines. In the face of the high vulnerability of forest soils and increased threads, e.g., through soil acidification, it is evident that active soil management strategies must be implemented with the aim to counteract the loss of soil functions or to recover them.

Keywords: forest soil characteristics; secondary soil structure; soil functions; ecosystem services; spatiotemporal integration level; forest soil management

Citation: Wilpert, K.v. Forest Soils—What's Their Peculiarity? *Soil Syst.* **2022**, *6*, 5. https://doi.org/10.3390/soilsystems6010005

Academic Editors: Andreas Fliessbach and Heike Knicker

Received: 18 October 2021
Accepted: 21 December 2021
Published: 6 January 2022

Publisher's Note: MDPI stays neutral with regard to jurisdictional claims in published maps and institutional affiliations.

1. Introduction—What Are Forest Soils Expected to Be and to Deliver?

The specific characteristic of forest soils is their long-term development under a more-or-less continuous vegetation cover. Trees as long-living organisms and through their magnitude shape soils in a specific way. A, compared to other land-use types, deeper-reaching rooting zone and high activity of microbes, soil fauna and plant roots result in high humus contents, as well as over-proportionally high porosity and continuity of the soil pore system [1]. Forest soils are the habitat of a high diversity of plants, macro-fauna and microbes [2]. Biological networks like the manifold symbioses between trees and mycorrhiza fungi optimize the supply of trees with nutrients and water and enhance the weathering of primary minerals and nutrient release from organic matter, as well as protect tree roots from toxic metal ions like Al^{3+} released through soil acidification [3]. Contamination with pesticides is comparably low in forest soils, since forests are nature-near ecosystems [4]. Moreover, the high demand of trees and soil biota for essential nutrients like phosphorous and nitrogen leads to low leaching rates of those elements in most forest soils [5]. Both the low load with pesticides and low leaching of phosphorous and nitrogen make forests sources of predominantly pure drinking water [6]. Luo et al. [7] found in the subtropical humid Chinese Hunan Province that permanent forest cover has a high potential for erosion prevention combined with a slight increase in water yield.

However, some of these functions of forest soils are endangered under the influence of environmental and climate change or even because of inadequate forest management measures under some circumstances, e.g., the high crown surface of forest combs out acids and nitrogen from air pollution, which leads to severe soil acidification in wide parts of Central Europe and other industrialized regions [8]. Additionally, the use of heavy forest machinery can cause soil compaction, leading to deficits in soil aeration, which can restrict the rooting space for forest trees to the uppermost soil layers [9–12]. Moreover, the optimization of the C-sequestration and greenhouse gas balances of forest soils through specific forest management practices is a topic of high actuality.

This study pursues the following aims and objectives:

1. To work out how physical, chemical and biological properties are interlinked in forest soils and how they define soil functions.
2. To clarify the scale levels of soil functions and ecosystem services.
3. Comparing soil properties under forests and other forms of land use to work out the peculiarity of forest soils.
4. To collect the specific threats on forest soil functions through environmental change and/or management.
5. To give hints for strategies to preserve forest soil functions.

The main emphasis of this study will be laid on making the specific quality and value of forest soils understandable (objectives 1–3). The specific threats and options for soil preservation in forests will be treated in a more exemplary way in the form of an outlook or discussion in order to not make the study too complex.

2. Materials and Methods—Perception of Forest Soils in the Scientific Literature

The literature research for this review was performed in an iterative procedure with stepwise refined and completed research criteria and exclusion of titles that either did not fit the objectives of this study or did not fulfil the quality criteria. A very general first overview on the scientific articles with a focus on forest soils was undertaken when beginning the work of this study on the specific functions of forest soils, their specific vulnerability and forest management options to preserve or restore them. For this purpose, 240 articles were collected with the search keys "forest" + "soil" and with a publication date not older than 5 years. The next step was to check the identified tiles if they fit the objectives of the study. Thus, 32 titles were excluded, mainly theoretical titles dealing with method development or with a too-local focus. According to the expectations on functions, threats and management options on forest soils, six thematic fields were drafted, and the remaining 208 articles assigned to them:

- Soil functions and silviculture: The effects of tree species and stand structures on soil chemical, soil physical and soil hydrological properties are dealt with in this field of interest. Since tree species selection and forest management systems, e.g., clear-cut vs. small-scaled harvesting regimes preserving ample crown cover over all stages of stand regeneration, these fundamental instruments of silviculture substantially influence soil processes [13] and soil characteristics. In this sense, silvicultural strategies can be taken as tools of long-term soil management [14].
- Forest and water: This field comprises the effect of forest soils on the quality and quantity of water yield. All over the world, forested areas are judged to be predominantly suited to provide high-quality drinking water [15]. The second important issue in this field is the function of forest soils as a store of plant-available water resources. This aspect is increasingly relevant under the actual increase of drought periods caused by climate change [16].
- Nutrient availability in forest soils: This item comprises the nutrient pools in forest soils, as well as processes governing the mobilization and availability of nutrients for forest trees.
- Climate change and forest soils: Forests and forest soils are concerned by climate change in two ways. On the one hand, forest soil functions are threatened by extreme weather events like droughts endangering continuous water and nutrient supplies for trees [17,18] or storms and storm floods causing wind throw and erosion damages. On the other hand, forest ecosystems and forest soils can contribute to lower greenhouse gas emissions through carbon sequestration or methane consumption in terrestrial forest soils [19].

- Soil compaction and erosion: Forest soils are in their natural stage over-proportionally unconsolidated and open-pored [12], and erosion is a seldom and subordinate process because of the coherent structure of the forest floor layer and the more-or-less continuous vegetation cover [20]. Therefore, soil compaction and erosion of forest soils are mainly manmade damages. They are caused by machine-bound harvesting techniques or inadequate management techniques like big clearcutting at steep slopes or forest roads and skidding tracks without sufficient water deduction facilities.
- Soil acidification and eutrophication: Soil processes caused by the deposition of acid compounds and nitrogen with precipitation seem to apparently be of minor relevance, since these problems have been somehow cursorily considered in the recent literature. This can be explained because, in the heavily industrialized regions, at least in Europe, the deposition of acidity was substantially reduced through effective filter techniques [21]. However, unnatural soil acidification and its after-effects remained as an inherited problem that still has to be counteracted by ecosystem-conforming measures aiming to rehabilitate the natural functions of forest soils [22,23].

The first three thematic fields deal predominantly with the functionality of forest soils and the last three ones with threats and management approaches for rehabilitation of the functionality of soils in forests.

The overview of 208 relevant articles from the first step of the review process that would potentially fit into the scope of this Special Issue revealed that the six thematic fields are represented with substantially differing intensities (Figure 1).

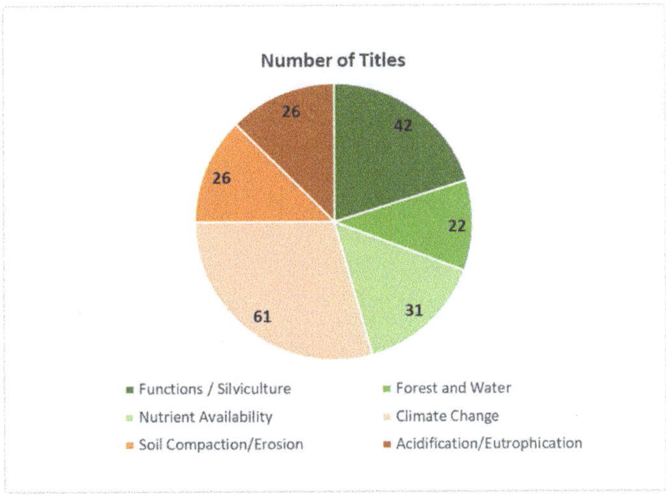

Figure 1. Number of articles in six thematic fields on forest soils and their functions, threats and management. Green colors show thematic fields focusing mainly on the specific functionality of forest soils; reddish colors indicate fields focused on threats to forest soil functions and measures to counteract them.

More than half (54%) of this first overview deals with threats on forest soils and only 46% with the functions of forest soils themselves. The by far dominating share of articles deals with climate change issues and, among them, the main part with the role of forest soils in greenhouse gas budgets. The thematic fields "forest and water", "nutrient availability", "soil compaction" and "soil acidification" are under-represented , respectively, by 15% or less of the titles. Moreover, the thematic focus of the articles in this overview seems, in many cases, not to distinguish between forest soil functions in the narrow sense and forest ecosystem services—the latter staying mainly in the foreground. This is understandable, because ecosystem services concern the effects of forest ecosystems as a whole. They address what mankind expects to receive as goods and services from forest ecosystems. Thus, ecosystem services are commonly perceived to be more relevant as the more abstract soil processes, even if most ecosystem services are mainly defined by soil properties.

Moreover, this first overview on the literature of the last five years revealed that a lot of contributions are focused on small-scaled detail processes like, e.g., microbial activity or the bio-chemical background of the nutrient acquisition of trees, which are, as a matter of course, important processes but provide a somehow scattered view on the specific and fundamental properties of forest soils defining the habitat characteristics, e.g., for microbes.

Therefore, in the second step of the review process, additional titles were included and checked. In order to represent specific characteristics and processes of forest soils and to get a more complete view on the interactions of soil physical, soil chemical and soil ecological processes confounding the specific values of forest soils, 77 additional titles were researched, including 59 older ones. Research-leading ideas were derived from contributions of the working group of Hildebrand on the theory of the "basic regulation unit", combining aspects of the physical soil structure, the chemical "climate" and the biological activity of soils [9] and the theory of Ulrich on the process hierarchy in forest ecosystems [24,25]. For all of these 285 titles, either PDF files were obtained or printed versions were available. Thus, an extensive check of their relevance for the objectives of the study, as well as a quality check, could be performed with a focus on the abstracts, keywords, conclusions and the whole text. An overview on the workflow and selection criteria of the literature research in this study is given in Figure 2.

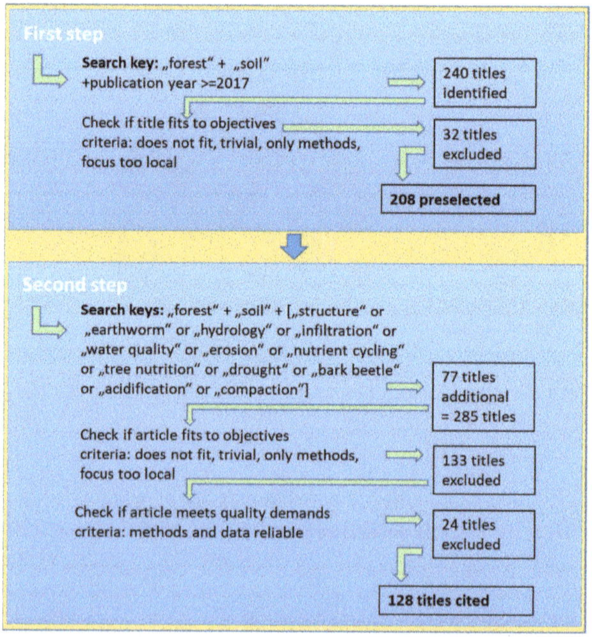

Figure 2. Iterative selection process of the literature research.

The publication years of the remaining 128 articles cited after the final selection are presented in Figure 3. The oldest title cited was published in 1990. Up to 2021, an exponentially increasing number of titles was cited, with 57% in the last five years.

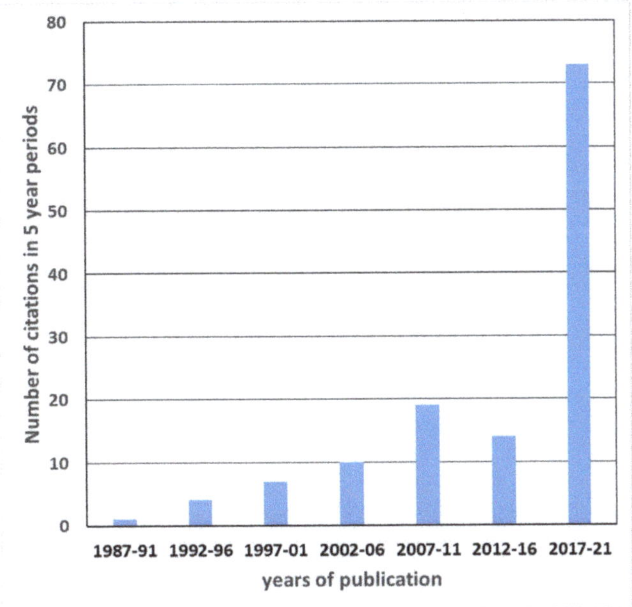

Figure 3. Distribution of 128 titles cited between 1987 and 2021 in 5-year periods.

Moreover, the regional and climatic contexts of the titles cited are given in Table A1. The regional and climatic context of all titles cited in this paper is presented in the Appendix A Table A1 at the end of the paper. The identification of the climatic context was done according to the updated climatic classification by Kottek et al. [26]. The main focus with about 43% of the citations was on temperate-humid and with about 25% on cold-humid climate conditions in Europe where soil genetic conditions are largely comparable and soil functions are not dominated by more-or-less extreme natural boundary conditions like, e.g., subpolar conditions resulting in a dominance of forest floor for soil functions or arid regions with their tendency toward salinification or tropic regions with their susceptibility to nutrient depletion. Those extreme natural conditions were included but in a more exemplary way and were represented with 5–10% of the citations, respectively.

3. Results—Forest Soils, the Basis for Multi-Functionality of Forest Ecosystems

"Natural forestlands are important to conserve soil and water, sequester C, and mitigate net emissions of greenhouse gases while providing wood, fuel, food, fodder, medicines, and other products (e.g., dyes, tannins, perfumes, ornamentals, exudates)" [27,28]. Forest soils are the thin, animated surface of earth where forest trees are rooting. However, they are not only the mechanical anchorage for tree stability rather than a consistent source for the supply of forests with nutrients and water, as well with fresh air for the high oxygen demand of growing roots. Forest soil does not serve only as the supplier for forest growth and, thus, for wood production. Moreover, it is the central "coordinating entity" for most of the ecosystem services that mankind expects from forest ecosystems. A study that was carried out in two contrasting regions in Europe (North–West of Belgium and North–East of Romania) on conditions and strategies to promote soil functions and soil biodiversity stated that "adaptive forest management is currently moving towards management for ecosystem functions and services", and therefore, "improved knowledge on functions delivered by soil biodiversity" is needed [29]. Complex interactions between chemical, physical and biological soil properties (e.g., nutrient availability, soil acidification and eutrophication, humus accumulation and soil structure) and environmental influences (e.g., atmospheric deposition) are judged by the scientific community to have a high "centrality" (or interconnectivity) to ecosystem services—predominantly to soil biodiversity—but are judged to be significantly less important by forest managers or the public. However, the importance of ground vegetation, tree species choice and nutrient cycling was judged by both communities synchronously to be comparably high, and the importance of climate conditions, recreation activities, timber production and harvesting was judged to be low [29]. This study revealed that strategies for preserving soil functions in a sustainable way need multivariate modeling approaches on a sound basis of quality-checked data (particularly from well-defined monitoring systems), as well as thorough transfer and communication of the results to practitioners and the public.

3.1. Soil Properties and Processes Founding Forest Soil Functions

The characteristics of forest soils are, on the one hand, shaped by long-lasting forest cover, the specific characteristics of matter cycles and dominated by high carbon input in forests, as well as by largely closed element cycles. On the other hand, the distribution of land use types in landscapes is not random. Agricultural land use types are preferentially situated on deeply developed soil types with high water holding capacities and high nutrient stocks, whereas for forests, the less fertile locations remain. Burst et al. [30] found in NW France that "forests were usually located slightly upslope of grasslands, and mainly because this non-random topographic position the topsoil texture was significantly more silty in forests, and clayey in grasslands", resulting in a higher soil porosity in forest soils that also persists after deforestation and land use changes to grasslands.

Forest soils are the overlapping space where the atmosphere, pedosphere and hydrosphere are closer interlinked among each other, as in soils of most other land use types, since the soil structure is predominantly fine-scaled in forest soils. They are the "reaction vessel", where the weathering of primary minerals to pedogenic substances, as well as the organic and nutrient matter cycles, take place, where the buffering of acids; storage of substances and their transformations between the solid, liquid and gaseous soil phases occur [9]. Thus, on-site effects like the filter, buffer and habitat functions of forest ecosystems are mainly directed by soil processes, as well as their function as sustainable breeding grounds of the climate-friendly raw material timber. Moreover, soil processes are the basis of off-site effects like the ability of forest ecosystems to mitigate climate change through their ability to minimize greenhouse gas concentrations by stable carbon storage predominantly in mineral soils, as well as by methane consumption and minimizing nitrous oxide efflux. Several studies demonstrated that, in forest soils, higher carbon pools are stored in comparison to arable or grassland soils [31–33]. A study in Central Poland revealed that the nitrous oxide emission was about three times lower from forest soils as from arable land [34], and Täumer et al. [35] derived from comparing 150 grassland/forest pairs that "reduction in grassland land-use intensity and afforestation has the potential to increase the CH_4 sink function of soils". Another crucial function of forest soils is to deliver high-quality drinking water and to equalize the landscape hydrology or even the resistance of forest soils against soil erosion [20,36].

3.1.1. Secondary Soil Structure—The Spatial Frame of Soil Functions

Forest soils are well-structured at different hierarchical levels. At the level of soil profiles, they are characterized by pronounced vertical layering. The specific characteristics are the high contents of organic materials in the humus layer and the upper A-horizon caused by the high litter input from the crown layer. The long-lasting forest cover allows for a more-or-less evolutionary development of a complex secondary aggregate- structure in mineral horizons of forest soils. Mainly, the activity of an abundant community of soil fauna does mix the mineral and organic components of the soil solid phase and forms secondary soil aggregates. Zangerle et al. [37] demonstrated with mesocosm experiments that "earthworms and plant roots, as ecosystem engineers, have large effects on biotic and abiotic properties of the soil system. They create biogenic soil macro-aggregates (i.e., earthworm casts and root macro-aggregates) with specific physical, chemical and microbiological properties". The initial formation of soil aggregates also results from abiotic processes like shrinking in drought periods, freezing or flocculation under the influence of multiple-charged cations. "Biologic activity is one of the main factors controlling the floating equilibrium between loosening and compacting forces in humic forest soils" [38]. Through the digging activity of soil fauna and growing roots, a high soil porosity with an over-proportional connectivity is created. Undamaged forest soils provide a significantly higher macro-porosity than cropland soils [39]. Sokołowska et al. [1] examined in the Carpathian region soil properties in a succession from meadow to mature forests. They found that "forest succession increased the soil porosity in the 10–20 cm layer, especially the volume of macro pores" and increased carbon sequestration on the long term. Damptey et al. [40] found after two decades of forest restoration on former mine areas in Ghana significantly increased carbon contents and decreased bulk density in the mineral soil. Zhang et al. [41] found in tropical forests in the Philippines that the bulk density was higher and porosity marginally lower in grasslands than in afforested areas, resulting in avoiding the Hortonian overland flow, which commonly occurred in the grassland areas.

The high porosity and the secondary aggregate structure is a dissipative and dynamic steady-state equilibrium that can only been maintained by continuous energy input from biologic activity against sagging forces [9]. The fundamental benefit of this complex soil structure is that it provides within elementary soil volumes a few mm^3 wide quasi-simultaneously and quasi-at the same locations water, nutrients and oxygen in plant-available forms. This seems to be contradictory on first sight but is realized by the close vicinity of meso-pores binding plant-available water, clay minerals storing exchangeable nutrients and nonwaterlogged macropores allowing for oxygen supply and carbon dioxide discharge (Figure 4). Thus, unlike sediments, the secondary structure is one of the most essential properties of forest soils, and it can serve to fulfill these quasi-contradictory needs of plant growth and the productivity of forests [9]. The distribution of roots is restricted to the surface zones of soil aggregates and macropores because of the high oxygen demand of growing roots.

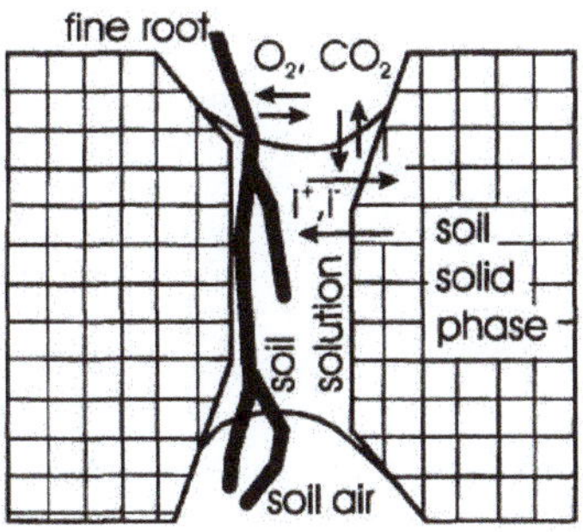

Figure 4. Basic regulation unit of soils in the range of a few mm^3 controlling the dissolution of minerals and organic matter, the transport of diluted compounds (i^+ and i^-) and simultaneously enabling the oxygen supply and carbon dioxide discharge. Thus, within these small, well-structured soil volumes, all essential demands for root growth are given under normal conditions: supply with water and nutrients, as well as sufficient aeration (from Reference [9]).

Since external mycorrhiza hyphae have diameters of 2–10 µm by one order of magnitude smaller than fine roots, they could potentially enter the mesopores of the intra-aggregate space. However, Schack-Kirchner et al. [42] and, also, Witzgall et al. [43] found in mesocosm experiments with naturally structured soil cores that hyphae also open the intra-aggregate space of forest soils very inefficiently, and thus, they do not show different behaviors than roots or aerobic microorganisms (Figure 5).

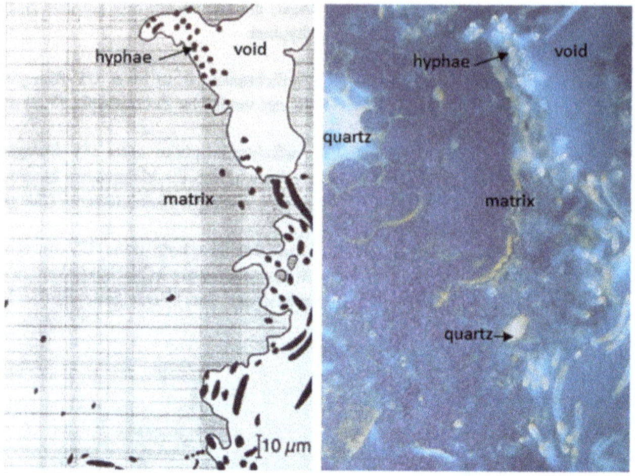

Figure 5. Evaluation of the soil preparations embedded in polyester resin polished and stained with acridine orange. Left: results of the object identification with the picture analysis system Leitz CBA 8000 at display windows of ca. 250 µm edge length. Hyphae and root fragments (black), soil matrix (grey-shaded) soil pores and voids (white). Right: micro-photo of the same display window reflecting light with UV activation (according to Reference [42]).

"The major part of hyphae grow within macro pores (>10 µm)" and "of the hyphae in the soil matrix, 70% were located in a 50 µm shell around the macro pores". It could be substantiated with geostatistical methods (e.g., pair correlation functions) that the hyphae in macropores and at the superficial shell of soil aggregates are heavily clustered (see the example in Figure 5). "Therefore, a considerable amount of chemically available nutrients is not directly accessible for the roots" [42]. This can cause deficiencies in the nutrient supply, especially in drought periods when the diffusive replenishment of aggregate surfaces with nutrients is interrupted when the waterlogged intra-aggregate pores dry out. "On the other hand, the inaccessibility of the intra-aggregate space by organisms can also be seen from a positive point of view. Storage of nutrients in intra-aggregate pores can be regarded as an efficient mechanism to prevent the highly mobile water-soluble ions from leaching" [42].

3.1.2. Soil Chemical Status

Most forests are not fertilized, disregarding artificial plantation forests. Therefore, the primary sources of chemical elements and nutrients are input fluxes with precipitation and the weathering of primary minerals, which is, in most forest ecosystems, the dominant input flux. Since weathering is a very slow process, it cannot equalize short-term fluctuations of the forest nutrition, e.g., caused by weather fluctuations like drought [24]. Clay minerals and clay–humus complexes provide negatively charged exchanger surfaces that adsorb cationic, basic nutrients (calcium, magnesium and potassium), as well as cation acids (aluminum, iron, manganese and ammonium). Thus, soils serve as short- to medium-term stores for plant-available nutrients, providing more-or-less constant nutrient availability for forest vegetation and trees, which can bridge such externally driven gaps in nutrient supply. An additional buffering function in the nutrient supply of forest stands can be fulfilled by the nutrient store in the humus layer and its mobilization by mineralization [24].

Soils naturally acidify in the course of soil development, but that is a very low process (e.g., in Central Europe under temperate–humid climate conditions, the mean soil acidification, since the last glaciation period, caused a pH decrease of 1 to 2 pH levels in about 10.000 years). This can be judged as a quasi-steady state [44].

If the saturation of exchanger surfaces with cation acids becomes dominant as a result of soil acidification, the selectivity for basic nutrients drops exponentially. Subsequently, essential basic nutrients are leached with seepage water accompanied with anions of carbonic acid or under the influence of air pollution and "acid rain" with the strong mineral acids sulfate and nitrate. This depletion of essential nutrients is highest for magnesium, since its bond strength to soil exchanger surfaces is predominantly low [45]. The transport pathway between the exchangeable ion pools and plants and/or seepage is the soil solution. Other relevant transport pathways of the element cycle are the mobilization of elements directly from organic materials or from inorganic minerals. This is the dominant source for tree nutrition with phosphorus, sulfur and, also, partially nitrogen, which is closely coupled with the cycle of organic substances [24,46]. The acid/base relation as expressed by the base saturation (relation between exchangeable cation bases and cation acids) is judged as an integrating indicator on the buffer function of soils, similar to the soil reaction (pH value). Both are constitutive parameters for the habitat quality for soil microbes and soil fauna [47]. Especially struck by soil acidification are earthworms, which are important as engineers of the soil structure, mixing organic material with mineral soil, enhancing the decomposition of organic material, forming stable soil aggregates, providing habitats for microorganisms [48] and creating highly continuous macropores enhancing soils aeration and water infiltration [49].

Since forest mineral soils are not a coherent paste or a slurry, the soil structure varies the mobilization and transport of chemical elements substantially like, e.g., the heterogeneous distribution of exchangeable nutrients in well-aggregated acid forest soils suggests, where concentrations of potassium, magnesium and calcium were found to be depleted at aggregate surfaces and high in the intra-aggregate space. The recharge of ion pools at the aggregate surfaces was substantially delayed, particularly for potassium [50]. This can be interpreted as an interaction between soil structure and chemical exchange kinetics and underlines the ecological value of the secondary soil structure and its predominant relevance for forest soil functions. The manifold interactions between soil chemical properties, soil structure and soil biological activity suggest that the full multifunctionality of forest soils requires the mutual optimization of those three aspects. This optimization process occurs in undisturbed forest ecosystems through evolutionary approximation to an attractor space being defined by climate, geology, soil development, soil texture, topographical position and species composition of the tree and ground vegetation layers. In managed forests, this approximation can either be supported or disturbed by management measures.

3.2. *Forest Ecosystem Services*

Soil processes confounding soil functions and, subsequently, ecosystem services of forests are defined by physical, chemical and biological soil properties and the interactions among them. Ecosystem services are related to higher spatiotemporal integration levels as soil functions. The typical spatial scale for ecosystem services is the pedon (area where soil characteristics are comparable) to catchment or landscape scales. The temporal scale for their generation is the mean life cycle of forest stands. Both are definitely macroscales compared to the meso-scaled dimension of the basic regulation unit being responsible for soil functions (Figure 4). The formulation of ecosystem services represents the benefits that mankind expects to be provided from forest ecosystems, which is a typically anthropocentric point of view.

3.2.1. Forest Soils as Basis for Growth and Existence of Forests

Forest soils are the basis for the growth intensity of forest trees and thus provide regenerative and climate friendly raw material timber. Between soil properties and forest trees, there do not exist simple and monocausal cause/effect relations rather than mutual influences and adaptations between trees and soil properties. On the one hand, soil properties define how tree species can grow and how they can assert themselves against the concurrence of other tree species. On the other hand, tree species are able to shape soil properties in an active way in order to optimize habitat conditions like, e.g., nutrient availability. The latter is the predominant factor under unfavorable environmental conditions. The supply with nutrients and water are the crucial variables of forest growth and are primarily determined by site characteristics like the geological provenience of bedrock, soil development and climate. Thus, site characteristics determine the ecological niches of forest trees. "Soil-sensitive tree species in temperate forests like ash or sycamore were found to be much more sensitive to soil variables than European beech. The most relevant soil variables for the competitive ability of the three species were found to be C/N-ratio, humus form, aluminum content, base saturation, magnesium content, and soil aeration. Shortage of nutrients limited the distribution of ash and sycamore and excess of toxic elements the distribution of ash" [51]. The supply of trees with mass nutrients like magnesium, potassium and calcium is closely linked to soil acidification when strongly adsorbing aluminum cations suppress adsorption of the nutrient cations at soil exchanger surfaces [50]. Furthermore, the secondary soil aggregation varies the plant availability of these nutrients [50,52]. This also applies for mycorrhiza hyphae, which are commonly described as spatial extensions of the reach of the rhizosphere of trees, since hyphae are, like roots, not able to enter the inner parts of soil aggregates (see Figure 5 and Reference [42]).

Since forest soils are, compared to soils of other land use forms, less fertile and mostly stonier [30], therefore, the potential contribution of the soil coarse fraction to the nutrition of trees and forest stands shall be considered in a short paragraph. Conventionally, nutrient pools in fine soil material (corn size < 2 mm) are taken as the dominant source of tree nutrition. However, under specific conditions like, e.g., in acidified and nutrient depleted podsols, the nutrient pool in the coarse soil fraction also seems to be accessible to trees, since "EcM fungi can" actively "dissolve mineral grains" [53]. Heisner et al. [54] found in heavily acidified forest soils of the Black Forest (SW Germany) that the skeletal fraction has a cation exchange capacity (CEC) within the same order of magnitude as fine earth. This finding was assumed to be attributed to fine material-filling fissures of skeleton grains. Koele et al. [55] substantiated this hypothesis and showed that "fine earth accumulated within the weathering fissures of the coarse-soil fraction (particles > 2 mm), so called stone-protected fine earth, can provide a high, short-term nutrient release by cation exchange". They could also demonstrate that "in the coarse-soil fraction of the BhBs horizon, the absolute hyphal length exceeded the hyphal length in the fine earth by factor 3" [56]. They concluded from their studies that "exchange processes were the main trigger of Ca and Mg mobilization and uptake rather than protolytic weathering by exudation of carboxylic acids", like van Schöll et al. [53] stated: "The exchange processes may be attributed to weathering cracks filled with fine material of high base saturation" [57]. These findings imply that the exchangeable nutrient pools in fine earth and "stone protected fine earth" in forest soils should be treated as a continuum when assessing the base cation supply of forest stands. However, since the mobilization of nutrients from primary minerals through weathering is a very slow process [58] that could recharge the exchangeable nutrient pool in stones, the use of these pools must be judged as a short-term emergency strategy to bridge a nutrient shortage, e.g., caused by the fast-developing soil acidification of the last decades in silicatic soils of Central Europe.

The supply of trees with nitrogen and phosphorous, which are predominantly essential nutrients for forest growth, is closely linked to the organic matter cycling in forest ecosystems and microbial activity. Waldner et al. [8] derived from the Europe-wide intensively monitored forest plots (Level II) that critical loads for inorganic nitrogen deposition were exceeded on about a third to one-half of the forest plots, which leads, on the one hand, to nutrient imbalances, such as low magnesium and potassium concentrations in foliage. On the other hand, a tendency toward elevated nitrate concentrations in the soil solution was observed at these plots, which propagates soil acidification and base cation export. Thus, nitrogen nutrition is, in Central Europe and other regions with high nitrogen deposition, no more a limited nutrient rather than an exuberance problem. In tropical forest ecosystems in Guinea, strong positive correlations were found between soil clay contents and total soil carbon stocks, as well as minerals associated C, N and P stocks, which were also correlated with mycorrhiza abundance, growth dynamics and the mortality of forest trees [59]. In stands of Chinese Fir, the microbial limitations of the mineralization of organic matter and organic phosphorous were found to be associated with the microbial demand of nitrogen [60]. The P-acquisition strategy was examined in stands of European beech in Germany. In P-rich soils, plants and soil organisms mobilize P mainly from primary and secondary minerals. In P-poor soils, roots and fungi seem to sustain their P demand more successfully than bacteria, mainly from the forest floor and soil horizons rich in organic matter. This underlines, in principle, a high adaptability of beech forest ecosystems to changing P supplies. Thus, "P deficiency is unlikely the result of a low P supply per se. More likely, sufficient P nutrition depends on supply-specific plant–microorganism–soil interactions" [46]. Moreover, it was found in this study that the phosphorous in particulate soil organic matter (SOM) within aggregates tends to increase with the decreasing soil P stock. That indicates that physically protected particulate SOM becomes increasingly relevant as a P cache in soils with a declining P status [61]. Additionally, Rodionov et al. [62] found "that P deficiency in the surface soil not only fosters microbial cycling of P in the organic and upper mineral soil layers but also causes the utilization of P from the deeper subsoil", and they concluded that "with continued weathering of the bedrock and mobilization of P from the weathered rocks, P cycling will proceed to greater depths, especially at sites characterized by P limitation". The complex interaction of chemical, microbial and physical processes in P cycling in forest soils is actually referenced for a large variety of ecosystems with a large number of publications (e.g., References [63–66]).

The second important factor of forest growth and health status besides nutrient supply is a widely continuous water supply. Compared to arable soils, the surface layers of forest soils (forest floor and upper A-horizon) have high humus contents and much more open-pored mineral soils, maximizing water infiltration and, thus, minimizing the occurrence of surface flow [20]. Soils act, as for nutrients, for water also like a "buffer store" enabling a high continuity in supply. Forest soils are able to maintain a site-specific water status caused by the nonlinearity of the unsaturated water conductivity function, which decreases exponentially with the decreasing pressure head. As a consequence of heavy rain events, water enters macropores, where the surplus water that gets not quickly stored in the meso- and micropores leaves the rooting zone quickly as seepage water. Thus, sufficient aeration in the rooting zone is also guaranteed in periods of heavy rainfall. The water in the meso- and micropore spaces gets retained against gravity over a long time and, thus, can, at least partially, sustain the water supply of trees during drought periods. Both the nutrient and water supply are directly related to the soil volume being opened up by roots and mycorrhiza hyphae and, thus, to the extent of the rhizosphere.

3.2.2. Secondary Ecosystem Services

Forest soils are the basis for manifold ecosystem services, besides growth and the existence of forest stands themselves. They serve as habitats of a broad variety of fauna [67,68], plants of the ground vegetation [2], fungi [69] and microbes [70]. Close interactions between the composition of the herbaceous ground vegetation and soil microbial diversity were found to drive forest ecosystem functioning in European temperate forests [2]. Giguère-Tremblay et al. [71] highlighted in boreal forests the "predominant role of soil organic matter on multi functionality . . . even though microbial diversity is important". Friggens et al. [72] found in boreal forests in Sweden "no trend in respiration with distance from trees, likely mediated by an extensive root and ectomycorrhizal network of the birch trees, which efficiently exploit resources throughout the forest". Strong correlations between dominant tree species and fungal communities were found on the local scale in mixed boreal stands (trembling aspen/black spruce) in Western Quebec [73]. Lots of soil functions and the ecosystem services linked to them are generated or at least influenced and shaped by these biota [37]. The adaption of these populations to the boundary conditions of site quality, climate conditions and phases of stand development is a slow process in the range of months to decades compared, e.g., to the very fast chemical exchange reactions. Thus, the integrity of the habitat function of forest ecosystems depends on the fact that, in not substantially disturbed cases, changes in boundary conditions are so slow that the biota can follow. This is, e.g., realized by the quasi-circular coevolution of animal communities, humus forms and mineralization processes along three stand phases of spruce in the Italian Alps [67]. Summarizing, it can be stated that forest ecosystems commonly provide a higher biodiversity and higher microbial activity [1] than agricultural land use types or forest succession on former agricultural land.

The high porosity and intrinsic surface in combination with their exchange capacity qualify forest soils as effective filters for water [74]. Especially along the passage through extensively rooted soil layers, potentially harmful substances for water quality are retained, adsorbed and/or transformed to less harmful compounds through chemical or biological soil processes. Phosphorous gets, e.g., in acid soils with high Al activity, quantitatively fixed by forming Al-phosphates and/or is quantitatively taken up by plants and microbes in the uppermost soil layers [46]. On the other hand, Missong et al. [75] indicated that 12 and 91% of the totally leached P from 20-cm-long soil columns were bound mainly to nano-colloids (0.6–29 nm) and fine particles (70–400 nm), depending on the type of the forest soil. They found that "size and composition was comparable to colloids present in acidic forest streams known from literature". Markowski et al. [5] observed during heavy rainfall events following dry periods a depth transport of P into the subsoil along preferential flow paths, especially for particle-bound P. Thus, evidence was obtained that P leaching from forest soils to the hydrosphere feasibly occurs, even if P–retaining soil processes are strong and effective. Nitrate is in anoxic soil layers subject to microbial denitrification and leaves the soil as gaseous nitrous oxides or elementary nitrogen [76]. Some tree species like European beech can quantitatively take up nitrate from the seepage water and, thus, act as effective nitrate sinks [14] and subsequently enhance the water quality even under the actual deposition conditions in Central Europe. Sucker et al. [77] explained the actual decreasing nitrate concentrations in headwater streams of the Ore Mountains with "a higher N uptake as a result of extensive reforestation and the continuous recovery and increasing vitality of damaged forests". Generally, forest soils provide an above-average infiltration capacity because of the very high porosity of the surface soil layer. Compared to agricultural land use, the infiltration rates in forests are 200–500% higher [41,78,79]. Thus, they prevent fast runoff on the soil surface or as surface-near interflow, which is predominantly relevant under tropical conditions with high precipitation intensities. This helps to minimize flood events in forest lands. Additionally, the protection of the soil surface by humus layers and ground vegetation is an effective security against soil erosion. Zhang et al. [80] found in a field experiment in the Loess Plateau (NW China) under grassland about 50 times and under arable farmland 100 times more eroded soil sediment after 30 min of heavy rainfall with 120 mm h^{-1} intensity as compared to forest land.

Recently, the role that forest ecosystems play in the context of climate change becomes more and more focused on by scientists and the public, as the high number of contributions to that thematic field imply (Figure 1). The most important contribution of forest ecosystems and forest soils for counteracting climate change is to sequester and/or metabolize greenhouse gases or components of them. The most important factor is carbon sequestration in forest stands as plant biomass or, subsequently, in the soil as organic matter on the forest floor and in the mineral soil (SOC). Witzgall et al. [43] stated that "The largest terrestrial organic carbon pool, carbon in soils, is regulated by an intricate connection between plant carbon inputs, microbial activity, and the soil matrix. This is manifested by how microorganisms, the key players in transforming plant-derived carbon into soil organic carbon, are controlled by the physical arrangement of organic and inorganic soil particles". This statement underlines the significance of the interaction of the structure, chemical status and microbial activity for the functionality of soils and ecosystem services. Caddeo et al. [33] modeled (Century5 model) for all of Italy the present soil carbon stocks and projections to the year 2095 under different agro-ecosystems and forests. They found that the current SOC stock estimates range from 51.3 in orchards to 129.5 Mg carbon ha^{-1} in coniferous forests. Projections under the influence of climate change (scenarios RCP4.5

and RCP8.5) "showed a moderate carbon loss suggesting that forest, grassland, and permanent crop soils could provide an important contribution to climate change mitigation". Pellis et al. [81] compared SOC and above-ground biomass in a 62-year-old forest afforested on former grassland, with the carbon pools in an adjacent grassland in the Italian Alps and the Apennines. They found under 62-year-old afforestation about two-times higher SOC stocks than under the correspondent grassland. In the Apennines with dryer and warmer climate conditions, the SOC increase was much higher than in the Alps. Additionally, the carbon stock in the above-ground biomass amounted in the old forest stands to 100–170% of the SOC stock. Moreover, this study highlighted the importance of considering the subsoil, since deep soil layers contributed 38% to the observed variations in the carbon stocks after land use change. Kalks et al. [82] found in three beech stands in Germany on sandy to loamy soils that 13C-labeled DOC injected at three soil depths was, after 17 months in the topsoil, largely lost (−19%), while DOC in the subsoil did not change much (−4.4%). The data indicated a high stabilization of injected DOC in the subsoils with no differences between the sites. This supports the significance of the subsoil carbon pool for long-term carbon sequestration. A study by Wordell-Dietrich et al. [83] supported this thesis, since they found in beech forests in Northern Germany that most of respired CO_2 (90%) was produced in the topsoil (<30 cm). However, the subsoil (>30 cm), which contained 47% of the SOC stocks, accounted for only 10% of the total soil respiration. Zachary et al. [84] determined SOC turnover rates by incubating trials with a silt loam-textured Luvisol from West Hungary. They determined the mean residence time (MRT) of four different SOC fractions. The particulate organic matter fraction was found to be the most labile C pool with a MRT of 3.6 years, and the most stable fraction was the chemically resistant soil organic carbon fraction associated with clay particles with MRT of 250 years. Forest continuity is obviously also an important factor for preserving high SOC pools. In NE Germany, significantly larger total SOC stocks were found in ancient forests (age > 200 years) as compared to 100–200 year-old afforestation. These differences were obtained partially in subsoils at depths of between 29 and 55 cm. Soils of "ancient" beech and pine forests stored, on average, twice as much SOC in the subsoils than did "old forests" [85]. In Denmark, also, a tendency toward increasing SOC stocks with increasing stand age after afforestation was found [86].

Besides carbon sequestration, the budget of methane (CH_4) and nitrous oxide (N_2O) of forest soils is highly relevant, because their greenhouse potential is much higher than carbon dioxide. According to the IPCC Fifth Assessment Report [87], the relative global warming potential (for a 100-year period) of CH_4 28- and for N_2O is 265 times as high as for CO_2. Undisturbed terrestric forest soils are a weak source of N_2O and a weak sink for CH_4 [88]. From wet soils, the emission of N_2O is much higher [89]. Schindler et al. [90] showed in a flooding experiment that soil water and nitrogen contents are the main controlling factors of stem and soil N_2O- and CH_4 fluxes. During flooding, CH_4 emission increased by a factor of 10, and the weak CH_4 sink turned to a strong source. The N_2O emission increased during flooding by 40%. Sosulski et al. [34] found in Central Poland that the N_2O-N emission from the arable soils is about 30% higher as compared to forest soils due to a greater amount of mineral nitrogen available for the nitrifying and denitrifying bacteria in the arable soils. They concluded that "conservation and sustainable management of forests would constitute an effective way to mitigate the N_2O-N emissions from the soil".

4. Outlook and Conclusions

After the detailed description of the functions of forest soils and ecosystem services of forests in the Results section, a short overview shall be given here on the actual threats on the integrity and functionality of forest ecosystems, as well as on the management options to counteract them. Some concluding remarks will summarize what forest soils differentiate from soils under other land use types.

4.1. Threats to Forest Soil Functions and Ecosystem Services

With respect to the high complexity and multifactorial boundary conditions of the functions of forest soils, as well as on the large scale of forest ecosystem services, it is intuitively understood that the vulnerability of them is high. Like their functionality, their threats are specifically branded by the characteristics of forests. A large crown surface, e.g., causes a high transpiration demand, which can provoke drought, and the height of trees is related to windthrow vulnerability. The main natural hazards are storms, insect calamities and their after-effects like increased fluxes of nitrate and phosphorous, which can endanger the water quality [91], and wildfire. Wasak et al. [92] found that windthrow does not only reduce the growth intensity of forest stands rather than also microbial activity. Reduced microbial activity after windthrow was predominantly attributed to a breakdown in fungal activity, which can be explained with a lack of substrate that feed trees to mycorrhiza fungi in undisturbed stands. These natural hazards are in the natural stage of site conditions to which tree species are evolutionarily adapted.

However, mankind accelerated soil processes like acidification and changed quasi-stable boundary conditions like climate characteristics to an amount that does not allow for the easy adaptation of trees. Additionally, forest management itself can contribute to manmade ecosystem damages, e.g. by the deformation and compaction of soils through moving heavy forest machines on unprotected soils. Kohler and Hildebrand [45] described this phenomenon as four unintended, large-scaled ecosystem experiments: the "titration and eutrophication experiment" with forest soils in Middle and Northern Europe yielded a drastic and self-accelerating depletion of exchangeable basic cations, "since bond strengths of exchangeable earth alkali ions decrease with increasing acidity". The drivers of soil acidification are the activity of strong and mobile acid anions—predominantly nitrate and sulfate. Even if, e.g., in Central Europe, the deposition of sulfate dropped in the last decade below the critical load threshold, the deposition of nitrogen remains high, causing, e.g., imbalances in tree nutrition and the ruderalization of ground vegetation [70]. Soil acidification, however, still persists as an inherited problem that results in a tendency towards flat rooting systems, thus increasing the susceptibility of trees for drought and disturbing tree nutrition when the potential rooting space is only partially exploited by roots like Matzner and Murach hypothesized [93]. The most threatening biological consequence of soil acidification is the drastic reduction of earthworm abundance at pH values below 4.5 [48], because earthworms are the main agents of the secondary soil structure [49], which is a key factor of soil function (see Section 3.1.1). Moreover, the progress of the acidification front towards the hydrosphere increases the risk of deterioration of the water quality and habitat characteristics of streams and lakes [77]. The "soil deformation experiment" [45] results in drastically reduced soil aeration and, thus, in a substantially reduced rooting intensity [10]. The recovery of compacted soils lasts decades [94]. Moreover, the shift from aerobic to anaerobic metabolism in compacted soils suggests that skid trails may be unconsidered hot spots of greenhouse gas balance because of substantially increased N_2O emissions and decreased CH_4 consumption, both due to the locally anaerobic soil conditions in skid trails [95,96], thus creating a link to the "greenhouse experiment" [45]. Climate change leads to warming and the increased frequency and intensity of extreme weather events threatening the existence and functionality of forests. Büntgen et al. [97] found in tree ring analyses that the sequence of recent European summer droughts since 2015 is unprecedented in the past 2110 years. Boden et al. [16] found in spruce forests in SW Germany that drought is an increasing threatening factor there and that the cumulation of drought events decreases the resilience of spruce to drought stress. Fleck et al. [98] derived from model projections that nitrate leaching from forest soils will increase because of increased organic matter decomposition. Hennings et al. [99] found that riparian areas in tropic rainforests in Sumatra have a high potential for C-sequestration and but a high C-loss potential if drained.

The impacts of environmental change on soil and ecosystem functions are complex and cannot be understood and managed in monocausal approaches. The complex interactions between climate change effects and soil functions and the large-scaled ecosystem services will be pointed out with the example of the actual forest dieback caused by climate change and its subsequent after-effects. This consideration will be focused on Europe, because there not only the direct effects of climate change are relevant rather than their interactions with the deposition history. Puhlmann et al. [100] modeled in Germany a significant increase of drought events since 1990 in terms of soil water availability. If "soil acidification and increased N availability decreased the fine root biomass of trees and shifted the rooting zone to upper soil layers" [93], this would aggravate drought stress for trees. In the Swiss Alps, a differential diagnostic study revealed that the actual tremendously increasing mortality through bark beetle attacks even in higher elevated areas (up to 1700 m a.s.l.) is not related to increasing bark beetle virulence or the raisin defense weakness of trees rather than to "drought-induced reduction in tree vigor ... under the ongoing climate warming" [101]. Rewald found that oak (*Quercus petraea*) and beech (*Fagus sylvatica*), as the two naturally dominating tree species in Central Europe, show very high vulnerability of fine roots to die off during drought events and thus prolong drought damage [102]. Additionally, in Switzerland, a study on the drought tolerance of trees stated that the "premature mortality of roots" leaves trees more vulnerable following drought years [103]. The same aspect was addressed by a modeling approach in the USA and Sweden that revealed that "host tree vulnerability plays an important role in bark beetle outbreak intensity" [104]. A study covering a climate gradient from Southern Sweden to Mediterranean Europe identified temperature warming, drought and storm effects as key climate drivers of the actual intensity of bark beetle calamities [105]. These studies provide evidence that an important determinant of tree mortality resulting from drought and subsequent bark beetle calamities is the predisposition of trees by deterioration of soil functions. Thus, it seems likely that the actual intensity of drought and bark beetle damages in Europe, Scandinavia and other industrialized regions is the result of the interrelation between predisposing stress factors arising from soil acidification/eutrophication and increasing climate stress. A study on long-term environmental monitoring data in Switzerland found that cation leaching losses actually mainly driven by nitrogen deposition are endangering forest sustainability. "Soil acidification has negative consequences for forest health, such as increased risk of windthrow on soils with low base saturation <40% or decreased rooting depth for soils with a base saturation <20%" [106]. Thus, the acidification legacy of former acid deposition and ongoing nitrogen deposition destabilizes forest ecosystems. It is reasonable to assume that predisposing and acute stress factors are acting together in an additive way.

4.2. Management Approaches for Protecting the Functionality of Forest Soils

In the face of the high vulnerability of forest soils and their functions and especially in industrialized regions, increased threats, e.g., through soil acidification, the loss of processes generating and maintaining the secondary soil structure and, thus, the loss of forest soil functions, it is evident that active soil management strategies must be implemented with the aim to counteract the loss of soil functions or to recover them as far as possible. At least the irreversible loss of soil functions like the destruction of clay minerals through heavy soil acidification must be avoided.

4.2.1. Silvicultural Management Options

The most important silvicultural management options are tree species selection and harvesting, respectively, thinning regimes. Several studies suggest that these management options would have different potentials for soil preservation besides the ostensible task of silviculture to optimize forest growth. The most fundamental silvicultural measure is to bring the tree species only to sites that meet their needs and thus maximizing the probability to get healthy stands, which can fulfill all services we expect from them. This was demonstrated with the example of the demands on soil properties of beech, ash and sycamore from 806 observation plots in Switzerland [51]. The study revealed that ash and sycamore are much more sensitive to soil characteristics than beech. "Shortage of nutrients limited the distribution of ash and sycamore and excess of toxic elements the distribution of ash". The authors concluded: "It is not advisable to plant ash or sycamore or to promote their natural regeneration beyond the critical values for soil acidity and nitrogen supply. A sound knowledge of the soil properties required by tree species is a prerequisite for addressing many practical and scientific issues such as forest management or the predictive mapping of tree species". A large number of studies have dealt with the effects of tree species and/or the harvesting regimes on ecosystem services. A monitoring study in mixed spruce/beech stands in the Czech Republic from subsequent observation campaigns in 1972, 1996 and 2010 revealed that beech dominated on dry terrestric soils and spruce on wetter and more acidic soils [107]. The authors conclude that the "current expansion of beech is expected to continue on terrestrial soils but will probably slow down with increasing soil wetness" under climate change conditions.

The effect of the admixture of evergreen and non-evergreen oaks in pine stands on microbial activity and the mineralization intensity of organic matter were examined in Southern France [108]. The study revealed an additive effect of oak admixture enhancing mineralization intensity and mobilization of nutrients from organic matter especially in stands with evergreen oak. The authors conclude that "admixtures of oaks and pines can potentially maximize the diversity of nutrient resources and consequently favor microbial diversity, biomass and catabolic potential, through complementary ecological niches". The dependence of mycorrhiza communities of tree species and nutrient availability was studied in Western Poland [69]. The study revealed that "Coniferous tree plots were characterized by lower pH values, plots with deciduous trees by higher concentrations of total Ca and exchangeable forms of Ca, K and Mg. Arbuscular mycorrhiza fungi abundance in soils and roots increased along with increasing soil alkalinity and macronutrient levels". Model scenarios assuming the whole forest area would be covered with spruce vs. beech were compared in a regionalization study in SW Germany based on data from regular soil monitoring. In the topsoil, at slope shoulders, no significant difference between the spruce and beech scenarios could be detected. At lower slope positions, the base saturation of the beech scenario was 0.3–2 times higher than that of the spruce scenario [109].

The deposition of acidity and nitrogen were substantially altered by tree species and stand structures in the Black Forest (SW Germany). The deposition load was in beech-dominated stands about 45–85% lower than in spruce stands. The leaching of nitrate out of the rooting zone (120cm depth) is equal in beech-dominated stands to the deposition and is, in spruce stands, about two to three times higher. It can be stated that the change from spruce stands to beech stands has a potential to reduce the impact of further deposition on the forest soil to about half the value in spruce stands. Moreover, beech has a strong water preservation potential in that region regarding nitrate leaching [13]. Zeller et al. [110] observed in 21 Douglas stands in France unexpectedly high nitrification rates and concluded "that even under optimal conditions for tree growth (high biomass increment) an excess of nitrate remains in the soil with a peak in autumn. As nitrate is highly mobile in the soil profile, leaching loss of nitrate and cations may affect surface and groundwater quality, as well as the sustainability of soils by an acidification process". Fleck et al. [98] suggested for the northern flatlands of Germany that the sink strength of forests for N should not be "additionally lowered by overly strong reductions of standing biomass, since they are already at the limit of their N retention capacity".

There have been several studies suggesting silvicultural approaches to support C-sequestration. Disturbance of the crown closure through clearcuts create long-lasting leaching of dissolved organic carbon (DOC) and, thus, decreases the SOM pool [111]. SOC stocks observed in oak-dominated stands in Denmark are not driven by decreased SOM decomposability. However, lower specific carbon mineralization in the 200-year-old forest suggests that the stability of C and retention of N may increase in a longer perspective [86]. In Poland, in beech and pine stands, it was observed that, under beech, much more organic matter was accumulated in mineral horizons than in organic horizons [112] and that beech stands after the removal of pine stands accumulated over 20% more organic carbon content.

To accumulate deadwood in forests is a usual measure to enhance the habitat value of forests. Wambganss et al. [113] examined if deadwood would contribute to the formation of stable SOM and found that, on silicate bedrock, deadwood increased the free light SOM fraction by 57% compared to the reference points. In contrast, on calcareous bedrock, deadwood decreased the free light fraction by 23%. Thus, it depends obviously on the chemical status of soils if deadwood contributes to stable or labile SOM forms, and the accumulation of deadwood cannot clearly be judged as a strategy to enhance long-term C-sequestration.

Besides tree species selection, the harvesting regime and disturbance of crown closure tends to increase the leaching rates, since plant uptake and crown interception are reduced in gaps or clearcuts. Moreover, a more open stand structure provokes increased mineralization of organic matter because of higher temperatures and water availability. Papaioannou et al. [114] found in spruce stands in Northern Greece that harvesting practices generally negative impact soil N and organic matter in mineral soil, as well as the C/N ratio and exchangeable Ca. The authors observed that, after 15 years, the nutrient availability and organic C accumulation recovered to similar levels to those of the unmanaged sites. In SW Germany, it was derived from extensive ecosystem monitoring plots (Level II) that, even under the influence of high deposition loads, permanent cover or gap-oriented harvesting regimes provided valuable options for the preservation of site sustainability in terms of equal or slightly positive balances of basic cations. In the opposite, "rough silvicultural management practices" like spruce monoculture with clearcut result in high losses of basic cations at the same site [14].

4.2.2. Technical Approaches for Forest Soil Preservation

The silvicultural management options through orienting tree species selection or harvesting regimes on preservation of soil functions and ecosystem services is the main part of sustainability strategies in forest management. However, some ecosystem disturbances are so heavy and natural recovery is so slow that technical management option must be used to accelerate recovery of soil functions and thus stabilize forest ecosystems which is strongly recommended in times when new strains and stresses are fast emerging, e.g., by climate change. Deposition driven soil acidification is such a fundamental and long-lasting damage on forest soil functions. Soil protection liming is an effective counter-measure, with low side effects, against unnatural soil acidification. By comparing the acidification status between the National Forest Soil Inventories of 1994 and 2008 in Germany it could be shown that on limed monitoring plots the base saturation increased by 88% more than on not limed plots [115]. Therefore, the authors conclude that "forest liming of soils with considerable acidification is furthermore recommended to balance negative impacts on soil functioning, the vitality, and growth of forests". A large-scale forest liming trial which was undertaken in SW Germany since 1983 represents with repeated liming after 20 years the liming intensity of a practical soil protective liming program. Natural recovery on the control plots in soil pH was in the time span 2003–2015, on average limited to an increase of 0.2–0.4 pH units in the forest floor and 0.1–0.3 pH units in the mineral soil. Exchangeable cations calcium and magnesium slightly increased also at the control plots, although the base saturation remained <20%. Lime treatment greatly accelerated the rise in pH by 1.2–1.3 units and base saturation by 40–70% in the organic layer, as well as 0.3–1.2 pH units and base saturation by 7–50% in mineral soil [116]. The authors conclude: "Liming of acidified forest soils significantly adds to natural recovery and therefore helps to establish greater buffering capacities and stabilize forest nutrition for the future". Berger et al. [117] found in beech stands in Austria that "the beech trees showed no sign of recovery from acidification although S deposition levels decreased". It is expected on the long-run that liming would lead to better exploitation of the potential rooting zone because of more favorable chemical and physical properties for root growth in the mineral soil. Thus, water and nutrient supply of tree should be enhanced. Kohler et al. [118] examined whether this would lead to an enhanced resistance, recovery or resilience of the growth rate of

Spruce against drought events. They found that "recovery and resilience of radial growth after severe drought events were generally better in spruce trees of limed treatments. This indicates a shorter stress period in spruce trees growing on limed soil, which may reduce their susceptibility to secondary, drought-related pests and pathogens".

If heavy forest machines are moving upon unprotected forest soils their deformation and compaction is inevitable. Therefore, the most effective strategy to minimize these damages is to establish soil preservation guidelines which restricts machine traffic to regular skidding trail systems with prescribed distances of the skid trails ranging from 20 to 40 m. Any wheeling of heavy machines between these more or less parallel tracks should be forbidden. Such guidelines already exist in most regions of Germany.

In order to focus counter strategies on sensitive sites, it would be helpful to know sites being tolerant against soil deformation. However, this differentiation, e.g., according to soil texture is difficult, because apparently, less susceptible textures like sand [11] or peat [119] also get damaged.

Technical measures like wide, low inflation tires or brush mats can alleviate the problem but generate no security because of a high uncertainty resulting from high variation of soil properties and dynamic machine impacts. A LIDAR–based study on rutting in skid trails revealed that low tire pressure may mitigate the impact of forwarders on soil deformation and the greater the number of passes, the greater the degree of soil disturbance [120]. Green et al. [121] could show that cable assisted, tethered harvesters and forwarders lowered the spatial distribution of machine influence on compaction.

The main problem of soil deformation and soil compaction is that natural recovery time is in any case very long. A controlled compaction trial on fine grained hydromorphic soils in France showed after seven years no sign of recovery [122]. In SW Germany skid trails of regular harvest operations with time delays up to 24 years between tracking and examination were investigated in order to characterize the status of recovery of essential soil functions. Gas diffusion coefficients and the fine root distributions of comparable sensitive silty loams were used to describe the disturbance of soil functions still detectable after decades. Up to 14 years after machine impact, gas diffusion coefficients and root densities in the upper mineral soil under wheel tracks showed no signs of restoration. In the subsoil, 24 years after machine impact, significantly reduced root densities occurred [94]. Therefore, a need of effective and not too cost-intensive measures for an active acceleration of soil deformation and their ecological effects is given. This applies especially for skid trails which should be abandoned because of technical reasons or in the case of irregular machine impact, e.g., when after windthrow regular skid trail systems are destroyed or not any more detectable. In a controlled wheeling experiment in SW Germany the recovery of soil structure on compacted skid trails, which had been treated with a combination of regeneration techniques (mulching, liming, planting alder trees or a combination of them) has been monitored. After four years, higher values of the diffusive gas permeability and macropores indicated significant improvement of soil aeration in the topsoil. In the topsoil, root density increased with increasing soil gas permeability, while in the deeper horizons only few macropores are occupied by fine roots [123]. The combination of technical treatments and planting of alder trees improves the circulation of air and water through the pore system. This leads to decreased CO_2 concentrations and increased root growth. Both are indicative of an initial recovery of soil structure. The planting of root-active trees showed a substantial regeneration effect. The root growth rate (cm cm^{-2}) in the mulched and planted variant approached after 4 years observation the range of the undisturbed control [124].

4.3. Conclusions

This literature review revealed that forest soils provide a predominantly differentiated soil structure being the basis for their high ecological functionality. Thus, forest soils are the favorable basis for manifold ecosystem services as FAO stated [27,28]. Since conventional soil descriptions provide only indicator variables and boundary conditions for soil functions and not data on the functions themselves, they have to be assessed from soil descriptions by means of so called pedotransfer functions (PTF) [125,126].

The literature cited in Section 3.1 support the conclusion that parameters like aggregate structure and connectivity of the soil pore system are crucial for soil functions. These parameters are not part of the conventional content of soil descriptions, e.g., of the World Reference Base [127] and can without additional and time-consuming measurements only indirectly and with high error probability been assessed by means of PTFs. Rabot et al. [128] suggested to derive information on aggregate structure and pore continuity by means of image processing techniques which is in line with the results discussed in Section 3.1.1 in this study. These approaches suggest that further research on quantification of the secondary soil structure is needed as a basis for modeling of soil functions like, e.g., water infiltration, plant available water storage, soil aeration and nutrient transport with the seepage water.

Ulrich stated that: "Forest ecosystems are characterized by a hierarchy of processes, differentiated according to spatial and temporal scales" [24,25]. Thus, natural forest ecosystems are in their natural status well buffered against external disturbance which can deflect them on the short run, but small scaled processes like, e.g., chemical acid-buffering reactions can soon bring them back to their specific attractor space. However, if the change of environmental conditions is too fast and too strong and thus over-ride the small-scaled buffering mechanisms, the signals of the disturbance reach the medium to large scale like die-back of the rooting system as reaction on deposition-driven soil acidification or by soil compaction through heavy forest machinery.

If forest management should optimize multi-functionality in a sustainable way, it must be based on "in-depth knowledge related to ecosystem processes and functions and soil state variables" [29]. Since scientists provide detailed quantitative information about soil functions but tend to "overlook practical, site-specific implications" which are common to practitioners [29], close cooperation between soil scientists and forest management practitioners seems to be the key to enable a holistic management approach comprising the relevant process scales up to the macro-scale where forest management takes place. Crucial precondition for the reliable transfer of point-related measuring data from environmental networks or scientific projects to the landscape level, are multivariate regionalization models which assess environmental and/or soil data on the whole forested area of landscapes with reliable error identification. "Base saturation could for example be predicted with an accuracy of 50–70% (in terms of the multiple R^2) using topographic variables, geologic substrate, stand characteristics and information about forest liming as predictor variables in multiple linear regression analyses. Thus, regionalization models achieve the role of decision support tools for planning of forest management at the landscape level" [109].

It is evident that active management measures must be set in action to preserve the vulnerable functional structures of forest soils under the actual fast changing environmental conditions. Doing so, always the complexity of physical, chemical and biologic agents contributing to build-up and sustain these prodigious structures has to be considered. Without keeping soil reaction in the range of living conditions of mycorrhiza and microbes, sufficient soil aeration for optimal root growth and planting tree species tolerating the expected climate conditions, the preservation of forest soil functions and forest ecosystem services, as we know them from experience, will fail.

Funding: This research received no external funding.

Informed Consent Statement: Not applicable.

Data Availability Statement: Nearly all cited articles are available from the author as PDF files. Some older articles are printed versions. The basis of the evaluations and the article are literal excerpts of the central statements of the cited literature.

Acknowledgments: The author is very grateful for the helpful discussions with my colleagues Heike Puhlmann, Peter Hartmann and Helmer Schack-Kirchner on the specific characteristics of forest soils.

Conflicts of Interest: The author declares no conflict of interest.

Appendix A

Table A1. Regional and climatic context of the 128 articles cited. Climate zones according to Reference [26].

Regional Context	Climate Zone Köppen-Geiger	Titles Cited	Titles	%
World wide	all except EF, ET, BW	[20,26–28,48,87,125,127,128]	9	7.0
Europe wide	Dfa, Dfb, Dfc, Cfa, Cfb, Csa, Csb, BSk	[8,15,17,21,31,47,49,70,76,97,103,106]	12	9.4
N-America, Canada, subpolar, no dry season	Dfa, Dfb, **Dfc**	[71,73,121]	3	2.3
Scandinavia, subpolar, no dry season	Dfa, Dfb, **Dfc**	[36,72,104,119,120]	5	3.9
Scandinavia, cold, no dry season	**Dfb**, Dfc	[3]	1	0.8
Europe, cold, no dry season	Cfa, Cfb, Dfa, **Dfb**, Dfc	[1,2,4,5,22,24,25,34,35,43,46,51,61,62,64,69,77,84,86,90,92,101,105,107,111,112,117]	27	21.1
Europe, temperate humid	Cfa, **Cfb**, Csb	[6,9–14,16,18,19,23,29,30,37,38,42,44,45,50,52–58,63,75,82,83,85,89,93–96,98,100,109,110,113,115,116,118,122–124,126]	48	37.5
Asia, N-America temperate humid	**Cfa**, Csb, Dfc	[60,88,91]	3	2.3
Europe, semi arid	BSk, **Csa**, Csb, Cfa	[32,33,65–68,81,102,108,114]	10	7.8
Asia, Africa, semi arid	**Bwk**, Cwa, Cfa	[7,74,79,80]	4	3.1
Asia, Africa, S-America tropic	**Af**, **Am**, As, Aw, Cfa	[39–41,59,78,99]	6	4.7

References

1. Sokołowska, J.; Józefowska, A.; Woźnica, K.; Zaleski, T. Succession from meadow to mature forest: Impacts on soil biological, chemical and physical properties—Evidence from the Pieniny Mountains, Poland. *Catena* **2020**, *189*, 104503. [CrossRef]
2. Ma, S.; De Frenne, P.; Boon, N.; Brunet, J.; Cousins, S.A.O.; Decocq, G.; Kolb, A.; Lemke, I.; Lemke, J.; Naaf, T.; et al. Plant species identity and soil characteristics determine rhizosphere soil bacteria community composition in European temperate forests. *FEMS Microbiol. Ecol.* **2019**, *95*, fiz063. [CrossRef] [PubMed]
3. Finlay, R.D. Ecological aspects of mycorrhizal symbiosis: With special emphasis on the functional diversity of interactions involving the extraradical mycelium. *J. Exp. Bot.* **2008**, *59*, 1115–1126. [CrossRef] [PubMed]
4. Orlinskiy, P.; Münze, R.; Beketov, M.; Gunold, R.; Paschke, A.; Knillmann, S.; Liess, M. Forested headwaters mitigate pesticide effects on macroinvertebrate communities in streams: Mechanisms and quantification. *Sci. Total Environ.* **2015**, *524*, 115–123. [CrossRef]
5. Makowski, V.; Julich, S.; Feger, K.-H.; Julich, D. Soil Phosphorus Translocation via Preferential Flow Pathways: A Comparison of Two Sites with Different Phosphorus Stocks. *Front. For. Glob. Chang.* **2020**, *3*, 48. [CrossRef]
6. Fiquepron, J.; Garcia, S.; Stenger, A. Land use impact on water quality: Valuing forest services in terms of the water supply sector. *J. Environ. Manag.* **2013**, *126*, 113–121. [CrossRef] [PubMed]
7. Luo, J.; Zhou, X.; Rubinato, M.; Li, G.; Tian, Y.; Zhou, J. Impact of Multiple Vegetation Covers on Surface Runoff and Sediment Yield in the Small Basin of Nverzhai, Hunan Province, China. *Forests* **2020**, *11*, 329. [CrossRef]
8. Waldner, P.; Thimonier, A.; Graf Pannatier, E.; Etzold, S.; Schmitt, M.; Marchetto, A.; Rautio, P.; Derome, K.; Nieminen, T.M.; Nevalainen, S.; et al. Exceedance of critical loads and of critical limits impacts tree nutrition across Europe. *Ann. For. Sci.* **2015**, *72*, 929–939. [CrossRef]
9. Hildebrand, E.E.; Schack-Kirchner, H. The influence of compaction on soil structure and functions in forest sites. In *Modern Trends in Applied Terrestrial Ecology*; Ambasht, N.K., Ambasht, R.S., Eds.; Springer-Science+Business Media: New York, NY, USA, 2002; pp. 1–11.
10. Gaertig, T.; Schack-Kirchner, H.; Hildebrand, E.E.; von Wilpert, K. The impact of soil aeration on oak decline in south-western Germany. *For. Ecol. Manag.* **2002**, *159*, 15–25. [CrossRef]
11. Ampoorter, E. Soil Compaction Due to Mechanized Forest Harvesting: Quantification of Ecosystem Effects and Exploration of Recovery Potential. Ph.D. Thesis, Ghent University, Ghent, Belgium, 2011; p. 182.
12. Schäffer, J.; von Wilpert, K.; Kublin, E. Analysis of fine rooting below skid trails using linear and generalized additive models. *Can. J. For. Res.* **2009**, *39*, 2047–2058. [CrossRef]

13. von Wilpert, K.; Zirlewagen, D.; Kohler, M. To what extent can silviculture enhance sustainability of forest sites under the immission regime in Central Europe? *Water Air Soil Pollut.* **2000**, *122*, 105–120. [CrossRef]
14. von Wilpert, K.; Zirlewagen, D. Forestry Management options to maintain sustainability—element budgets at Level II sites in South—West Germany. In *Forests in a Changing Environment—Results of 20 years ICP Forests Monitoring*; Eichhorn, J., Ed.; Schriften aus der Forstlichen Fak. University Göttingen: Göttingen, Germany, 2007; Volume 142, pp. 170–179.
15. Aubin, D.; Varone, F. The evolution of European water policy. In *The Evolution of National Water Regimes in Europe. Transitions in Water Rights and Water Policies*; Kissling-Näf, I., Kuks, S., Eds.; Kluwer Academic Publishers: Dordrecht, The Netherlands; Boston, UK, 2004; p. 4986.
16. Boden, S.; Kahle, H.P.; von Wilpert, K.; Spiecker, H. Resilience of Norway spruce (*Picea abies* (L.) Karst) growth to changing climatic conditions in Southwest Germany. *For. Ecol. Manag.* **2014**, *315*, 12–21. [CrossRef]
17. Mellert, K.H.; Lenoir, J.; Winter, S.; Kölling, C.; Čarni, A.; Dorado-Liñán, I.; Gégout, J.C.; Göttlein, A.; Hornstein, D.; Jantsch, M.; et al. Soil water storage appears to compensate for climatic aridity at the xeric margin of European tree species distribution. *Eur. J. For. Res.* **2018**, *137*, 79–92. [CrossRef]
18. Chakraborty, T.; Reif, A.; Matzarakis, A.; Saha, S. How Does Radial Growth of Water-Stressed Populations of European Beech (*Fagus sylvatica* L.) Trees Vary under Multiple Drought Events? *Forests* **2021**, *12*, 129. [CrossRef]
19. Maier, M.; Paulus, S.; Nicolai, C.; Stutz, K.P.; Nauer, P.A. Drivers of Plot-Scale Variability of CH_4 Consumption in a Well-Aerated Pine Forest Soil. *Forests* **2017**, *8*, 193. [CrossRef]
20. Blanco-Canqui, H.; Lal, R. Soil Erosion Under Forests. In *Principles of Soil Conservation and Management*; Springer: Dordrecht, The Netherlands, 2010. [CrossRef]
21. Banzhaf, S.; Schaap, M.; Kerschbaumer, A.; Reimer, E.; Stern, R.; van der Swaluw, E.; Builtjes, P. Wet Deposition: Model Development and Evaluation. In Air Pollution Modeling and its Application. *Springer* **2011**, *21*, 459–465.
22. Navratil, T.; Kurz, D.; Kram, P.; Hofmeisteer, J.; Hruska, J. Acidification and recovery of soil at a heavily impacted forest catchment (Lysina, Czech Republic)—SAFE modeling and field results. *Ecol. Model.* **2007**, *205*, 464–474. [CrossRef]
23. Johnson, J.; Graf Pannatier, E.; Carnicellt, S.; Cecchini, G.; Clarke, N.; Cools, N.; Hansen, K.; Meesenburg, H.; Nieminen, T.M.; Pihl-Karlsson, G. The response of soil solution chemistry in European forests to decreasing acid deposition. *Glob. Chang. Biol.* **2018**, *24*, 3603–3619. [CrossRef]
24. Ulrich, B. Process hierarchy in forest ecosystems: An integrating ecosystem theory. In *Effects of Acid Rain on Forest Processes*; Godbold., D.L., Hüttermann., A., Eds.; Wiley-Liss: New York, NY, USA, 1994; pp. 353–397.
25. Ulrich, B.; Bigham, J. The role of soil processes in the process hierarchy of forest ecosystems. In *Soils and Environment—Soil Processes from Mineral to Landscape Scale*; Auerswald, K., Stanjek, H., Eds.; Advances in Geoecology, Catena Verlag: Reiskirchen, Germany, 1997; Volume 30, pp. 11–22.
26. Kottek, M.; Grieser, J.; Beck, C.; Rudolf, B.; Rubel, F. World Map of Köppen-Geiger Climate Classification updated. *Meteorol. Z.* **2006**, *15*, 259–263. Available online: http://koeppen-geiger.vu-wien.ac.at (accessed on 11 September 2021). [CrossRef]
27. Food and Agriculture Organization of the United Nations (FAO). *Global Forest Resources Assessment 2000. FAO Main Report. Forestry Paper 140*; Rome, Italy, 2001. Available online: http://www.fao.org/forestry/site/7949/en/ (accessed on 19 April 2021).
28. Food and Agriculture Organization of the United Nations (FAO). Soils Deliver Ecosystem Services that Enable Life on Earth. *International Year of Soils (Poster) 2015.* Available online: http://www.fao.org/soils-2015/en/ (accessed on 17 October 2021).
29. Vanermen, I.; Muys, B.; Verheyen, K.; Vanwindekens, F.; Bouriaud, L.; Kardol, P.; Vranken, L. What do scientists and managers know about soil biodiversity? Comparative knowledge mapping for sustainable forest management. *For. Policy Econ.* **2020**, *119*, 102264. [CrossRef]
30. Burst, M.; Chauchard, S.; Dambrine, E.; Dupouey, J.L.; Amiaud, B. Distribution of soil properties along forest-grassland interfaces: Influence of permanent environmental factors or land-use after-effects? *Agric. Ecosyst. Environ.* **2020**, *289*, 106739. [CrossRef]
31. Hernández, L.; Jandl, R.; Blujdea, V.N.B.; Lehtonen, A.; Kriiska, K.; Alberdi, I.; Adermann, A.; Cañellas, I.; Marin, G.; Moreno-Fernandez, D.; et al. Towards complete and harmonized assessment of soil carbon stocks and balance in forests: The ability of the Yasso07 model across a wide gradient of climatic and forest conditions in Europe. *Sci. Total Environ.* **2017**, *599*, 1171–1180. [CrossRef]
32. Charro, E.; Moyano, A.; Cabezón, R. The Potential of Juniperus thurifera to Sequester Carbon in Semi-Arid Forest Soil in Spain. *Forests* **2017**, *8*, 330. [CrossRef]
33. Caddeo, A.; Marras, S.; Sallustio, L.; Spano, D.; Sirca, C. Soil organic carbon in Italian forests and agroecosystems: Estimating current stock and future changes with a spatial modelling approach. *Agric. For. Meteorol.* **2019**, *278*, 107654. [CrossRef]
34. Sosulski, T.; Szara, E.; Szymańska, M.; Stępień, W.; Rutkowska, B.; Szulc, W. Soil N_2O emissions under conventional tillage conditions and from forest soil. *Soil Tillage Res.* **2019**, *190*, 86–91. [CrossRef]
35. Täumer, J.; Kolb, S.; Boeddinghaus, R.S.; Wang, H.; Schöning, I.; Schrumpf, M.; Ulrich, T.; Marhan, S. Divergent drivers of the microbial methane sink in temperate forest and grassland soils. *Glob. Change Biol.* **2021**, *27*, 929–940. [CrossRef]
36. Räty, M.; Järvenranta, K.; Saarijärvi, E.; Koskiaho, J.; Virkajärvi, P. Losses of phosphorus, nitrogen, dissolved organic carbon and soil from a small agricultural and forested catchment in east-central Finland. *Agric. Ecosyst. Environ.* **2020**, *302*, 107075. [CrossRef]
37. Zangerlé, A.; Hissler, C.; Lavelle, P. Effects of earthworms and plants on the soil structure, the physical stabilization of soil organic matter and the microbial abundance and diversity in soil aggregates in a long term study. In Proceedings of the EGU General Assembly, Vienna, Austria, 27 April–2 May 2014; Geophysical Research Abstracts. Volume 16, p. 16293.

38. Schack-Kirchner, H.; Hildebrand, E.E. Changes in soil structure and aeration due to liming and acid irrigation. *Plant Soil* **1998**, *199*, 167–176. [CrossRef]
39. Saputra, D.D.; Sari, R.R.; Hairiah, K.; Roshetko, J.M.; Suprayogo, D.; van Noordwijk, M. Can cocoa agroforestry restore degraded soil structure following conversion from forest to agricultural use? *Agrofor. Syst.* **2020**, *94*, 2261–2276. [CrossRef]
40. Damptey, F.G.; Birkhofer, K.; Nsiah, Ü.K.; de la Riva, E.G. Soil Properties and Biomass Attributes in a Former Gravel Mine Area after Two Decades of Forest Restoration. *Land* **2020**, *9*, 209–227. [CrossRef]
41. Zhang, J.; Bruijnzeel, L.A.; Quiñones, C.M.; Tripoli, R.; Asio, V.B.; van Meerveld, H.J. Soil physical characteristics of a degraded tropical grassland and a 'reforest': Implications for runoff generation. *Geoderma* **2019**, *333*, 163–177. [CrossRef]
42. Schack-Kirchner, H.; von Wilpert, K.; Hildebrand, E.E. The spatial distribution of soil hyphae in structured spruce-forest soils. *Plant Soil* **2000**, *224*, 195–205. [CrossRef]
43. Witzgall, K.; Vidal, A.; Schubert, D.I.; Höschen, C.; Schweizer, S.A.; Buegger, F.; Pouteau, V.; Chenu, C.; Mueller, C.W. Particulate organic matter as a functional soil component for persistent soil organic carbon. *Nat. Commun.* **2021**, *12*, 4115. [CrossRef]
44. von Wilpert, K. Chemical deposition and seepage water quality in forests. In *Forest Hydrology—Results of Research in Germany and Russia, Part 1*; Puhlmann, H., Schwarze, R., Eds.; IHP/HWRP: Koblenz, Germany, 2007; Volume 6, pp. 23–35.
45. Kohler, M.; Hildebrand, E.E. New Aspects of Element Cycling and Forest Nutrition. In *Towards the Sustainable Use of Europe's Forests—Forest Ecosystem and Landscape Research: Scientific Challenges and Opportunities*; Andersson, F., Birot, Y., Päivinen, R., Eds.; European Forest Institute: Joensuu, Finland, 2004; EFI Proceedings; Volume 49, pp. 171–180.
46. Lang, F.; Krüger, J.; Amelung, W.; Willbold, S.; Frossard, E.; Bünemann, E.K.; Bauhus, J.; Nitschke, R.; Kandeler, E.; Marhan, S.; et al. Soil phosphorus supply controls P nutrition strategies of beech forest ecosystems in Central Europe. *Biogeochemistry* **2017**, *136*, 5–29. [CrossRef]
47. Rusek, J.; Marshall, V.G. Impacts of airborne pollutants on soil fauna. *Annu. Rev. Ecol. Syst.* **2000**, *31*, 395–423. [CrossRef]
48. Ojha, R.B.; Devkota, D. Earthworms: Soil and Ecosystem Engineers—A Review. *World J. Agric. Res.* **2014**, *2*, 257–260. [CrossRef]
49. Le Bayon, R.C.; Bullinger-Weber, G.; Schomburg, A.C.; Turberg, P.; Schlaepfer, R.; Guenat, C. Earthworms as ecosystem engineers: A review. In *Earthworms. Types, Roles and Research*; Clayton, H.G., Ed.; Nova Science Publishers: New York, NY, USA, 2017; pp. 129–177. ISBN 978-1-53612-176-6.
50. HILDEBRAND, E.E. The Heterogeneous Distribution of Mobile Ions in the Rhizosphere of Acid Forest Soils: Facts, Causes, and Consequences. *J. Environ. Sci. Health* **1994**, *A29*, 1973–1992. [CrossRef]
51. Walthert, L.; Graf Pannatier, E.; Meier, E.S. Shortage of nutrients and excess of toxic elements in soils limit the distribution of soil-sensitive tree species in temperate forests. *For. Ecol. Manag.* **2013**, *297*, 94–107. [CrossRef]
52. Hildebrand, E.E. The spatial heterogeneity of chemical properties in acid forest soils and its importance for tree nutrition. *Water Air Soil Pollut.* **1990**, *54*, 183–191. [CrossRef]
53. van Schöll, L.; Kuyper, T.W.; Smits, M.M.; Landeweert, R.; Hoffland, E.; van Breemen, N. Rock-eating mycorrhizas: Their role in plant nutrition and biogeochemical cycles. *Plant Soil* **2008**, *303*, 35–47. [CrossRef]
54. Heisner, U.; Raber, B.; Hildebrand, E.E. The Importance of the Skeleton in Forest Sites of the Southern Black Forest. *Eur. J.For. Res.* **2004**, *123*, 249–257. [CrossRef]
55. Koele, N.; Hildebrand, E.E.; Schack-Kirchner, H. Effects of weathering state of coarse soil fragments on tree-seedling nutrient uptake. *J. Plant Nutr. Soil Sci.* **2010**, *173*, 245–251. [CrossRef]
56. Koele, N.; Storch, F.; Hildebrand, E.E. The coarse-soil fraction is the main living space of fungal hyphae in the BhBs horizon of a Podzol. *J. Plant Nutr. Soil Sci.* **2011**, *174*, 750–753. [CrossRef]
57. Koele, N.; Hildebrand, E.E. The ecological significance of the coarse soil fraction for *Picea abies* (L.) Karst. seedling nutrition. *Plant Soil* **2008**, *312*, 163–174. [CrossRef]
58. Kohler, M.; von Wilpert, K.; Hildebrand, E.E. The Soil Skeleton as a Source for the short-term Supply of "Basic Cations" in Forest Soils of the Black Forest (Germany). *Water Air Soil Poll.* **2000**, *122*, 37–48. [CrossRef]
59. Soong, J.L.; Janssens, I.A.; Grau, O.; Margalef, O.; Stahl, C.; Van Langenhove, L.; Urbina, I.; Chave, J.; Dourdain, A.; Ferry, B.; et al. Soil properties explain tree growth and mortality, but not biomass, across phosphorus-depleted tropical forests. *Sci. Rep.* **2020**, *10*, 2302. [CrossRef]
60. Liu, J.; Chen, J.; Chen, G.; Guo, J.; Li, Y. Enzyme stoichiometry indicates the variation of microbial nutrient requirements at different soil depths in subtropical forests. *PLoS ONE* **2020**, *15*, e0220599. [CrossRef]
61. Stahr, S.; Graf-Rosenfellner, M.; Klysubun, W.; Mikutta, R.; Prietzel, J.; Lang, F. Phosphorus speciation and C:N:P stoichiometry of functional organic matter fractions in temperate forest soils. *Plant Soil* **2018**, *427*, 53–69. [CrossRef]
62. Rodionov, A.; Bauke, S.L.; von Sperber, C.; Hoeschen, C.; Kandeler, E.; Kruse, J.; Lewandowski, H.; Marhan, S.; Mueller, C.W.; Simon, M.; et al. Biogeochemical cycling of phosphorus in subsoils of temperate forest ecosystems. *Biogeochemistry* **2020**, *150*, 313–328. [CrossRef]
63. Baumann, K.; Glaser, K.; Mutz, J.E.; Karsten, U.; MacLennan, A.; Hu, Y.; Michalik, D.; Kruse, J.; Eckhardt, K.U.; Schall, P. Biological soil crusts of temperate forests: Their role in P cycling. *Soil Biol. Biochem.* **2017**, *109*, 156–166. [CrossRef]
64. Zederer, D.P.; Talkner, U. Organic P in temperate forest mineral soils as affected by humus form and mineralogical characteristics and its relationship to the foliar P content of European beech. *Geoderma* **2018**, *325*, 162–171. [CrossRef]

65. Yang, F.; Magh, R.K.; Ivanković, M.; Lanšćak, M.; Haberstroh, S.; Du, B.; Dannenmann, M.; Rennenberg, H.; Herschbach, C. Foliar P nutrition of European beech (*Fagus sylvatica* L.) depends on the season but remains unaffected by co-cultivation with silver fir (Abies alba Mill.). *Eur. J. For. Res.* **2020**, *139*, 853–868. [CrossRef]
66. Bueis, T.; Bravo, F.; Pando, V.; Kissi, Y.A.; Turrión, M.B. Phosphorus availability in relation to soil properties and forest productivity in Pinus sylvestris L. plantations. *Ann. For. Sci.* **2019**, *79*, 69102. [CrossRef]
67. Salmon, S. Changes in humus forms, soil invertebrate communities and soil functioning with forest dynamics. *Appl. Soil Ecol.* **2018**, *123*, 345–354. [CrossRef]
68. Landi, S.; d'Errico, G.; Binazzi, F.; Di Salvatore, U.; Gardin, L.; Marchi, M.; Mazza, G.; Roversi, P.F.; Simoncini, S.; Torrini, G.; et al. The Short-Term Impact of Different Silvicultural Thinnings on Soil Nematode and Microarthropod Biodiversity in Artificial Black Pine Stands. *Forests* **2020**, *11*, 1212. [CrossRef]
69. Rożek, K.; Rola, K.; Błaszkowski, J.; Leski, T.; Zubek, S. How do monocultures of fourteen forest tree species affect arbuscular mycorrhizal fungi abundance and species richness and composition in soil? *For. Ecol. Manag.* **2020**, *465*, 118091. [CrossRef]
70. Galluzzi, M.; Giannetti, F.; Puletti, N.; Canullo, R.; Rocchini, D.; Bastrup-Birk, A.; Chirici, G. A plot-level exploratory analysis of European forest based on the results from the BioSoil Forest Biodiversity project. *Eur. J. For. Res.* **2019**, *138*, 831–845. [CrossRef]
71. Giguère-Tremblay, R.; Laperriere, G.; de Grandpré, A.; Morneault, A.; Bisson, D.; Chagnon, P.L.; Germain, H.; Maire, V. Boreal Forest Multifunctionality Is Promoted by Low Soil Organic Matter Content and High Regional Bacterial Biodiversity in Northeastern Canada. *Forests* **2020**, *11*, 149–167. [CrossRef]
72. Friggens, N.L.; Aspray, T.J.; Parker, T.C.; Subke, J.A.; Wookey, P.A. Spatial patterns in soil organic matter dynamics are shaped by mycorrhizosphere interactions in a treeline forest. *Plant Soil* **2020**, *447*, 521–535. [CrossRef]
73. Nagati, M.; Roy, M.; Manzi, S.; Richard, F.; Desrochers, A.; Gardes, A.; Bergeron, Y. Impact of local forest composition on soil fungal communities in a mixed boreal forest. *Plant Soil* **2018**, *432*, 345–357. [CrossRef]
74. Suh, J.; Kim, S.M.; Yi, H.; Choi, Y. An Overview of GIS-Based Modeling and Assessment of Mining-Induced Hazards: Soil, Water, and Forest. *Int. J. Environ. Res. Public Health* **2017**, *14*, 1463. [CrossRef]
75. Missong, A.; Holzmann, S.; Bol, R.; Nischwitz, V.; Puhlmann, H.; von Wilpert, K.; Siemens, J.; Klumpp, E. Leaching of natural colloids from forest topsoils and their relevance for phosphorus mobility. *Sci. Total Environ.* **2018**, *634*, 305–315. [CrossRef]
76. Ambus, P.; Zechmeister-Boltenstern, S. Denitrification and N-Cycling in Forest Ecosystems. In *Biology of the Nitrogen Cycle*; Bothe, H., Ferguson, S.J., Newton, W.E., Eds.; Elsevier: Amsterdam, The Netherlands, 2007; pp. 343–358.
77. Sucker, C.; von Wilpert, K.; Puhlmann, H. Acidification reversal in low mountain range streams of Germany. *Environ. Monit. Assess.* **2011**, *174*, 65–89. [CrossRef] [PubMed]
78. Ilstedt, U.; Malmer, A.; Verbeeten, E.; Murdiyarso, D. The effect of afforestation on water infiltration in the tropics: A systematic review and meta-analysis. *For. Ecol. Manag.* **2007**, *251*, 45–51. [CrossRef]
79. Mireille, N.M.; Mwangi, H.M.; Mwangi, J.K.; Gathenya, J.M. Analysis of Land Use Change and Its Impact on the Hydrology of Kakia and Esamburmbur Sub-Watersheds of Narok County, Kenya. *Hydrology* **2019**, *6*, 86. [CrossRef]
80. Zhang, X.; Yu, G.Q.; Li, Z.B.; Li, P. Experimental Study on Slope Runoff, Erosion and Sediment under Different Vegetation Types. *Water Resour. Manag.* **2014**, *28*, 2415–2433. [CrossRef]
81. Pellis, G.; Chiti, T.; Rey, A.; Yuste, J.C.; Trotta, C.; Papale, D. The ecosystem carbon sink implications of mountain forest expansion into abandoned grazing land: The role of subsoil and climatic factors. *Sci. Total Environ.* **2019**, *672*, 106–120. [CrossRef]
82. Kalks, F.; Liebmann, P.; Wordell-Dietrich, P.; Guggenberger, G.; Kalbitz, K.; Mikutta, R.; Helfrich, M.; Don, A. Fate and stability of dissolved organic carbon in topsoils and subsoils under beech forests. *Biogeochemistry* **2020**, *148*, 111–128. [CrossRef]
83. Wordell-Dietrich, P.; Wotte, A.; Rethemeyer, J.; Bachmann, J.; Helfrich, M.; Kirfel, K.; Leuschner, C.; Don, A. Vertical partitioning of CO_2 production in a forest soil. *Biogeosciences* **2020**, *17*, 6341–6356. [CrossRef]
84. Zacháry, D.; Filep, T.; Jakab, G.; Molnár, M.; Kertész, T.; Király, C.; Hegyi, I.; Gáspár, L.; Szalai, Z. Carbon Isotope Measurements to Determine the Turnover of Soil Organic Matter Fractions in a Temperate Forest Soil. *Agronomy* **2020**, *10*, 1944. [CrossRef]
85. Nitsch, P.; Kaupenjohann, M.; Wulf, M. Forest continuity, soil depth and tree species are important parameters for SOC stocks in an old forest (Templiner Buchheide, northeast Germany). *Geoderma* **2018**, *310*, 65–76. [CrossRef]
86. Rahman, M.M.; Bárcena, T.G.; Vesterdal, L. Tree species and time since afforestation drive soil C and N mineralization on former cropland. *Geoderma* **2017**, *305*, 153–161. [CrossRef]
87. IPCC. *Climate Change 2014: Synthesis Report. Contribution of Working Groups I, II and III to the Fifth Assessment Report of the Intergovernmental Panel on Climate Change*; Core Writing Team, Pachauri, R.K., Meyer, L.A., Eds.; IPCC: Geneva, Switzerland, 2014; p. 151.
88. Morishita, T.; Sakata, T.; Takahashi, M.; Ishizuka, S.; Mizoguchi, T.; Inagaki, Y.; Terazawa, K.; Sawata, S.; Igarashi, M.; Yasuda, H.; et al. Methane uptake and nitrous oxide emission in Japanese forest soils and their relationship to soil and vegetation types. *Soil Sci. Plant Nutr.* **2007**, *53*, 678–691. [CrossRef]
89. Jungkunst, H.F.; Flessa, H.; Scherber, C.; Fiedler, S. Groundwater level controls CO_2, N_2O and CH_4 fluxes of three different hydromorphic soil types of a temperate forest ecosystem. *Soil Biol. Biochem.* **2008**, *40*, 2047–2054. [CrossRef]
90. Schindler, T.; Mander, Ü.; Machacova, K.; Espenberg, M.; Krasnov, D.; Escuer-Gatius, J.; Veber, G.; Pärn, J.; Soosaar, K. Short-term flooding increases CH_4 and N_2O emissions from trees in a riparian forest soil-stem continuum. *Sci. Rep.* **2020**, *10*, 3204. [CrossRef] [PubMed]

91. Arango, C.; Ponette-González, A.; Neziri, I.; Bailey, J. Western spruce budworm effects on throughfall N, P, and C fluxes and soil nutrient status in the Pacific Northwest. *Can. J. For. Res.* **2019**, *49*, 1207–1218. [CrossRef]
92. Wasak, K.; Klimek, B.; Drewnik, M. Rapid effects of windfall on soil microbial activity and substrate utilization patterns in the forest belt in the Tatra Mountains. *J. Soils Sediments* **2020**, *20*, 801–815. [CrossRef]
93. Matzner, E.; Murach, D. Soil changes induced by air pollutant deposition and their implication for forests in central Europe. *Water Air Soil Pollut.* **1995**, *85*, 63–76. [CrossRef]
94. von Wilpert, K.; Schäffer, J. Ecological effects of soil compaction and initial recovery dynamics: A preliminary study. *Eur. J. For. Res.* **2006**, *125*, 129–138. [CrossRef]
95. Teepe, R.; Brumme, R.; Beese, F.; Ludwig, B. Nitrous Oxide Emission and Methane Consumption Following Compaction of Forest Soils. *Soil Sci. Soc. Am. J.* **2004**, *68*, 605–611. [CrossRef]
96. Warlo, H.; von Wilpert, K.; Lang, F.; Schack-Kirchner, H. Black Alder (*Alnus glutinosa* (L.) Gaertn.) on Compacted Skid Trails: A Trade-off between Greenhouse Gas Fluxes and Soil Structure Recovery? *Forests* **2019**, *10*, 726. [CrossRef]
97. Büntgen, U.; Urban, O.; Krusic, P.J.; Rybníček, M.; Kolář, T.; Kyncl, T.; Ač, A.; Koňasová, E.; Čáslavský, J.; Esper, J.; et al. Recent European drought extremes beyond Common Era background variability. *Nat. Geosci.* **2021**, *14*, 190–196. [CrossRef]
98. Fleck, S.; Ahrends, B.; Sutmöller, J.; Albert, M.; Evers, J.; Meesenburg, H. Is Biomass Accumulation in Forests an Option to Prevent Climate Change Induced Increases in Nitrate Concentrations in the North German Lowland? *Forests* **2017**, *8*, 219. [CrossRef]
99. Hennings, N.; Becker, J.N.; Guillaume, T.; Damris, M.; Dippold, M.A.; Kuzyakov, Y. Riparian wetland properties counter the effect of land-use change on soil carbon stocks after rainforest conversion to plantations. *Catena* **2021**, *196*, 104941. [CrossRef]
100. Puhlmann, H.; Schmidt-Walter, P.; Hartmann, P.; Meesenburg, H.; von Wilpert, K. Soil Water Budget and Drought Stress. In *Status and Dynamics of Forests in Germany. Results of the National Forest Monitoring*; Wellbrock, N., Bolte, A., Eds.; Springer Open, Ecological Studies: New York, NY, USA, 2019; Volume 237, Chapter 3; pp. 55–91.
101. Wermelinger, B.; Rigling, A.; Schneider Mathis, D.; Kenis, M.; Gossner, M.M. Climate Change Effects on Trophic Interactions of Bark Beetles in Inner Alpine Scots Pine Forests. *Forests* **2021**, *12*, 136. [CrossRef]
102. Rewald, B. Impact of Climate Change Induced Drought on Tree Root Hydraulic Properties and Competition Belowground. Ph.D. Thesis, Georg-August-Universität, Göttingen, Germany, 2008; p. 172.
103. Brunner, I.; Herzog, C.; Dawes, M.A.; Arend, M.; Sperisen, C. How tree roots respond to drought. *Front. Plant Sci.* **2015**, *6*, 547. [CrossRef] [PubMed]
104. Bentz, B.J.; Jönsson, A.M. Modeling Bark Beetle Responses to Climate Change. In *Bark Beetles*; Vega, F.E., Hofstetter, R.W., Eds.; Elsevier: Amsterdam, The Netherlands, 2015. [CrossRef]
105. Marini, L.; Økland, B.; Jönsson, A.M.; Bentz, B.; Carroll, A.; Forster, B.; Grégoire, J.C.; Hurling, R.; Nageleisen, L.M.; Netherer, S.; et al. Climate drivers of bark beetle outbreak dynamics in Norway spruce forests. *Ecography* **2017**, *40*, 1–10. [CrossRef]
106. Braun, S.; Tresch, S.; Augustin, S. Soil solution in Swiss forest stands: A 20 year's time series. *PLoS ONE* **2020**, *15*, e0227530. [CrossRef]
107. Daněk, P.; Šamonil, P.; Vrška, T. Four decades of the coexistence of beech and spruce in a Central European old-growth forest. Which succeeds on what soils and why? *Plant Soil* **2019**, *437*, 257–272. [CrossRef]
108. Brunel, C.; Gros, R.; Ziarelli, F.; Da Silva, A.M.F. Additive or non-additive effect of mixing oak in pine stands on soil properties depends on the tree species in Mediterranean forests. *Sci. Total Environ.* **2017**, *590*, 676–685. [CrossRef]
109. Zirlewagen, D.; von Wilpert, K. Using model scenarios to predict and evaluate forest management impacts on soil base saturation at landscape level. *Eur. J. For. Res.* **2004**, *123*, 269–282. [CrossRef]
110. Zeller, B.; Legout, A.; Bienaimé, S.; Gratia, B.; Santenoise, P.; Bonnaud, P.; Ranger, J. Douglas fir stimulates nitrification in French forest soils. *Sci. Rep.* **2019**, *9*, 10687. [CrossRef]
111. Mayer, M.; Matthews, B.; Rosinger, C.; Sanden, H.; Godbold, D.L.; Katzensteiner, K. Tree regeneration retards decomposition in a temperate mountain soil after forest gap disturbance. *Soil Biol. Biochem.* **2017**, *115*, 490–498. [CrossRef]
112. Łabęda, D.; Kondras, M. Influence of forest management on soil organic carbon stocks. *Soil Sci. Annu.* **2020**, *71*, 165–173. [CrossRef]
113. Wambsganss, J.; Stutz, K.P.; Lang, F. European beech deadwood can increase soil organic carbon sequestration in forest topsoils. *For. Ecol. Manag.* **2017**, *405*, 200–209. [CrossRef]
114. Papaioannou, E.; Chatzistathis, T.; Menexes, G. The Impact of Management Practices on Soil Fertility and Foliar Nutrient Concentrations in a Spruce (*Picea abies*) Forest Ecosystem of Rodopi Mountainous Area, in Northern Greece. *Not. Bot. Horti Agrobot.* **2018**, *46*, 301–308. [CrossRef]
115. Bolte, A.; Block, J.; Eichhorn, J.; Sanders, T.G.M.; Wellbrock, N. Sustainable Use and Development of Forests and Forest Soils: A Resume. In *Status and Dynamics of Forests in Germany Results of the National Forest Monitoring*; Wellbrock, N., Bolte, A., Eds.; Springer, Ecological Studies: New York, NY, USA, 2019; Volume 237, Chapter 12; pp. 355–374.
116. Jansone, L.; von Wilpert, K.; Hartmann, P. Natural Recovery and Liming Effects in Acidified Forest Soils in SW-Germany. *Soil Syst.* **2020**, *4*, 38. [CrossRef]
117. Berger, T.W.; Türtscher, S.; Berger, P.; Lindebner, L. A slight recovery of soils from Acid Rain over the last three decades is not reflected in the macro nutrition of beech (Fagus sylvatica) at 97 forest stands of the Vienna Woods. *Environ. Pollut.* **2016**, *216*, 624–635. [CrossRef] [PubMed]

118. Kohler, M.; Kunz, J.; Herrmann, J.; Hartmann, P.; Jansone, L.; Puhlmann, H.; von Wilpert, K.; Bauhus, J. The Potential of Liming to Improve Drought Tolerance of Norway Spruce [*Picea abies* (L.) Karst.]. *Front. Plant Sci.* **2019**, *10*, 382. [CrossRef] [PubMed]
119. Lepilin, D.; Laurén, A.; Uusitalo, J.; Tuittila, E.S. Soil deformation and its recovery in logging trails of drained boreal peatlands. *Can. J. For. Res.* **2019**, *49*, 743–751. [CrossRef]
120. Marra, E.; Cambi, E.; Fernandez-Lacruz, R.; Giannetti, F.; Marchi, E.; Nordfjell, T. Photogrammetric estimation of wheel rut dimensions and soil compaction after increasing numbers of forwarder passes. *Scand. J. For. Res.* **2018**, *33*, 613–620. [CrossRef]
121. Green, P.Q.; Chung, W.; Leshchinsky, B.; Belart, F.; Sessions, J.; Fitzgerald, S.A.; Wimer, J.A.; Cushing, T.; Garland, J.J. Insight into the Productivity, Cost and Soil Impacts of Cable-assisted Harvester-forwarder Thinning in Western Oregon. *For. Sci.* **2019**, *66*, 82–96. [CrossRef]
122. Bonnaud, P.; Santenoise, P.; Tisserand, D.; Nourrisson, G.; Ranger, J. Impact of compaction on two sensitive forest soils in Lorraine (France) assessed by the changes occurring in the perched water table. *For. Ecol. Manag.* **2019**, *437*, 380–395. [CrossRef]
123. Flores Fernández, J.L.; Rubin, L.; Hartmann, P.; Puhlmann, H.; von Wilpert, K. Initial recovery of soil structure of a compacted forest soil can be enhanced by technical treatments and planting. *For. Ecol. Manag.* **2018**, *431*, 54–62. [CrossRef]
124. Flores Fernández, J.L.; Hartmann, P.; Schäffer, J.; Puhlmann, H.; von Wilpert, K. Initial recovery of compacted soil—planting and technical treatments decrease CO_2 concentrations in soil and promote root growth. *Ann. For. Sci.* **2017**, *74*, 73–85. [CrossRef]
125. Reynolds, C.A.; Jackson, T.J.; Rawls, W.J. Estimating soil water-holding capacities by linking the Food and Agriculture Organization soil map of the world with global pedon databases and continuous pedotransfer functions. *Water Resour. Res.* **2000**, *36*, 3653–3662. [CrossRef]
126. Puhlmann, H.; von Wilpert, K. Pedotransfer functions for water retention and unsaturated hydraulic conductivity of forest soils. *J. Plant Nutr. Soil Sci.* **2012**, *175*, 221–235. [CrossRef]
127. IUSS Working Group WRB. *World Reference Base for Soil Resources 2014, Update 2015, International Soil Classification System for Naming Soils and Creating Legends for Soil Maps*; Rome World Soil Resour. Rep. No 106; Food and Agriculture Organization of the United Nations (FAO): Rome, Italy, 2015; 203p.
128. Rabot, E.; Wiesmeier, M.; Schlüter, M.; Vogel, H.J. Soil structure as an indicator of soil functions: A review. *Geoderma* **2018**, *314*, 122–137. [CrossRef]

soil systems

Article

Recovery of Soil Structure and Fine Root Distribution in Compacted Forest Soils

Jürgen Schäffer

Department of Soil Science and Site Ecology, University of Applied Forest Sciences, Schadenweiler Hof, D-72108 Rottenburg, Germany; schaeffer@hs-rottenburg.de

Abstract: Soil compaction is a major concern in the context of ensuring sustainable forest and agricultural management practices. Productivity gains during the last decades were also achieved by increasing mechanization. This change was associated with growing machine weights and impacts on trafficked soils. Up to now, knowledge about the resilience of compaction phenomena is still poor. At 11 forest sites on compaction-sensitive silt and loam substrates in the federal state of Baden-Wuerttemberg (Germany), fine root distribution, macroporosity, and apparent gas diffusion coefficients were used to characterize the status of recovery from former soil compaction by machine impact. The time span of recovery at the investigated sites ranged from 6 to 37 years. The investigated soil physical parameters indicate the beginning of soil structure recovery in the wheel tracks, comprising the first 10 cm of soil depth at most sites that were trafficked 10 years or more before the investigation. Synchronously with this restructuring, fine root propagation has started to recover in the topsoils. However, a high persistence of damage was observed below that depth. A synoptic interpretation of the data led to the conclusion that time spans up to almost four decades are not sufficient for the restoration of soil functionality in formerly compacted soils characterized by silt loam texture and low activity of soil biota. In view of the long-lasting persistence of the negative compaction effects, soil protection strategies combined with monitoring of their strict compliance must be implemented into forest practice. In case of an insufficient natural recovery potential, active measures to accelerate the regaining of soil functionality in compacted soils at irregular wheel tracks should be considered in order to shorten the time spans of disturbed soil conditions.

Keywords: soil compaction; soil aeration; soil structure recovery; soil protection strategy

Citation: Schäffer, J. Recovery of Soil Structure and Fine Root Distribution in Compacted Forest Soils. *Soil Syst.* **2022**, *6*, 49. https://doi.org/10.3390/soilsystems6020049

Academic Editor: Klaus von Wilpert

Received: 20 February 2022
Accepted: 20 May 2022
Published: 24 May 2022

1. Introduction

Beginning in the 1960s in Northern America and Scandinavia, mechanization in forest harvesting and logging operations started to replace manual systems [1,2]. This transition was almost complete within the next three decades. Following this trend, specialized forest machines also became more and more common in Central Europe [3].

More than 40 years of trafficking history led to the alarming accumulation of soil structure damage in forest stands [4], which endangers the fulfillment of the multiple functions and services of forest soils and forest ecosystems. The increase in soil disturbances can be attributed to the long-lasting persistence of soil deformation effects and to the poor compliance with soil-protection regulations, which aim to concentrate damage at skid trails and prohibit trafficking away from these lanes [5].

Up to now, only a few investigations have given valuable information about the persistence of soil structure damage and the impedance of related soil functions. This information is essential to judge the necessity of active countermeasures in order to support the restructuring of compacted soils. In most of these studies, soil physical parameters such as bulk density and penetration or shear resistance were used to judge the status of soil structure recovery [6–10]. Hildebrand [11] showed that comparisons of compacted and uncompacted soils based solely on bulk density might lead to misinterpretation. The same applies to parameters linked directly to bulk density, such as penetration or shear resistance. Hildebrand and Schack-Kirchner [12] postulated that the gas budget in the rhizosphere of forest soils is a sensitive indicator and closely linked to soil structure and root growth. Ampoorter et al. [13] confirmed this finding. Von Wilpert and Schäffer [14] proved that fine root propagation is an ecologically sensitive indicator for the evaluation of enduring soil compaction effects.

Investigations using sensitive parameters such as soil gas concentrations, gas exchange properties or fine root propagation for the detection of recovery stages are rather rare or do not cover long time periods of natural regeneration. Goutal et al. [15] reported a decreased air-filled porosity and elevated CO_2 concentrations in compacted soils 3 to 4 years after a logging operation in France. Schack-Kirchner [16] found a highly significant reduction in the gas diffusion coefficients in the upper soil layer at a loamy investigation site 12 years after machine passage in Baden-Wuerttemberg. Summarizing the results from six study sites in Southern Germany spanning from sandy loams to silty clays, Kremer [17] concluded that within the considered time span of 35 years, no significant recovery of the sensitive parameters took place. Von Wilpert and Schäffer [14] found a rather transient stage and not an overall recovery of soil aeration properties and root propagation 24 years after machine impact at a luvisol in Baden-Wuerttemberg. Ebeling et al. [18] showed a significantly reduced gas diffusivity 40 years after the last machine impact for a loamy sandy podzol in Lower Saxony. For cambisols on limestone with a high activity of soil biota and high clay content, the authors could find evidence of a structure recovery within a time span of 10–20 years.

In this paper, investigations of macroporosity, gas exchange properties, and fine root propagation are presented from 11 study sites in Baden-Wuerttemberg, characterized by time spans of soil structure recovery from 6 to 37 years. The evaluation of the results is performed in the sense of pattern recognition rather than a classical "space for time" substitution with statistical testability. The expectation was that the longer the time span for the restructuring of the formerly compacted soils lasted, the more closely the soil structure and fine rooting should approximate towards natural site-specific levels.

2. Materials and Methods

2.1. Investigation Sites

In Central Europe, there is a lack of long-term wheeling experiments conducted under controlled experimental conditions (accurately reported machine configurations, soil water content, and soil physical preconditions) that cover long time ranges between impact and investigation of recovery. For this reason, the material of this study had to be extracted from well-documented practical harvesting campaigns, where wheeling activity at the formerly passed skid trails could be excluded unequivocally. With an inquiry at the competent forest authorities, the technical and operational aspects of the harvesting and logging operations were clarified for the presented sites.

Figure 1 and Table 1 give an overview of the locations of the investigation sites and their characteristics. At Biberach (BIB), Ettenheim (ETT 1 and ETT 2), Mengen (MEN), Rottweil (ROT), Stockach (STO), Weil im Schoenbuch (WIS 1 and WIS 2), and Wolfegg (WOL), luvisols developed on loess loam covered Triassic or Pleistocene deposits. The transition zones between the Ah/E and the clay enriched Bt or Bv/Cv horizons varied in depth between 40 and 50 cm. The sites at BIB, WIS 2, and MEN showed stagnic properties beginning at depths of more than 40 cm. The parent material of the stagnosol at Ravensburg (RAV) was a freshwater molasse sediment. There, rooting below 30 cm was limited due to pseudogleyization. The soil type at Todtmoos (TOD) was a cambisol developed on a para-gneiss periglacial coverbed.

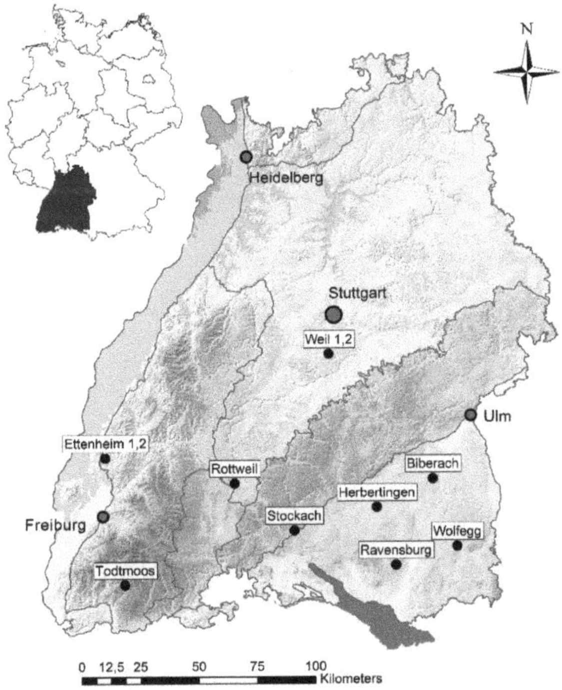

Figure 1. Location of the 11 investigation sites in Baden-Wuerttemberg (Germany).

Table 1. Site conditions, logging operations, and time span of soil structure recovery.

Site	Soil Type Texture Stone Content	Forest Operation (Species Composition; Age)	Harvesting and Skidding Equipment(Total Mass in Mg)	Time Since Vehicle Movement
Weil im Schoenbuch WIS 1 (6)	Luvisol SiL F	Thinning (spruce [a] with beech [b] and larch [c]; 35–40)	*Timberjack 1270 & Ponse HS 10* (~15 Mg, 600 mm tyres), *Valmet Forwarder* (>15 Mg incl. load, 500 mm tyres)	6
RavensburgRAV (9)	Stagnosol SiL (SiCL) C	Thinning (spruce [a]; 120)	*Manually* *MB-Track, Timberjack* (>15 Mg)	9
Weil im Schoenbuch WIS 2 (11)	Luvisol SiL F	Thinning (ash [d] with sycamore [e] and beech [b]; 125)	*Manually* *Small sized Fendt truck* (~ 8 Mg incl. load)	11
Wolfegg WOL (12)	Luvisol SL C	Thinning (spruce [a]; 50–55)	*FMG 746/250 ÖSA Super Eva* (14 Mg, 600 mm tyres) *FMG 678 Mini-Bruunett* (~16 Mg incl. load, 600 mm tyres)	12
Todtmoos TOD (14)	Cambisol SL M	Thinning (beech [b], spruce [a], and fir [f]; 5–70)	*FMG 746/250 ÖSA Super Eva* (14 Mg, 600 mm tyres) *FMG 678 Mini-Bruunett* (~16 Mg incl. load, 500 mm tyres)	14
Stockach STO (16)	Luvisol SL F	Thinning (spruce [a]; 56)	*ÖSA 250 Eva* (12.6 Mg, 600 mm tyres) *FMG 678 Mini-Bruunett* (~16 Mg incl. load, 500 mm tyres)	16

Table 1. *Cont.*

Site	Soil Type Texture Stone Content	Forest Operation (Species Composition; Age)	Harvesting and Skidding Equipment(Total Mass in Mg)	Time Since Vehicle Movement
Ettenheim ETT 1 (18)	Luvisol SL F	Stripwise clearcut (beech [b] with oak [g], spruce, and pine [h] ;18)	*Manually* *Welte Ökonom/Unimog U90* (~10 Mg, 300 to 500 mm tyres)	18
Biberach BIB (21)	Stagnic Luvisol SiL (L) C	Stripwise clearcut (spruce [a], pole-sized stand; 21)	*Manually* *Unimog U90* (~10 Mg)	21
Ettenheim ETT (24)	Luvisol SL F	Stripwise clearcut (beech [b] with oak [g]) spruce, and pine [h]; 18)	*Manually* *Welte Ökonom/Unimog U90* (~10 Mg)	24
Mengen MEN (35)	Stagn. Luvisol SiL (SiCL) C	Clear cut (spruce [a] pole-sized stand; 35)	*Manually* *Welte "Forstmann" and small agricultural trucks* (~8 Mg)	35
Rottweil ROT (37)	Luvisol L (CL) C (M)	Clear cut (silver fir [f], spruce [a], and beech [b]; 37)	*Manually* *"Eicher Königstiger"* (~5 Mg)	37

Texture class
L: Loam
SiL: Silt loam
SL: Sandy loam
SiCL: Silty clay loam

Coarse fraction content (Ø > 2 mm):
F: Very few (2–5 vol%)
C: Common (5–15 vol%)
M: Many (15–40 vol%)

Texture and coarse fraction classified according to FAO guideline for soil description [19]

Tree species:
[a] Norway spruce (*Picea abies* (L.) Karst.)
[b] European beech (*Fagus sylvatica* L.)
[c] European larch (*Larix decidua* Mill.)
[d] European ash (*Fraxinus excelsior* L.)
[e] Sycamore (*Acer pseudoplatanus* L.)
[f] Silver fir (*Abies álba* Mill.)
[g] Sessile oak (*Quercus petraea* Liebl.)
[h] Scots pine (*Pinus sylvestris* L.)

The two skid trail situations at MEN and ROT resulted from forest operations after storm events in 1967. We have to assume that at these two sites the machine impact spread over the whole forest stand. At the other sites, the investigated skid trails were part of regular skid trail systems that were set up for the first mechanized thinning campaigns (WIS 1, STO, and TOD) or had been in use during several harvesting campaigns (RAV, WIS 2, WOL, BIB, ETT 1 and ETT 2). The weights of the machines exceeded 10 Mg with the exception of WIS 2, MEN (~8 Mg at both sites), and ROT (~5 Mg). The time spans elapsed since vehicle movement ranged from 6 to 37 years.

2.2. Soil Physical Analysis and Root Counting

Persisting disturbances of soil structure and fine root propagation were assessed by examination of macroporosity, soil gas diffusivity, and fine root density densities.

At each investigation site, two profile pits were dug across the skid trails (up to 4.8 m wide and 0.6–1.0 m deep) and one in the un-trafficked forest close to the skid trails (varying in width from 0.4–2.0 m and depth from 0.6–1.0 m). The widths of the skid trail pits were defined according to the locations of wheel ruts, which were visible at the soil surface as depressions of several centimeters. Because side-effects close to the wheel track locations were expected, the width of the pits was extended into the right and left transition zones to the undisturbed forest stand. As strata for the analysis of the soil physical parameters and rooting densities, the skid trail trench walls were categorized into "wheel tracks", "median strips" (in between the two-wheel track ruts), and "margin zones" (outside the wheel track, transition to the adjacent untrafficked forest stand).

At each of these skid trail strata and at a control plot five replicates of undisturbed soil columns (100 cm^3) were sampled in four depth layers down to a maximum depth of 34 cm (0–4 cm, 10–14 cm, 20–24 cm, and 30–34 cm measured from the top of the mineral soil surface). At STO (16), BIB (21), MEN (35), and ROT (37) sampling depths were 0–4 cm, 5–9 cm, 15–19 cm, and 25–29 cm and 200 cm^3 soil columns were used. In the lab, total pore volume was determined using vacuum pycnometry [20] and macroporosity (volume of the pores with diameters larger than 10 μm) was calculated by taking the weight after dehydration of the soil cores on a kaolin bed at a water suction of −300 hPa [21]. The gas diffusion properties were analysed using a one-chamber method [22,23] after dehydration at a water suction of −300 hPa. The gas flux of the inert gas neon from the inside of the chamber through the soil cores (Ds) was analysed by measuring the decrease in neon concentration in subsequent time intervals with a Chrompack Micro GC CP-2002 (molsieve 5 Å column). Helium was used as carrier gas. By dividing the measured D_s of neon by its binary diffusion coefficient (D_0) in free atmosphere, a proportionality factor is gained that characterizes the reduction of the diffusion velocity through the soil pore system in relation to the one in free atmosphere (relative apparent diffusion coefficient D_s/D_0).

At all investigation sites, fine roots (∅ < 2 mm) were counted using the profile wall method [24,25]. The grid cell dimension was 4 × 4 cm. Distribution of fine roots was determined across the whole profile walls at the skid trail strata and at the control pit.

2.3. Modelling of Root Densities

A methodological approach for modelling of fine root densities is the use of Generalized Additive Models (GAMs). In contrast to Linear or Generalized Linear Models (LMs, GLMs), GAMs offer much flexibility in modelling continuous root density surfaces without integrating uncertain stratification information [26,27].

For the prediction of the root densities, a two-dimensional smoothing function where S is the smoothing term in the x and y directions was used.

$$h(\mu_{ij}) = \eta_{ij} = S\left(s_{x_{ij}}, s_{y_{ij}}\right)$$

In the model formula, we integrated a log link, and a quasi-Poisson family assumption. Smoothing in the horizontal and vertical directions was restricted to the trench wall boundaries. An exponential covariance function was used to cope with the spatial autocorrelation between root counts of neighbouring grid cells.

First, we used the modelled fine root data of the GAM to calculate the depth gradients at the wheel track situations based on the field classification of the wheel rut location. In a second approach, we did not use this strata information, but instead looked up the skid trail positions with the lowest fine root densities. Therefore, the x-coordinates of the minima for the first five horizontal grid cell layers were calculated. The detection algorithm was limited to a distance of 150 cm to each side of the skid trail centre. For the assignment of the horizontal extent of the most impacted area, the means of the minima were determined as centre locations. The spreading in the horizontal direction was defined based on the horizontal extent of the wheel rut depression detected at the topsoil of the skid trail transects. This spreading was assumed to provide a suitable approximation for the re-positioning of the area with a maximum impact concerning fine root propagation. Figure 2 shows this approach for the depth layer from 8–12 cm of skid trail 2 at WIS 1 (6).

With this algorithm, 28 locations characterized by local minima of fine rooting at 14 transects could be detected. At 12 locations, the calculated minima were located at the same horizontal position in all depth layers; at the remaining 16, they spread over 3 or fewer neighbouring grid cells. For 8 transects, no local minima could be derived. An overview of the modelled minima at the investigation sites is given in Appendix A Table A1. Figure 3 shows the two principal shapes of the root density curves that did not portray local minima. Although wheel rut depressions were visible in the field, the recalculation didn´t reveal corresponding minima of fine root counts at these locations. For these transects, the original field stratification of the wheel track location was used to extract the GAM model result for this stratum for calculation of inference statistics.

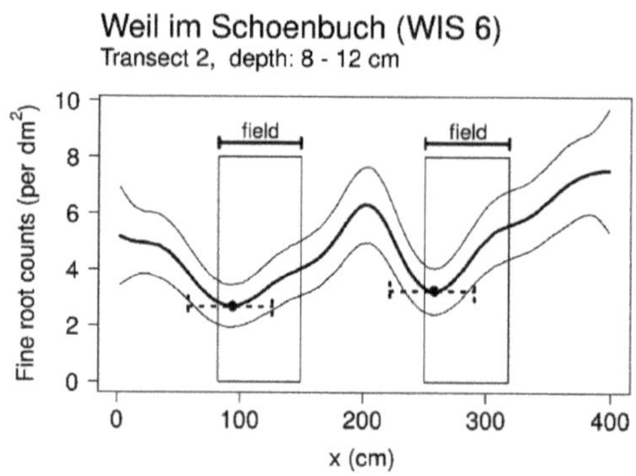

Figure 2. Modelled fine root densities for the depth of 8–12 cm at transect 2 at (WIS 1). Mean root density (bold line) is shown together with the confidence interval (± twofold standard deviations; narrow lines). The transparent-filled bar represents the field classification of the wheel track location. The filled dots show the two minima of fine rooting for this depth layer, which was used as a centre location for the reallocation of the horizontal spreading (dotted line).

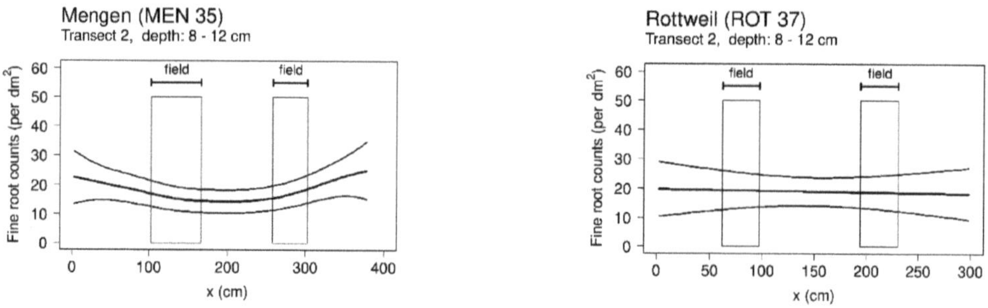

(**a**) (**b**)

Figure 3. (**a**) The modelled fine root densities decrease from the margin zones towards the centre of skid trail transect 2 at MEN (35). (**b**) The root density curve at ROT (37) follows a straight line across the whole transect.

In 52.4% of the tests for the investigated depths layers based on the field stratification of the wheel track situation, significant differences were detected. Using the reallocated skid trail locations, 54.8% of the comparisons showed a significant result. Thus, the results for the reallocated locations are presented in the following chapters as "wheel track" situations.

2.4. Statistical Analysis

The statistical computations were performed using the open-source software R 3.2.2 [28]. Groupwise testing of macroporosity at the different skid trail strata is presented in Figure 4; the post hoc Kruskal–Dunn test from the package "PMCMR" was used. The non-parametric Wilcoxon rank test (package "exactRankTests") was applied for the pairwise comparisons of macropore volume and apparent gas diffusion coefficients (Ds/Do) at the wheel track strata and control situations. Differences between wheel track and control situations were tested at significance levels of $p < 0.05$ and $p < 0.01$. With the package "mgcv" the GAMs were fitted and the predicted standard deviations and confidence limits were calculated. For the graphical presentations of the rooting depth gradients, the cubic smoothing spline function "sreg" from the package "fields" was applied.

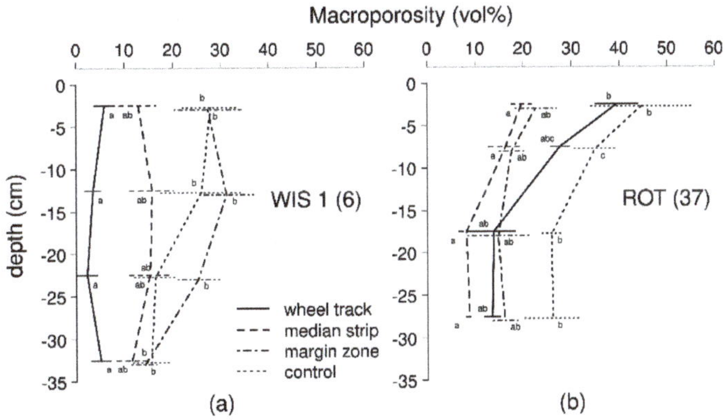

Figure 4. Macroporosities for the skid trail strata and control situation at (**a**) WIS 1 (6) and (**b**) ROT (37). Statistical distinctness between the strata was tested with the multiple non-parametric Kruskal–Dunn test at a significance level of $p < 0.05$ and is marked by small Arabic letters. The horizontal lines represent the two-fold standard deviations.

3. Results

In this paper the macropore volume (diameter of pores > 10 µm) and relative apparent diffusion coefficient measured at −300 hPa are presented for the characterization of the status of soil structure recovery. In addition to these two soil structural parameters, comparisons of fine root densities in wheel tracks and control situations are shown in order to judge the persistence of reduced root propagation in the trafficked soils.

3.1. Macropore Volume

In Figure 4, macroporosities for all investigated skid trail strata are shown for WIS 1 (6) and ROT (37). These two sites represent the longest and shortest time-period of soil structure recovery for which different patterns of the persisting loss of macroporosity at the skid trail strata can be observed exemplarily.

At the WIS 1 (6), the depth gradients are ranked clearly according to the expected impact at the skid trail strata. The most persistent loss of macropore volume (Figure 4a) was found at the wheel track strata (less than 5.2 vol%) in all the investigated depth layers compared with the control (15–29 vol%). The median strip represents an intermediate status: the groupwise testing did not reveal any significant difference between this and the other strata. Macroporosity at the margin zones and at the control situation were significantly higher compared with the wheel track strata (with the exception of the depth layer of 20–25 cm at the control site).

In contrast to the findings at the site WIS 1 (6), at ROT (37), the clarity of the ranked relationship between the skid trail strata was lost (Figure 4b). As expected, the macropore volume was highest in all depth layers at the control site. The lowest macroporosities were found at the median strip situation. The depth profiles at the margin zone, and even more pronounced at the wheel track location, represent an intermediate status of persisting macropore volume loss.

In Figure 5, the ratios of macroporosity between wheel tracks and controls are presented for all investigation sites categorized in three-time spans of soil structure recovery. At the study sites with recovery time spans of up to 12 years (Figure 5a), all differences are significant down to a depth of 24 cm (with the exception of WOL (12)). At all sites, a pronounced but significant only for the WIS 1 (6)-persistence of macroporosity loss is still visible in 30–34 cm. At TOD (14), STO (16), and ETT 1 (18); the ratios are grouped above 48% (Figure 5b). In the upper soil, the levels of significance are lower compared with the sites characterized by shorter time spans of soil structure recovery. In the upper soil of the sites with recovery time spans of more than 20 years, the variation in ratios is high compared with the sites with recovery time spans of less than 20 years: the ratios extend from 62% up to 167%. In the depth below 15 cm, all wheel track locations show a high and, in most cases, significant persistence of reduced macroporosity (Figure 5c) compared with the corresponding controls. The depth gradient at the site ETT 1 shows unexpected discontinuity in the depth layer of 10–14 cm. This may have been caused by trafficking away from the regular skid trail system during or after the clear-cutting operations, which was overlooked during the selection procedure of the site.

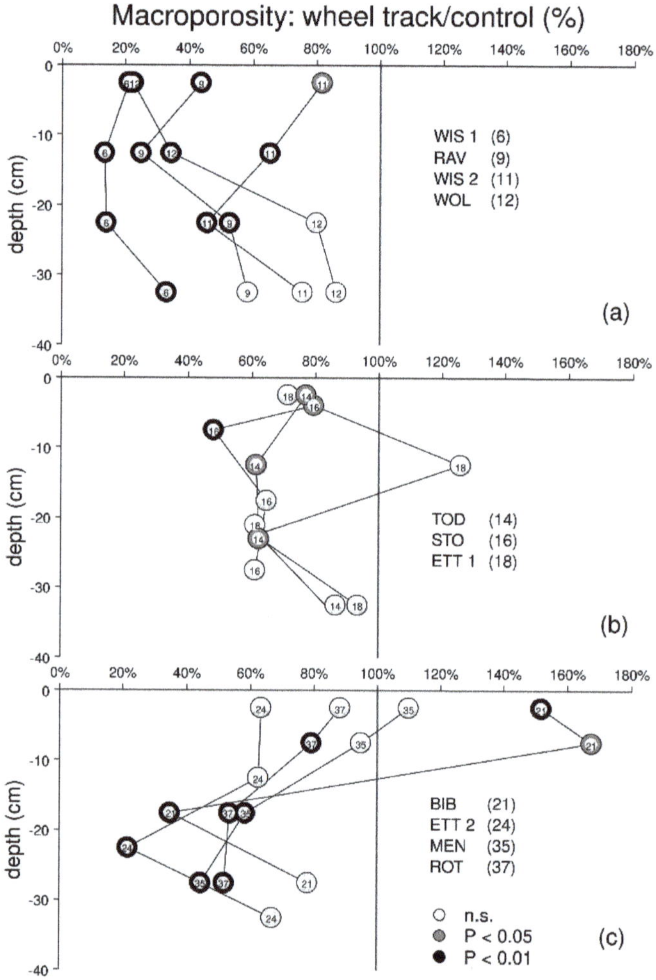

Figure 5. Ratios of macroporosity (wheel track/control×100) for sites with (**a**) recovery time of up to 12 years, (**b**) recovery time from 12 to 18 years, and (**c**) recovery time from 21 to 37 years. The significance levels of the pairwise Wilcoxon rank test are highlighted by differing grey shading intensities (● $p < 0.01$, ● $p < 0.05$, and ○ = not significant).

3.2. Diffusive Permeability

The relations between the wheel track strata and control for the relative apparent diffusion coefficients measured at −300 hPa (Figure 6) show a pattern comparable to that presented for macroporosity.

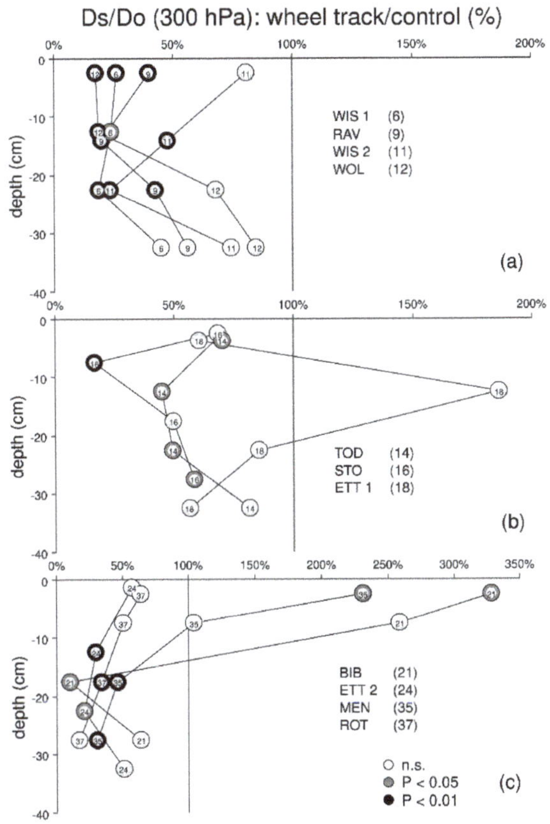

Figure 6. Ratios of relative apparent diffusion coefficients D_s/D_o measured at −300 hPa (wheel track/control×100) for sites with (**a**) recovery time of up to 12 years, (**b**) recovery time from 12 to 18 years, and (**c**) recovery time from 21 to 37 years. The significance levels of the pairwise Wilcoxon rank test are highlighted by differing grey shading intensities (● $p < 0.01$, ● $p < 0.05$, and ○ = not significant).

At the study sites with recovery time spans of up to 12 years (Figure 6a), the diffusion velocities at the wheel tracks are substantially lower in all depth layers compared with the corresponding controls. In most cases, the ratios lie below 50%. However, significant differences could only be proved down to a depth of 24 cm (with the exception of WIS 2 (11) in 0–5 cm and WOL (12) in 20–24 cm). Not as substantial as presented for the sites with time periods for structure recovery of up to 11 years, but still obvious, the ratios at the sites TOD (14) and STO (16) reveal the persistence of reduced diffusion properties in the wheel tracks (Figure 6b). Besides the outlying values in the 5–10 cm depth in STO (16) and ETT 1 (18) in 10–14 cm, the ratios span from 44% to 82%. In correspondence with the findings for macroporosity, the range of proportions for Ds/Do in the topsoil is wide for the sites with recovery time spans of more than 20 years (Figure 6c). In the uppermost soil layer at the wheel tracks at BIB (21) and MEN (35), 3.3-fold and 2.3-fold higher relative apparent diffusion coefficients were detected at the wheel tracks compared with the control situations, respectively. An outlier exceeding the diffusion velocities at the control was also present at the depth of 5–9 cm (BIB 21). At the other sites, this effect cannot be observed. There, the ratios in the topsoil and in the deeper soil layers still reveal a substantial and most often highly significant disturbance of gas exchange properties.

At the study site BIB (21), the original wheel rut was filled with loose mineral and organic matter (Figure 7). Due to this filling, in which a high activity of pore-generating soil biota was found, the microporosity, as well as the gas exchange properties, were much higher compared with the control strata. The abrupt boundary to the bleached mineral soil indicates that no spreading of pore volume generation took place into the compacted area below and to the sides of the wheel rut. This finding is consistent with the sharp transition of diffusion velocities as well as the reduced macropore volumes at the wheel tracks below 14 cm at this site and at MEN (35) in 0–4 cm.

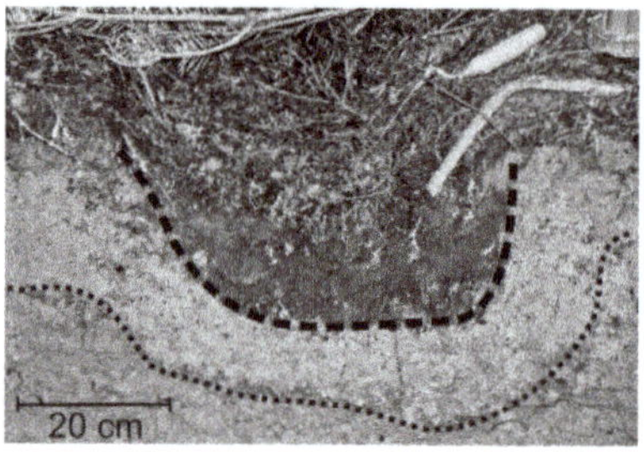

Figure 7. Filling of wheel rut with loose organic and mineral material at the study site BIB (21). The bold dashed line indicates the boundary of the former depression zone. The transition to the unbleached mineral soil is marked by the dotted line (photograph: H. Buberl).

3.3. *Recovery of Fine Rooting*

In Figure 8, the modelled fine root densities are presented for the study sites WIS 1 (6) and ROT (37) in order to illustrate the different patterns of fine rooting after a short and long timespan for the regaining of site-specific rooting conditions.

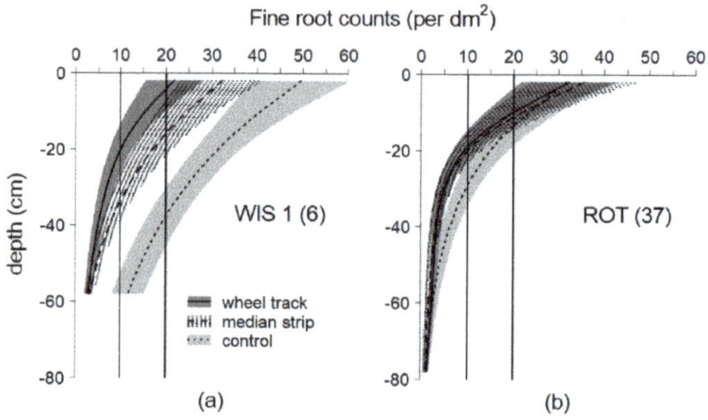

Figure 8. Depth gradients of the mean fine root densities at WIS 1 (**a**) and ROT (**b**) for wheel track, median strip, and control strata. Confidence intervals were calculated using the standard deviations of the GAM predictions (±2 sd). The bold vertical reference lines indicate the boundaries for intense rooting (20 fine roots × 100 cm^{-2}); the finer reference lines the threshold for weak rooting (10 fine roots × 100 cm^{-2}).

As it was presented for the macroporosity parameter, the depth gradients of fine rooting for the site WIS 1 (6) are arranged according to the expected intensity of impact (Figure 8a). In contrast to the control strata, where intense fine rooting (more than 20 fine roots × 100 cm^{-2}) reached down to approximately 40 cm, this intensity was only achieved in the uppermost grid cell layer at the wheel track. The intensively rooted zone at the median strip spread down to 16 cm. Throughout the whole control profile, fine rooting did not fall below the threshold of 10 fine roots × 100 cm^{-2}, which is classified as a weak rooting density. This threshold was only surpassed in the upper 20 cm at the wheel track and above 36 cm at the median strip. As for the macroporosity presented in Figure 4, the median strip represents an intermediate status of recovery concerning the persistence of reduced fine root propagation.

At the investigation site ROT (37) with the longest period of soil restructuring, fine root densities for the three strata were comparable in the upper 10 cm soil depth and below 60 cm (Figure 9b). In between, the mean fine root densities at the control profile exceeded the densities found at the wheel track and median strip. A statistical significance ($p < 0.05$) between the control situation and wheel track strata could be proved from 20 cm down to 36 cm.

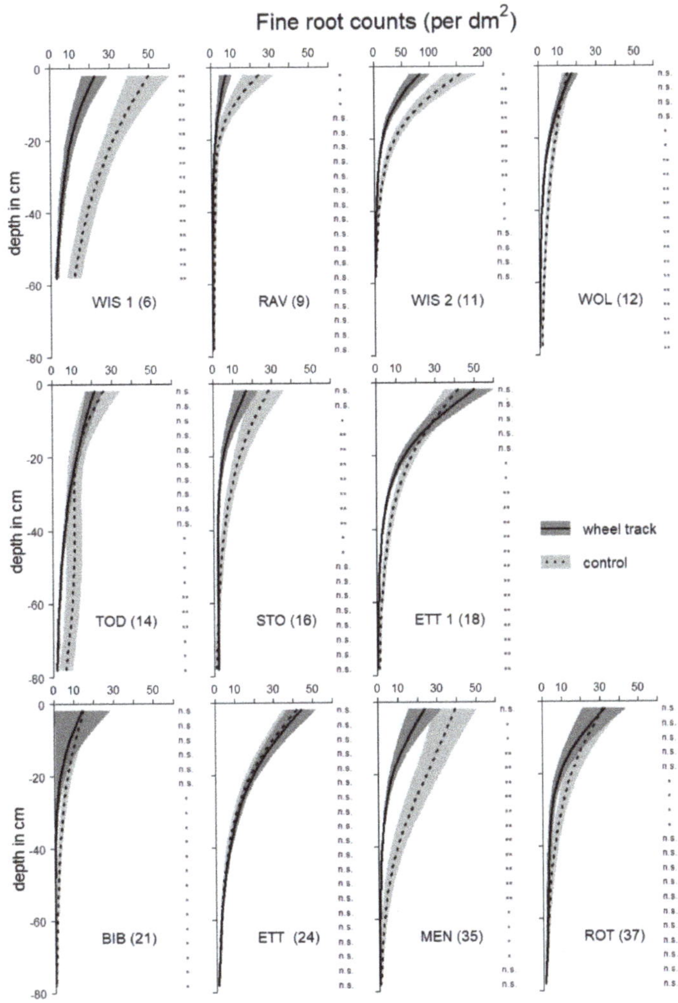

Figure 9. Depth gradients for the mean fine root densities of the wheel track and control strata. Confidence limits were derived based on the GAM predictions. Significance levels were assigned on the basis of three (** $p < 0.01$) and two standard deviations (* $p < 0.05$), n.s.= not significant).

Figure 9 presents the fine root depth gradients for the wheel track strata and control situations for all investigation sites in a chronological manner.

At WIS 1 (6), WIS 1 (12), and RAV (12), all the upper soil layers reveal a significant persistence of reduced fine rooting (Figure 9a). Fine rooting in the deeper soil layers at RAV (9) was limited by stagnic properties beginning at a soil depth of 25 to 30 cm. Thus, at that site no deeper-reaching effects caused by wheeling activity can be detected, as is possible for the other representations within this time span of recovery. The site WOL (12) exhibits reduced fine root propagation below 16 cm down to 80 cm soil depth.

For the sites shown in Figure 9b,c, no consistent chronology concerning the recovery of fine root propagation according to the time span of recovery can be derived. As a common characteristic for both groups, overlapping of the confidence bands in the uppermost layer(s) can be observed. With the exception of ETT (24), reduced root densities in the wheel tracks persisted in different depths and vertical extents. Even at the sites where the investigations took place more than 3 decades after the last trafficking, root propagation was still reduced in relation to the controls. This can be seen in particular at MEN (35).

As a characteristic for the sites BIB (21), MEN (35), and ROT (37), the confidence bands for the wheel track locations are wide, reflecting a high variability within these strata. At the sites with a shorter time span of recovery, the confidence bands at the wheel tracks are narrow compared with the corresponding control strata.

At WIS 2 (11) the intensity of fine rooting and at TOD (14) the shape of the depth gradients differ from the other investigation sites. At WIS 2 (11), ash and sycamore were the dominant tree species, which are characterized by a more intense fine root exploration in the top soil. The fine root density curve at the control site at TOD (14) shows hardly any decrease below 20 cm down to 60 cm of soil depth, whereas the other sites portrayed continuously declining density curves. There, a higher macroporosity and diffusion velocities in the cambisol allowed for intense rooting even in deeper soil horizons.

4. Discussion

In forest stands, nutrition and water supply can only be maintained if fine roots have sufficient access to these essential resources. This is ensured by a hierarchical pore system in naturally structured forest soils. Disturbance of gas exchange processes caused by pore volume loss and pore continuity in compacted forest soils impedes the survival and propagation of the fine roots, which have a high oxygen demand [29–31]. Parameters closely linked to soil aeration status were shown to be suitable to characterize the ecological damage [11,12,32]. Thus, macroporosity, relative apparent diffusion velocity, and fine root propagation were used for the evaluation of the recovery status of the formerly compacted soils at 11 sites in Baden-Wuerttemberg, with time spans reaching from 6 to 37 years of soil structure recovery.

4.1. Soil Physical Parameters Indicate Restructuring of the Pore System at the Topsoils

The comparisons for the macropore volume and apparent diffusion coefficient between wheel track strata and controls revealed a high persistence of pore volume loss and decreased gas diffusion velocity down to a 35 cm soil depth, even more than 3 decades after the trafficking events. Closer to the soil surface, the still persisting impedances differed in intensity depending on the time spans of recovery. Extremely lower macroporosities and diffusion velocities compared with the controls (ratios considerably smaller than 50%) were found at the sites with time spans of recovery of up to 12 years. At the investigation sites with longer periods of recovery, the ratios exceed 50%. This might be cautiously interpreted as an indication of topsoil structure recovery.

At two investigation sites, even higher topsoil macroporosity and diffusion coefficients compared with the control sites were found. There, the former wheel ruts obviously acted as traps for organic and mineral matter. Loosely packed organic and mineral filling of the wheel ruts in combination with a sufficient water supply led to favourable conditions for soil-structuring biota that in one case were furthermore promoted by a liming campaign. This effect was reported in earlier studies [33–35] but its ecological relevance was questioned [34]. This doubt can be confirmed by the presented results: The "apparent recovery effect" was obviously limited to the organically enriched uppermost part of the wheel track situation. A significant spreading of the biological restructuring into the compacted soil below the wheel tracks, attributed to, e.g., earthworm burrowing as reported by Capowiez et al. [36], cannot be deduced from our soil physical measurements and was also not confirmed by Ampoorter et al. [32].

The high persistence of a reduced macroporosity and gas diffusion velocity in the deeper soil characterized by time spans of recovery of more than 20 years seems to be contradictory. Two aspects might explain this long persistence of damage: In contrast to specialized forest machinery equipped with wide tyres, small-tyred machines may sink deep into the mineral soil when used in wet soil conditions, causing intense subsoil damage. Furthermore, wheeling at sites with storm damage was in former times not restricted to equidistant skid trails, with the consequence of areal soil structure damage. Both aspects might have led to severe inhibition of the restructuring soil biota due to the widespread destruction of habitats.

In the soil depth below 30 cm, the macropore volume and gas diffusion coefficient at the wheel tracks were still markedly lower compared with those of the control strata. However, the statistical testing did not reveal significant differences due to the higher variability of the measured parameters. This depth can be cautiously interpreted as a transition zone to the subsoil, where pore volume loss and gas diffusion properties had been less affected by the wheeling activity.

4.2. Recovery of Root Propagation

Based on the extended material of this study, the principal findings of von Wilpert and Schäffer [13], who found hardly any recovery of fine rooting within recovery time spans of up to 12 years, are substantiated. This can be explained by the "bottleneck effect" of a disturbed gas exchange at the topsoil: loss of pore volume and lack of pore continuity in the uppermost soil layers limit the oxygen supply and carbon dioxide disposal in the whole root zone [29,30].

At the sites with longer time delays between trafficking and root counting, no clear chronological ordering according to the hypotheses of a step-by-step approximation of fine root densities towards site-specific levels can be detected. Even at sites with time spans of soil structure recovery of more than 30 years, the depth gradients of fine rooting at the wheel tracks exhibit significantly lower densities compared with the control. Although the statistical testing of the macroporosity and diffusion velocity reveal a restructuring in the topsoil, this effect is not sufficient to overcome the impedance of fine root propagation in the deeper soil. If limitations of gas exchange persist in the deeper soil (as realized at the sites with long periods of a potential structure recovery), fine rooting might still be affected substantially in the deeper soil layers.

One of the study sites was characterized by a higher content of coarse fragments and another by tree species with a higher rooting capacity [37]. Soils with higher contents of sand or coarse fragments are assumed to be less prone to machine impacts [38,39]. At both sites, site-specific soil structure properties and fine root propagation should be achieved in shorter time periods. In contrast to this expectation, the fine root propagation and the soil structural parameters do not indicate an outstanding recovery dynamic.

4.3. Methodological and Statistical Aspects

4.3.1. Space for Time Substitution

The space for time substitution is a frequently used approach in ecosystem research when continuous measurements over longer time spans are lacking [40]. Its use is not reliable, when the comparability of sites is not given or unrecognised effects in the past were of a large magnitude. We tried to ensure a high comparability between the sites and operations through a detailed inquiry at the forest districts responsible for forest management and a verification process in the field. Vossbrink and Horn [41] proved that even lighter machines used in forest operations might cause soil stress comparable to that from heavy weighted machines due to dynamic force components. Thus, comparable impacts at the investigated sites can be assumed. However, uncertainty concerning the comparability of the original disturbance intensity remains. Besides this aspect, the unknown dynamic and velocity of the soil re-structuring processes at the different sites is critical for the evaluation of the data in the sense of a space for time substitution. Comparability concerning this aspect can only be achieved when study sites are located close together and site conditions do not differ among the considered representations. Neither of the mentioned aspects can be fully complied with in a retrospective study, looking back for decades. Nevertheless, the presented results give valuable information for judging the persistence of soil compaction phenomena in the sense of pattern recognition.

4.3.2. Modelling of Fine Root Surfaces

Fine root patterns derived at trench walls across skid trails give valuable information about the vertical and horizontal spreading of damaged soil compartments [14]. Generalized Additive Models (GAMs) [26,27] are a powerful tool to analyse root surfaces without integrating the uncertain field information on skid trail stratification. In this study, inference statistics between control situations and skid trail locations with the maximum impedance of rooting (reallocated "wheel track") were performed after detecting the local minima of root densities in the first five topsoil grid layers across the skid trails. A high matching rate detected for the horizontal locations of the minima confirms the suitability of this approach for an objective derivation of the most impacted zone below skid trails concerning fine root propagation. In addition, the failure of the algorithm at the sites with longer time spans of recovery indicates that field stratification involves a risk of misinterpretation, e.g., when sampling of soil cores at formerly passed skid trails is stratified according to the visual depression at the soil surface in the field. Furthermore, prevention of wheel rutting during harvesting and logging operations has become a major concern in forest machinery construction during the last decades. This fact also underpins the value and need of flexible statistical tools that are able to cope with uncertain or even missing information for stratification purposes and pattern recognition.

5. Conclusions

In the assessment of soil structure damage caused by heavy machinery and the judgment of their ecological relevance, the persistence of soil functionality disturbance is of great importance. Although numerous case studies have proved the negative effects on soil physical properties persisting over decades, there is still a high uncertainty concerning the persistence of the damage for different soil types and site conditions.

For the investigated sites with a silt loam texture and low activity of soil biota, long-lasting negative effects on the soil structure and fine root propagation exceeding time spans of more than three decades have been proved. The prognosis concerning soil disturbances caused by the much heavier machines that are used in practical forestry today is even more alarming. The lack of compliance with existing regulations in addition to the lack of awareness concerning long-term effects induced by soil compaction, as well as the long-lasting persistence of damage, are the main reasons for the historical accumulation of soil structure damage. The results presented in this study underpin the necessity for an implementation of soil protection strategies that ensure the preservation of naturally structured soils as a prerequisite for fulfilling the diverse ecosystem services. This is becoming more and more relevant in the face of the upcoming threats to forest ecosystems initiated by climate change.

If time spans of more than three decades are necessary for at least a partial restoration of soil functionality, we have to discuss whether it is justifiable or not to leave irregularly deformed soils in their critical physical conditions. If the latter is answered affirmatively, active measures have to be considered in order to restore soil structure deficits. Although soil tilling by ploughing or milling is a well-known strategy in agricultural management practices, its use in forestry is subject to several restrictions and success is uncertain [42,43]. Thus, methods that are based on biological pore structure generation and stabilization promise more success. Planting or sowing compaction-tolerant herbal or woody plant species might be a realistic option to accelerate the natural recovery processes and to promote a deeper soil restructuring [44,45].

More investigations on this topic are essential in order to develop practicable and feasible strategies for the activation and stabilization of soil-structuring processes in compacted soils. This would be in accordance with the EU-Soil Strategy 2030 [46], which aims to reduce the number of compacted soils, take active measures in the case of low natural recovery potential, and create greater awareness of the vulnerability of soils concerning compaction threads.

Funding: This work was funded by the Federal Ministry of Education and Research (Germany) (BMBF project n 0339767).

Informed Consent Statement: Not applicable.

Data Availability Statement: Data available on request.

Conflicts of Interest: The author declares no conflict of interest.

Appendix A

Table A1. Location of modelled fine root density; the five depth layers were located at the same horizontal grid cell position. Minima: The interval represents the range of horizontal grid cell positions for the minima in the first five grid layers at the transects of the investigation sites. Skid trail locations with identical minima in all of these layers are printed in italics.

	Transect 1		**Transect 1**	
Site	**Wheel Track 1**	**Wheel Track 2**	**Wheel Track 1**	**Wheel Track 2**
	Min–max (cm)		Min–max (cm)	
Ravensburg (RAV)	118–130	282–286	–	–
Weil im Schoenbuch (WIS 2)	50–54	250–258	118–122	*242*
Todtmoos (TOD)	*146*	358–362	*102–114*	*318*
Stockach (STO)	94–102	298–306	*166*	*346*
Biberach (Bib)	–	–	–	–
Ettenheim (ETT 2)	–	–	–	–
Rottweil (ROT)	*62*	206–210	–	–

References

1. Macdonald, P.; Clow, M. What a difference a skidder makes: The role of technology in the origins of the industrialization of tree harvesting systems. *Hist. Technol.* **2003**, *2*, 127–149. [CrossRef]
2. Nordfjell, T.; Öhman, E.; Lindroos, O.; Ager, B. The technical development of forwarders in Sweden between 1962 and 2012 and of sales between 1975 and 2017. *Int. J. For. Eng.* **2019**, *1*, 1–13. [CrossRef]
3. Hamberger, J. Wie Mechanisierung und Umweltvorsorge die Forstwirtschaft Veränderten. *LWFaktuell* **2003**, *39*, 33–39. Available online: https://www.lwf.bayern.de/service/publikationen/lwf_aktuell/076550/index.php (accessed on 23 April 2022).
4. Schäffer, J.; Buberl, H.; von Wilpert, K. Deformation damages in forest topsoils—An assessment based on Level-I soil monitoring data from Baden-Württemberg (SW Germany). *J. Plant Nutr. Soil Sci.* **2012**, *1*, 24–33. [CrossRef]
5. Gaertig, T.; Ebeling, C.; Riggert, R. Bodenschutz beim Forstmaschineneinsatz. *Bodenschutz* **2018**, *1*, 9–15. [CrossRef]
6. Hatchell, G.E.; Rolston, C.W. Natural recovery of surface soils disturbed in logging. *Tree Plant. Notes* **1971**, *2*, 5–9. Available online: https://www.srs.fs.usda.gov/pubs/ja/1971/ja_1971_hatchell_001.pdf (accessed on 23 April 2022).
7. Froehlich, H.A.; Miles, D.W.R.; Robbins, R.W. Soil bulk density recovery on compacted skid trails in central Idaho. *Soil Sci. Soc. Am. J.* **1985**, *49*, 1015–1017. [CrossRef]
8. Webb, R.H.; Steiger, J.W.; Wilshire, H.G. Recovery of compacted soils in Mojave Desert ghost towns. *Soil Sci. Soc. Am. J.* **1986**, *5*, 1341–1344. [CrossRef]

9. Labelle, E.R.; Jaeger, D. Soil Compaction Caused by Cut-to-Length Forest Operations and Possible Short-Term Natural Rehabilitation of Soil Density. *Soil Sci. Soc. Am. J.* **2011**, *6*, 2314–2329. [CrossRef]
10. Klaes, B.; Struck, J.; Schneider, R.; Schüler, G. Middle-term effects after timber harvesting with heavy machinery on a fine-textured forest soil. *Eur. J. For. Res.* **2016**, *6*, 1083–1095. [CrossRef]
11. Hildebrand, E.E. Der Einfluss der Bodenverdichtung auf die Bodenfunktionen im Forstlichen Standort. *Forstwiss. Cent.* **1983**, *102*, 111–125. Available online: https://www.freidok.uni-freiburg.de/fedora/objects/freidok:5830/datastreams/FILE1/content (accessed on 23 April 2022). [CrossRef]
12. Hildebrand, E.E.; Schack-Kirchner, H. The Influence of Compaction on Soil Structure and Functions in Forest Sites. In *Modern Trends in Applied Terrestrial Ecology*; Ambasht, R.S., Ambasht, N.K., Eds.; Springer Science + Business Media (Springer eBook Collection): Boston, MA, USA, 2002; pp. 1–11. [CrossRef]
13. Ampoorter, E.; van Nevel, L.; de Vos, B.; Hermy, M.; Verheyen, K. Assessing the effects of initial soil characteristics, machine mass and traffic intensity on forest soil compaction. *For. Ecol. Manag.* **2010**, *10*, 1664–1676. [CrossRef]
14. Von Wilpert, K.; Schäffer, J. Ecological effects of soil compaction and initial recovery dynamics. A preliminary study. *Eur. J. For. Res.* **2006**, *125*, 129–138. [CrossRef]
15. Goutal, N.; Renault, P.; Ranger, J. Forwarder traffic impacted over at least four years soil air composition of two forest soils in northeast France. *Geoderma* **2013**, *193–194*, 29–40. [CrossRef]
16. Schack-Kirchner, H. Struktur und Gashaushalt von Waldböden Ber. *Forsch. Waldökosyst.* **1994**, *112*, 145.
17. Kremer, J. *Befahrungsbedingte Strukturveränderungen von Waldböden und ihre Auswirkungen auf das Wachstum von Fichten, Kiefern und Buchen auf ausgewählten Standorten*; GCA-Verlag: Herdecke, Germany, 1998; p. 177.
18. Ebeling, C.; Lang, F.; Gaertig, T. Structural recovery in three selected forest soils after compaction by forest machines in Lower Saxony, Germany. *For. Ecol. Manag.* **2016**, *359*, 74–82. [CrossRef]
19. Food and Agriculture Organization (FAO). *Guidelines for Profile Description*, 4th ed.; Food and Agriculture Organization (FAO): Rome, Italy, 2006; p. 97. Available online: https://www.ipcinfo.org/fileadmin/user_upload/soils/docs/FAO_guidelines_soil_description__20063.pdf (accessed on 23 April 2022).
20. Danielson, R.E.; Sutherland, P.L. Porosity. In *Methods of Soil Analysis. Part 1. Physical and Mineralogical Methods*, 2nd ed.; Klute, A., Ed.; Agronomy monograph; American Society of Agronomy: Madison, WI, USA, 1986; Volume 9, pp. 443–450.
21. Hartge, K.H.; Horn, R. *Die Physikalische Untersuchung von Böden*, 4th ed.; E. Schweizerbart'sche Verlagsbuchhandlung: Stuttgart, Germany, 2009; p. 178.
22. Frede, H.G. *Der Gasaustausch des Bodens*; Göttinger Bodenkundliche Ber: Göttingen, Germany, 1986; Volume 87, p. 130.
23. Kühne, A.; Schack-Kirchner, H.; Hildebrand, E.E. Gas diffusivity in soils compared to ideal isotropic porous media. *J. Plant Nutr. Soil Sci.* **2012**, *1*, 34–45. [CrossRef]
24. Böhm, W. *Methods of Studying Root Systems*; Ecological Studies 33; Springer: Berlin, Germany, 1979; p. 208.
25. van Nordwijk, M.; Brouwer, G.; Meijboom, F.; Oliveira, M.; do Rosario, G.; Bengough, A.G. Trench profile techniques and core break methods. In *Root methods—A handbook*; Springer: Berlin, Germany, 2000; pp. 212–233.
26. Wood, S.N. *Generalized Additive Models: An Introduction with R*; Chapman & Hall; CRC: Boca Raton, FL, USA, 2006; p. 391.
27. Schäffer, J.; von Wilpert, K.; Kublin, E. Analysis of fine rooting below skid trails using linear and generalized additive models. *Can. J. For. Res.* **2009**, *39*, 2047–2058. [CrossRef]
28. R Core Team. *R: A Language and Environment for Statistical Computing*; R Foundation for Statistical Computing: Vienna, Austria. Available online: https://www.R-project.org/ (accessed on 10 November 2015).
29. Gaertig, T. *Bodengashaushalt, Feinwurzeln und Vitalität von Eichen*; Freiburger Bodenkundliche Abhandlungen: Freiburg, Germany, 2001; Volume 40, p. 157.
30. Gaertig, T.; Schack-Kirchner, H.; Hildebrand, E.E.; Wilpert, K.V. The impact of soil aeration on oak decline in southwestern Germany. *For. Ecol. Manag.* **2002**, *159*, 15–25. [CrossRef]
31. Qi, J.; Marshall, J.D.; Mattson, K.G. High soil carbon dioxide concentrations inhibit root respiration of Douglas fir. *New Phytol.* **1994**, *128*, 435–442. [CrossRef]
32. Ampoorter, E.; de Schrijver, A.; de Frenne, P.; Hermy, M.; Verheyen, K. Experimental assessment of ecological restoration options for compacted forest soils. *Ecol. Eng.* **2011**, *11*, 1734–1746. [CrossRef]
33. Benthaus, M.; Matthies, D. Regeneration befahrener Waldböden. *Allg. Forst Z.* **1993**, *9*, 448–451.
34. Schack-Kirchner, H.; Hildebrand, E.E.; von Wilpert, K. Bodensauerstoffhaushalt unter Fahrspuren. *Allg. Forst Z.* **1993**, *3*, 118–121.
35. Eklkofer, E.; Matthies, D. Das Wuchsverhalten eines gepflanzten Fichtenbestandes auf einem vorverdichteten Standort—30 Jahre nach Befahrung. *Allg. Forst Z.* **1995**, *22*, 1222–1226.
36. Capowiez, Y.; Cadoux, S.; Bouchard, P.; Roger-Estrade, J.; Richard, G.; Boizard, H. Experimental evidence of the role of earthworms in compacted soil regeneration based on field observations and results from a semi-field experiment. *Soil Biol. Biochem.* **2009**, *4*, 711–717. [CrossRef]
37. Horn, A.; Murach, D. Vertikale Feinwurzelverteilung und Hinweise auf interspezifische Wurzelkonkurrenz in Eschen/Buchen-Naturverjüngungen. *Forstarchiv* **2003**, *74*, 46–52.
38. Greacen, E.L.; Sands, R. Compaction of forest soils. A review. *Soil Res.* **2002**, *2*, 163–189. [CrossRef]
39. Cambi, M.; Certini, G.; Neri, F.; Marchi, E. The impact of heavy traffic on forest soils: A review. *For. Ecol. Manag.* **2015**, *338*, 124–138. [CrossRef]

40. Pickett, S.T.A. Space-for-Time Substitution as an Alternative to Long-Term Studies. In *Long-Term Studies in Ecology*; Likens, G.E., Ed.; Springer: New York, NY, USA, 2012; pp. 110–135. [CrossRef]
41. Vossbrink, J.; Horn, R. Modern forestry vehicles and their impact on soil physical properties. *Eur. J. For. Res.* **2004**, *2*, 259–267. [CrossRef]
42. Horn, R.; Lebert, M. Möglichkeiten und Grenzen der Bodenbearbeitung der physikalischen Bodenrekultivierung in der Forstwirtschaft. *Allg. Forst Z.* **1992**, *47*, 998–1004.
43. Gaertig, T.; Hildebrand, E.E.; Schäffer, J.; von Wilpert, K. Wirkung mechanischer Bodenlockerung auf Bodenbelüftung und Durchwurzelung. *Allg. Forst Z.* **2000**, *55*, 1124–1126.
44. Meyer, C.; Lüscher, P.; Schulin, R. Enhancing the regeneration of compacted forest soils by planting black alder in skid lane tracks. *Eur. J. For. Res.* **2014**, *3*, 453–465. [CrossRef]
45. Flores Fernández, J.L.; Rubin, L.; Hartmann, P.; Puhlmann, H.; von Wilpert, K. Initial recovery of soil structure of a compacted forest soil can be enhanced by technical treatments and planting. *For. Ecol. Manag.* **2018**, *431*, 54–62. [CrossRef]
46. EU-COM. EU Soil Strategy for 2030—Reaping the Benefits of Healthy Soils for People, Food, Nature and Climate. 2021. Available online: https://eur-lex.europa.eu/legal-content/EN/TXT/?uri=CELEX%3A52021DC0699 (accessed on 23 April 2022).

soil systems

MDPI

Article

Nutrient Storage and Stoichiometry of the Forest Floor Organic Matter in Japanese Forests

Masamichi Takahashi [1,2]

[1] Japan International Forestry Promotion and Cooperation Center, Rinyu Building, 1-7-12 Kouraku, Bunkyo-ku, Tokyo 112-0004, Japan; masamichi@jifpro.or.jp

[2] Forestry and Forest Products Research Institute, 1 Matsunosato, Tsukuba 305-8687, Japan

Abstract: Nutrient storage in the forest floor is regulated through litter decomposition and nutrient cycling. Stoichiometry of nutrients can provide characterization of the forest floor. To quantify nutrient storage in the forest floor and to determine stoichiometry among different forest types, available data on nutrients were meta-analyzed. The data on nutrients—nitrogen, phosphorus, potassium, calcium, and magnesium—were collected from published reports and original data on Japanese forests. The relationship between nutrient storage and forest floor mass was also examined. Japanese cypress and cedar plantations had small N and P storage in the forest floor with high C:N and C:P ratios, whereas subalpine conifers had large N and P storage in the forest floor with low C:N and C:P ratios; cedar plantations showed large Ca-specific storage in the forest floor. The stoichiometry of the forest floor varied between different forest types, namely C:N:P ratios were 942:19:1 for cedar and cypress plantations, 625:19:1 for broad-leaved forests, and 412:13:1 for subalpine conifers and fir plantations. N storage was closely correlated; however, P and other mineral storages were weakly correlated with the forest floor mass. Nutrient storage and stoichiometry can provide a better perspective for the management of forest ecosystem.

Keywords: C:N:P ratio; dead organic matter; forest ecosystem; forest soil; humus layer; nutrient cycling; nutrient reservoir; organic layer

Citation: Takahashi, M. Nutrient Storage and Stoichiometry of the Forest Floor Organic Matter in Japanese Forests. *Soil Syst.* **2021**, *5*, 51. https://doi.org/10.3390/soilsystems5030051

Academic Editor: Klaus von Wilpert

Received: 14 July 2021
Accepted: 24 August 2021
Published: 29 August 2021

1. Introduction

The forest floor developed over mineral soils includes organic residues—such as leaves, branches, bark, and stems—in various stages of decomposition. Forest floor organic matter forms an important carbon pool within the forest ecosystem since it makes up the organic humus layers, such as the L (Oi), F (Oe), and H (Oa) layers, and fine woody debris. The Inter-Governmental Panel on Climate Change defined this carbon pool as a "litter", separating it from soil and deadwood in the generalized category for dead organic matter [1]. Since then, there has been a large amount of data summarized on "litter" carbon stock in the recent decades [2–4]. Apart from providing this vital carbon pool for forest ecosystems, the forest floor serves as critical habitats and food sources for soil fauna and microbes, which play vital roles in litter decomposition [5]. The thickness and density of the forest floor affect the abundance and diversity of soil fauna [6,7]. Conversely, the forest floor quality, i.e., its nutrient contents and energy sources, further affects the activities of decomposers [8]. For nutrient cycling in forest ecosystems, the forest floor functions as a nutrient reservoir and chemical buffering layer against acid deposition [9–11]. Although it plays such a pivotal role in ecosystem functioning, nutrient storage within the forest floor is given much less attention in research than that of litter decomposition.

Numerous studies have been conducted on litter decomposition processes that have revealed changes in litter decomposition rate following nutrient release and/or nutrient immobilization [12–15]. The litter bag method is widely used to track the time course of nutrient dynamics within litter during the early stages of decomposition. However, the method has limitations in the later decomposition stages [16,17], meaning that the litter bag study does not provide accurate and suitable information on the storage of nutrients in the forest floor. Because the forest floor comprises a mixture of organic materials going through both early and late decomposition, the stratification of organic layers can enable the accurate estimation of nutrient storage in thick humus forms—moder and mor humus [18,19]; however, the subdivision is not applicable for the mull humus form. Root development in the F and H layers also affect nutrient dynamics in the forest floor [20]. Although nutrient storage in the forest floor may provide a snapshot of nutrient cycling in the ecosystem, this labile nutrient pool is important for soil fertility and soil food web. Thus, compiling data on nutrient storage in the overall forest floor of different forest types in Japan could enable further research for evaluating the importance of the forest floor as a nutrient reservoir and its chemical buffering capacity for the ecosystem.

The stoichiometry of the forest floor is a good indicator of the conditions of decomposing organic materials. Fresh fallen litter shows a large variation in nutrient contents and carbon-to-nutrient ratios among tree species gradually converge at a particular range by progressing decomposition and humification [2,21,22]. Since a large divergence in stoichiometry exists between fresh litter and soil [23], the stoichiometry of the forest floor, which serves as an intermediary body between them, may assist in understanding the decomposition processes and element dynamics at the soil surface boundary. In Canada, long-term litterbag experiments using 10 tree species were conducted over a six-year period [24]. The results showed a convergence in litter stoichiometry of the carbon-to-nitrogen (C:N), carbon-to-phosphorus (C:P), and nitrogen-to-phosphorus (N:P) ratios to 30, 450, and 16, respectively, which is equivalent to a C:N:P ratio of 450:16:1, based on P content. Ma et al. [25] studied the C:P:N ratio variations in decomposed forest floor litter at different successional stages in eastern China and found that C:N:P varied with successional vegetation stages, ranging from 377:26.5:1 to 782:57:1. Tree species and climate conditions were noted to affect the stoichiometry of the forest floor. Because Japan has a wide variety of forest types and climate conditions even in small archipelagic nations [26,27], evaluating the range of forest floor stoichiometry will be valuable to understand litter quality in various forest conditions.

The amount of nutrient storage that the forest floor can provide is greatly affected by its mass. The accumulation of this forest floor mass varies depending on site condition factors, such as soil, topography, and forest management; this is even so in a monoculture plantation [28–30]. Using the variation of the forest floor mass, carbon concentration and storage in the forest floor have been determined by the regression with dry weight of the forest floor mass adjusted by the ash content [31]. Similarly, it may be possible to develop regression equations to estimate the nutrient storage within the forest floor. Because nutrient concentrations in fresh litter varies with tree species [32], correlations may differ based on predominant tree species in a particular forest.

The aims of this study were to: (1) summarize dry weight mass and nutrient storage in the forest floor of different forest types in Japan, (2) analyze stoichiometry of organic materials in the forest floor to evaluate their quality, and (3) examine regression equations between forest floor mass and nutrient storage in the forest floor. To achieve the study aims, a meta-analysis was performed using existing data collected from scientific papers and project reports, as well as data based on the author's original research conducted in Japan.

2. Materials and Methods

Data on dry weight, as well as carbon and nutrient storage of the forest floor organic layer, were obtained from the existing research articles and project reports. This literature review included both English and Japanese manuscripts, and the data from the literature was combined with original data of the author to undertake the meta-analysis. Some of the data were used for compiling forest soil carbon stock in varying forests in Japan [33]. Nutrient mass within the forest floor was calculated by using the forest floor dry weight multiplied by the total concentration of nutrients: nitrogen (N), phosphorus (P), potassium

(K), calcium (Ca), and magnesium (Mg). Analytical methods and equipment varied between the different researchers and their respective studies. When the studied report did not provide the amount of carbon storage of the forest floor, a conversion equation was used in its place [31]. Data for all nutrients was not available for every study; therefore, the sample size for each nutrient varied among forest types.

Data were grouped based on predominant forest types, plantation, or natural and semi natural forests. The predominant forests are shown in Table 1. Plantation forestry mainly consists of Japanese cedar (*Cryptomeria japonica*, 44% of the total plantation area) and Japanese cypress (*Chamaecyparis obtusa*, 25%). Well-drained and productive moist hills and mountain slopes in Japanese forests are largely occupied by these two species [34]. Japanese red pine (*Pinus densiflora*) is a representative planted and naturally established species on xeric ridges. Japanese black pine (*Pinus thunbergia*) is often seen planted on coastal sandy soils [35,36]. Subalpine coniferous forests are often develop on podzolic soil with thick organic layers [37,38]. Japanese larch (*Larix kaempferi*) and Todo fir (*Abies sachalinensis*) plantations are mostly distributed in cool northern prefectures [39,40]. Subalpine coniferous forests include cypress and fir old-growth natural forests. For broad-leaved deciduous forests, Japanese beech (*Fagus crenata*) covers cool temperate areas, whereas oak (*Quercus* spp.) widely occurs in cool and warm temperate forests [27]. *Castanopsis* spp. are evergreen broad-leaved species found in warm southern parts of Japan [41]. Because of the limited number of available nutrient data on broad-leaved forests, the broad-leaved species were grouped into deciduous and evergreen forests. The groups based on forest types included those with more than 10 stands and those older than 20 years. Distribution maps of sample size in a prefecture are shown in Figure 1 and Table S1. The data on Todo fir were collected only from the Hokkaido prefecture, North Island.

Table 1. Forest types and their predominant tree species in Japan.

Forest Type	Scientific Names of Predominant Species
Cedar	*Cryptomeria japonica*
Cypress	*Chamaecyparis obtusa*
Larch	*Larix kaempferi*
Pine	*Pinus densiflora*, *P. thunbergii*
Todo fir	*Abies sachalinensis*
Subalpine coniferous	*Abies veitchii*, *A. mariesii*, *A. sachalinensis*, *Picea jezoensis* var. *hondoensis*, *P. jezoensis*, *P. glehnii*, *Tsuga diversifolia*
Deciduous broad-leaved	*Fagus* spp., *Quercus* spp., *Betula* spp., *Acer* spp., *Alnus* spp., *Carpinus* spp., *Pterocarya rhoifolia*, *Aesculus turbinata*, *Fraxinus* spp.
Evergreen broad-leaved	*Castanopsis* spp., *Lithocarpus* spp., *Quercus acuta*, *Machilus thunbergii*, *Cinnamomum camphora*, *Camellia* spp.

Most of the group data on nutrient storage were not normally distributed and had positive skewness. Hence, the medians and quartiles were used for the representative values. The nonparametric Steel–Dwass test was performed for multiple comparisons of all pairs of the forest types. For comparing nutrient concentrations among forest types, Turkey–Kramer test was used because of normal distributions of the data. The linear regression between the dry weight of forest floor mass and the nutrient storage was conducted by using logarithm transformation. Statistical analysis was conducted using JMP software (SAS Institute Inc., Cary, NC, USA).

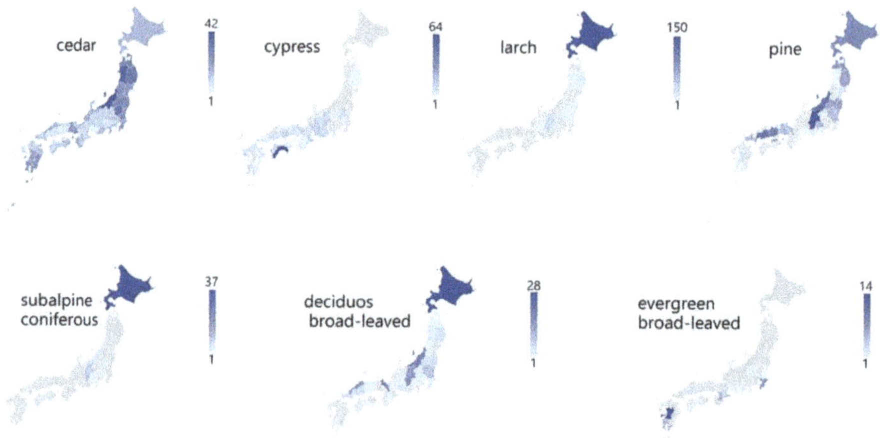

Figure 1. Distribution of the sample size of forest floor mass data in the prefecture level in Japan. Nutrient storage data are shown in Table S1.

3. Results

3.1. Dry Mass and Nutrient Storage in Predominant Forest Type

Among coniferous forest types, minimum accumulation of forest floor mass occurred in cypress plantations (6.6 Mg ha^{-1}), followed by cedar plantations (9.3 Mg ha^{-1}; Figure 2A, Table S2). Cypress and cedar were significantly different from the other forest types, except evergreen broad-leaved forests. The largest mass was represented by subalpine coniferous forests (37.0 Mg ha^{-1}), and fir natural forests (32.5 Mg ha^{-1}) next. Other coniferous forests ranged from 12.1 to 14.5 Mg ha^{-1} were mostly no significant difference. As for broad-leaved forests, deciduous forests (12.2 Mg ha^{-1}) had relatively larger forest floor mass than evergreen forests (9.2 Mg ha^{-1}), despite of insignificance.

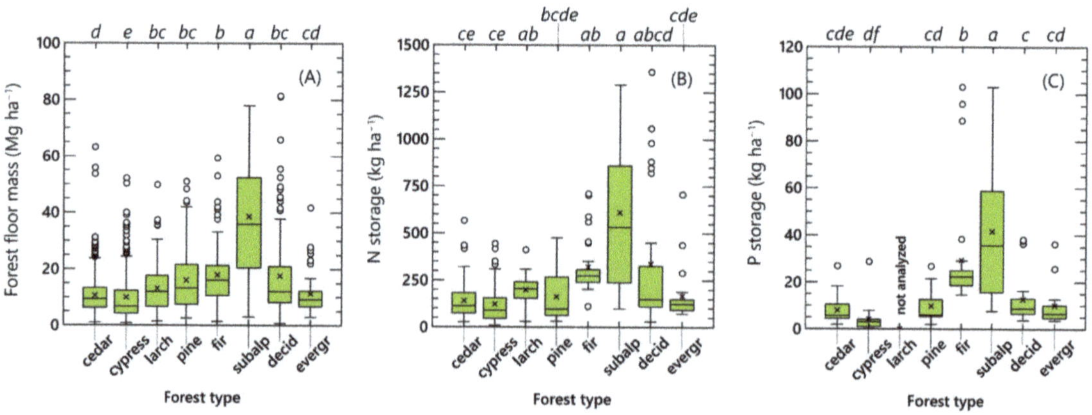

Figure 2. Box and whisker plots of forest floor mass (**A**), nitrogen (**B**), and phosphorus (**C**) storage in the forest floor among Japanese forest types. Subalpine coniferous forests, deciduous broad-leaved forests, and evergreen broad-leaved forests are represented by "subalp", "decid", and "evergr" respectively. The box covers the 75th and 25th percentiles. The horizontal line within the box marks the median, and the cross indicates the mean. The length of the whisker is 1.5 times the interquartile range. Outliers are indicated by open circles. Different lowercase letters on the figure denote significant differences between forest types at $p < 0.05$ based on the Steel–Dwass test.

Nitrogen storage in the forest floor showed significant variation among forest types (Figure 2B, Table S3). Among coniferous forests, subalpine coniferous forests and fir plantations had significantly larger N storage, at 533 and 261 kg ha^{-1}, respectively. Cedar and cypress plantations and evergreen broad-leaved forest showed significantly lower N storage, ranging from 89 to 123 kg ha^{-1}. Phosphorus storages were significantly small in cypress (3.10 kg ha^{-1}) and cedar (5.66 kg ha^{-1}) plantations and large in subalpine coniferous forests (35.6 kg ha^{-1}) (Figure 2C, Table S4). Evergreen and deciduous broad-leaved forests also had small P storages (6.55 kg ha^{-1} and 8.75 kg ha^{-1}, respectively).

Regarding mineral stocks, K storage was the lowest (10.9 kg ha^{-1}) in cypress and deciduous broad-leaved forests (Figure 3A, Table S5). Cedar plantations also showed small K storage in the forest floor. Compared to cedar, cypress, and deciduous broad-leaved forests, significantly higher K storages occurred in subalpine coniferous forests (51.6 kg ha^{-1}) and fir plantations (35.7 kg ha^{-1}). Calcium storage was significantly higher (147 kg ha^{-1}) in cedar than in cypress plantations (57.3 kg ha^{-1}) (Figure 3B, Table S6). The highest occurred in the subalpine coniferous forest (233 kg ha^{-1}). Evergreen broad-leaved forests had significantly higher Ca storage than deciduous broad-leaved forests. The range of Mg storage in the forest floor was narrow and no significant differences were found among the forest types (Figure 3C, Table S7).

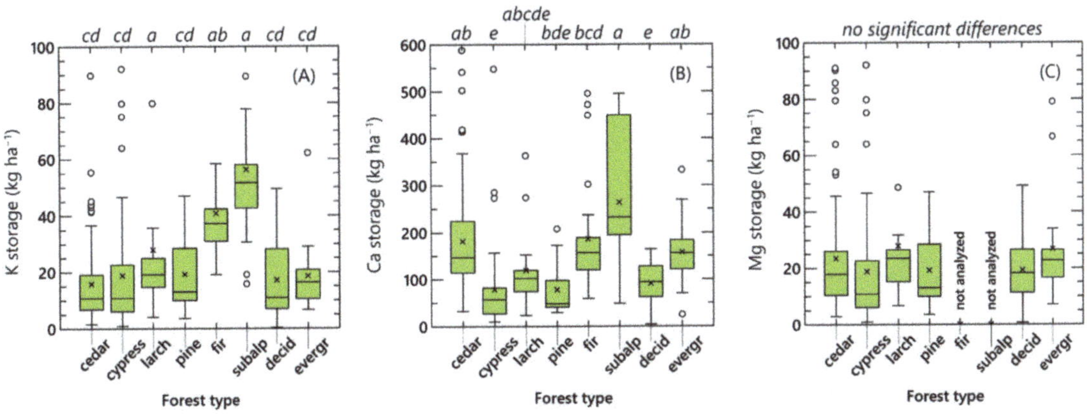

Figure 3. Box and whisker plots of potassium (**A**), calcium (**B**), and magnesium (**C**) storage in the forest floor among Japanese forest types. The definitions of box and whisker are the same as those used in Figure 2. Different lowercase letters on the figure denote significant differences between forest types at $p < 0.05$ based on the Steel–Dwass test.

3.2. Total Nutrient Concentrations in the Forest Floor

Table 2 shows the nutrient concentrations of each forest type. The N concentration in the forest floor of broad-leaved forests was significantly higher than that of coniferous forests. The range of P concentration was large; the low values (about 0.55 g kg^{-1}) occurred in cedar and cypress plantations, while high values (about 1.1 g kg^{-1}) were observed in fir plantations and subalpine coniferous forests. Broad-leaved forests showed intermediate P concentrations. Ca concentrations in cedar plantations and evergreen broad-leaved forests were almost twice as high as the other forest types. No significant differences were detected in K and Mg concentrations of the forest floor among forest types.

Table 2. Mean and standard deviation (s.d.) of total nutrient concentrations (g kg^{-1}) in the forest floor of predominant forest types.

Nutrient Forest Type	N Mean [(2)]	s.d.	P Mean [(2)]	s.d.	K Mean	s.d.	Ca Mean [(2)]	s.d.	Mg Mean	s.d.
Coniferous										
Cedar	10.0 d	2.86	0.542 d	0.265	1.24	0.870	13.9 a	4.27	1.65	1.05
Cypress	10.0 d	3.03	0.552 cd	0.270	1.70	1.46	7.11 b	3.72	1.37	1.11
Larch	12.5 bcd	2.55	n.a [(1)]	n.a	1.55	0.815	7.37 b	3.34	1.64	0.64
Pine	11.1 cd	2.67	0.770 bcd	0.195	1.51	0.597	6.42 b	2.53	1.84	1.58
Todo fir	13.4 bc	1.90	1.19 ab	0.446	1.73	0.360	7.96 b	3.20	n.a.	n.a.
Subalpine coniferous	14.0 ab	3.46	1.09 a	0.624	1.48	0.574	7.47 b	4.27	n.a.	n.a.
Broad-leaved										
Deciduous	15.5 a	4.83	0.813 bc	0.340	1.13	0.870	7.85 b	6.53	1.33	0.888
Evergreen	13.5 abc	4.48	0.633 cd	0.270	1.37	0.647	13.7 a	10.9	2.16	1.58

[(1)] n.a.: not analyzed due to small sample size. [(2)] Values with different letters indicate significant differences between forest types at $p < 0.05$ based on the Turkey–Kramer test.

3.3. Stoichiometry of Carbon and Nutrients in the Forest Floor

C:N ratios were significantly high in the cedar and cypress plantations—49.8 and 47.2, respectively (Figure 4A, Table S8). The C:N ratio of other coniferous forests ranged from 32.4 to 40.7. The lowest C:N ratio occurred in the deciduous broad-leaved forests (28.2). For C:P ratio, cedar and cypress plantations again showed significantly high values—957 and 914, respectively (Figure 4B, Table S9). The lowest C:P ratio occurred in the fir plantations (394). The N:P ratios of cedar and cypress plantation and broad-leaved forests were significantly higher than that of fir plantations (Figure 4C, Table S10). Fir plantations, pine forests, and subalpine coniferous forests had low N:P ratios with no significant differences. Based on P concentration, C:N:P ratios were 943:19.5:1 for cedar and cypress plantations, 625:19.3:1 for broad-leaved forests (deciduous and evergreen), and 412:13.0:1 for subalpine coniferous forests and fir plantations, on average.

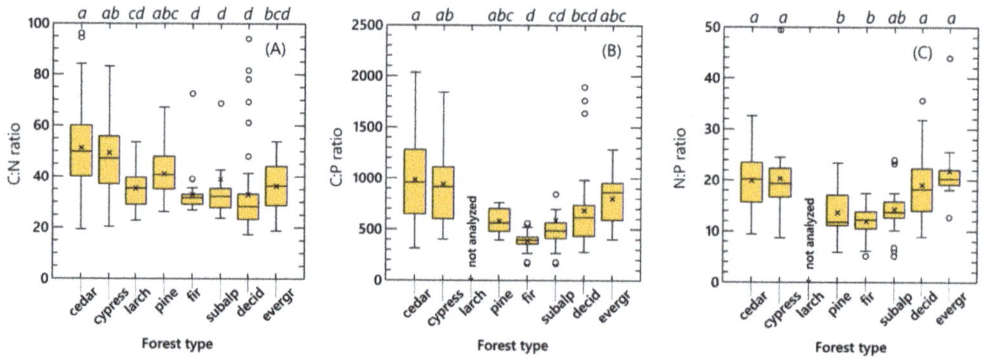

Figure 4. Box and whisker plots of stoichiometry (C:N (**A**), C:P (**B**), and N:P (**C**) ratios) of the forest floor among Japanese forest types. The definitions of box and whisker are the same as those used in Figure 2. Different lowercase letters on the figure denote significant differences between forest types at $p < 0.05$ by the Steel–Dwass test.

Having the lowest C:Ca (34.8) was considered a special characteristic of cedar plantations. This was significant when compared with the C:Ca ratio of other forest types, which ranged from 59.9 to 72.6 (Table S11). No significant difference among forest types was found for C:Mg ratios, that is, from 267 to 444 (Table S12). Carbon-to-potassium (C:K) ratios were significantly lower in fir plantations (233) than in the cedar plantations (510) (Table S13).

3.4. Relationship between the Dry Weight of the Forest Floor and Nutrient Storage

Nutrient storage in the forest floor was significantly correlated with its dry weight for all elements ($p < 0.001$). However, the coefficient of determination (R^2) varied with the different elements and forest types (Table A1). The relationships in cedar plantations and those in deciduous broad-leaved forests are shown in Figures 5 and 6, respectively, as examples. The storage of N was most closely correlated with the dry weight of the forest floor mass in most forest types: the R^2 of N ranged from 0.697 (evergreen broad-leaved forests) to 0.924 (pine forests). However, except for N, nutrient storage was weakly correlated with forest floor mass in most of elements. Particularly, the variation of the R^2 was large in P, ranging from 0.443 (cedar) to 0.903 (pine). In coniferous plantations and forests, the storages of mineral elements had moderate correlations with their forest floor mass, approximately 0.6–0.7 in the R^2, irrespective of elements and forest types. Low R^2 occurred in broad-leaved forests particularly for Ca and Mg, which ranged from 0.102 to 0.381.

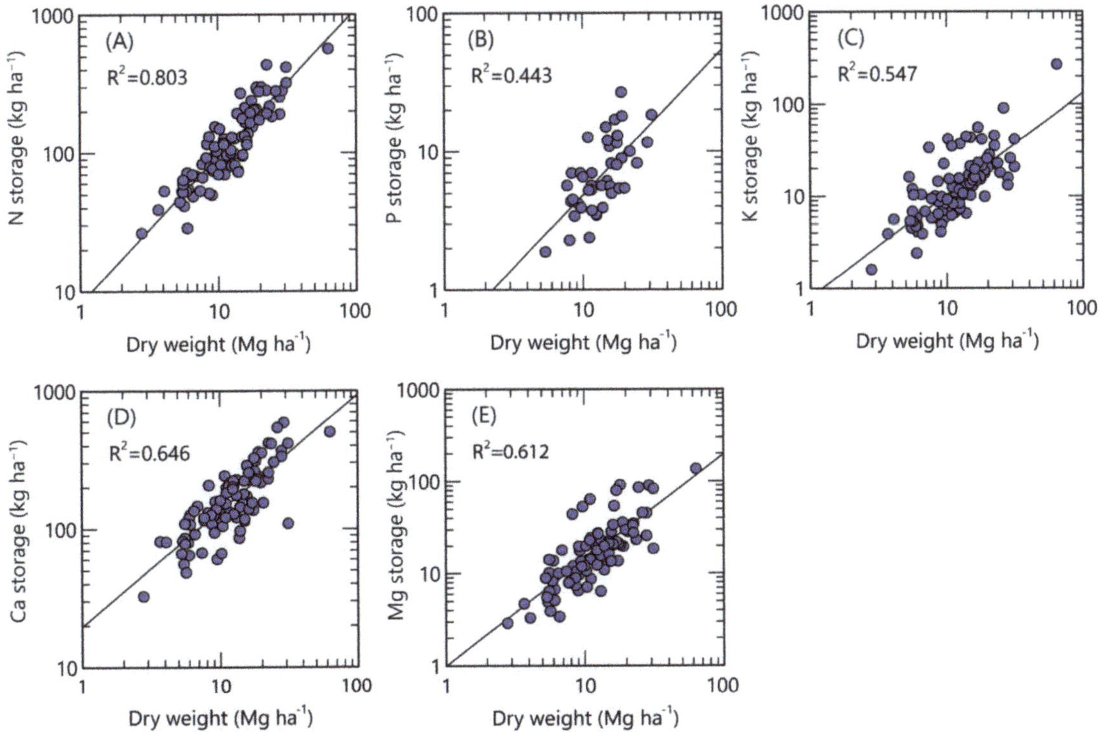

Figure 5. Relationship between nutrient storage (nitrogen (**A**), phosphorus (**B**), potassium (**C**), calcium (**D**), and magnesium (**E**)) and dry weight of the forest floor in Japanese cedar plantations. Regression equations are shown in Table A1.

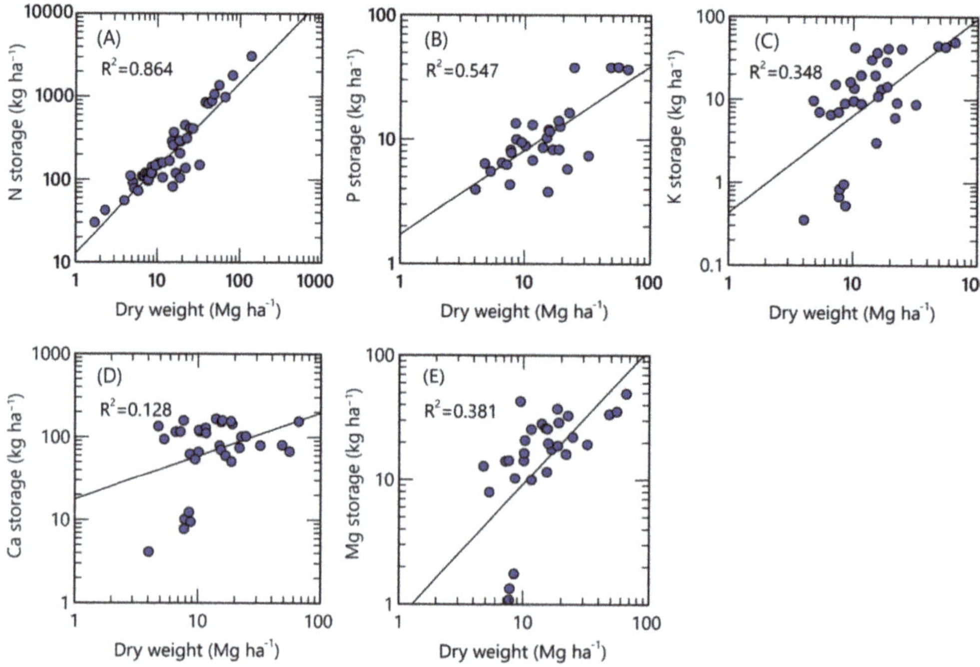

Figure 6. Relationship between nutrient storage (nitrogen (**A**), phosphorus (**B**), potassium (**C**), calcium (**D**), and magnesium (**E**)) and dry weight of the forest floor in deciduous broad-leaved forests in Japan. Regression equations are shown in Table A1.

4. Discussion

4.1. Forest Floor Mass

Nutrient storage is calculated by multiplying the forest floor nutrient concentration by its dry weight. Factors that play a role in developing forest floor mass should be discussed, especially for cedar and cypress plantations. According to the well-applied litter bag method, the decomposition rate of cedar and cypress leaf litter is not high when compared with other representative tree species in Japan [13,16]. In cypress plantations, it has often been suggested that the rapid physical fragmentation of cypress leaf litter progresses in the earliest stage of decomposition. Fresh fallen cypress leaves were fragmented quickly—within half a year from the litter fall season—into small leaflets, which are more susceptible to erosion or migration into mineral soil [42–44]. Hence, twigs and small branches most commonly remained on the forest floor. It has been emphasized that the forest floor is generally sparsely distributed in cypress plantations with a partly uncovered surface soil [42,45–47], which might explain the small mass of its forest floor.

In cedar plantations, soil macro fauna, such as earthworms and soil crustaceans, are abundant [48,49]. A higher population of soil fauna is reported in cedar plantations compared to cypress plantations [16,48,50]. These faunal activities seem to be responsible for the rapid decomposition of cedar litter. Mesofauna was also likely to contribute to cedar litter decomposition. Kaneko et al. [16] found that oribatid mites invade the inside of the cedar needle litter to ingest it, creating a hollow within leaves that could be detached from rachis in early decomposition stages. Consequently, rachis, petioles, twigs, and cones most commonly remained on the forest floor, likewise cypress plantations. Fallen branches were also scattered as bundles of foliage in cedar plantations [16]. Although the forest floor consisted of a mixture of organic materials going through both early and late decomposition stages, in cedar and cypress plantations, easily decomposable litter parts disappeared quickly, with recalcitrant woody organic materials dominating the forest floor, which is consistent with our findings that cedar and cypress forest floors had high C:N and C:P ratios.

The forest floor of old-growth natural cypress forests developed a thick mass at 44.8 Mg ha^{-1} (Table S2), which was different from their plantations. The large forest floor mass was supposed to resulted from specific site conditions, such as dense understory dwarf bamboo and slow litter decomposition in cool climate regions [51,52]. Besides cool climate conditions, such large amounts of the forest floor in old-growth forests may be responsible for a long steady-state period for accumulating organic matter in forest floor without disturbances and harvesting. Similarly, the forest floor mass in natural fir forests (32.5 Mg ha^{-1}) is larger than those in fir plantations (14.5 Mg ha^{-1}) in Hokkaido prefecture.

4.2. Nutrient Storage of Nitrogen and Phosphorus

The lowest and second-lowest N storage occurred in cypress and cedar plantations at 89 and 111 kg ha^{-1}, respectively. Phosphorus storage was also low in these forest floors, at 3.10 kg ha^{-1} for cypress and 5.66 kg ha^{-1} for cedar plantations. Furthermore, high C:N and C:P ratios within these forest floors suggest that net N and P mineralization does not progress during organic matter decomposition. These findings suggest that the forest floor, in situ, rarely functions as an available nutrient reservoir in the ecosystem. In addition, it is known that cedar litter has high N-fixing activity levels during decomposition [53] but that the high C:N ratio and low N storage suggest that N fixation contributes little toward N storage. Fragmented litter that physically migrated to the A horizon, as discussed above, may then serve as available N and P to vegetations.

On the contrary, in subalpine and cool temperate plantations of larch and fir, large amounts of N and P were stored in the forest floor, which was probably due to the slow litter decomposition taking place in the cool climate. Among broad-leaved forests, beech forests distributed in cool regions also stored large N amounts (450 kg ha^{-1}) in the forest floor (Table S3). Moreover, C:N and C:P ratios were low in these forest floor. Thus, the forest floor could function as an actual nutrient pool for these ecosystems, despite slow decomposition of litter. The observation that fine roots systems were developed in the F and H layers [18,19] indicates the tight relationships between tree growth and forest floor nutrient in the ecosystems. These findings inspire fragile ecosystem of subalpine forests, such that if the site conditions were changed—through climate change and forest disturbances like harvesting and typhoon damage—the nutrient pool will be directly disturbed and would then release large amounts of N and P from the forest floor. Recovery from the disturbances may then require time to again form the necessary thick forest floor.

4.3. Mineral Storage in the Forest Floor

It is a special characteristic of cedar plantations that a considerably high amount of Ca can be stored even within the thin forest floor. The total Ca concentration was significantly higher in the forest floor of cedar plantations. Several studies reported that exchangeable Ca in the A horizon was higher in cedar plantations than in other forest types [49,54–56]. The abundance of faunal decomposers, such as soil crustaceans and earthworms, seemed to be related to Ca richness in the forest floor [48,49]. The Ca concentrations in the fresh fallen cedar leaves were relatively higher among the tree species in Japan, averaging at 17.9 mg g^{-1} [32]. Harada et al. [57] reported that the Ca concentration of the bark of Japanese cedar was considerably higher at 9.5–23.1 mg g^{-1}, when compared to its other parts. Considering the high C:N ratio in the forest floor, the high Ca storage in the forest floor may partly be responsible for the recalcitrant bark of cedar trees.

High Ca storage of the forest floor is likely to play a large buffering capacity for acid deposition. Baba et al. [55] observed that a large amount of exchangeable Ca was leaching from the cedar forest floor, which alleviated the Ca loss by proton load. Takahashi [58] demonstrated that the Ca present in needle litters became water soluble when the decomposing needles became black, suggesting that an increase in the water solubility contributes to Ca mobility from the litter to A horizon. However, such Ca behavior does not contribute Ca accumulation in the forest floors. Further studies are still needed to understand how high Ca concentration and storage develop in the forest floor of cedar plantations.

Significantly high K and P storage were found in subalpine and cool temperate coniferous forests. It was reviewed that the strong correlation between P and K concentrations observed in the fungal component of symbiotic ectomycorrhizal and arbuscular mycorrhizal fungi suggests that P transport to plants was accompanied by K ion [59]. K is a typical ion that is quickly released from fresh litter in initial decomposition stages [60]. However, it might be caught by fungal hyphae developed in the F layers of forest floors [19,61], resulting in large K storage in the forest floor of cool ecosystems. Additionally, some mushrooms are known to act as a K accumulator in forest floors [62–64]. Both K and P storage may be partly controlled by fungal activity in thick forest floors. Mg storage in the forest floors did not clearly show any special tendency among the forest types despite the small numbers of samples. On the basis of this, Mg may have little biological interaction with decomposers in the forest floor.

4.4. Stoichiometry

The interquartile ranges (IQRs) of C:N, C:P, and N:P ratios among the forest types were narrow when compared with those of nutrient storage elements themselves (Tables S8–S10). Although data were collected from varying prefectures with varying site conditions, the stoichiometry of the forest floor ranged within narrow ratios depending on the specific forest type. Cedar and cypress plantations showed high C:N, C:P, and N:P ratios, with broad-leaved forest types showing low C:N and C:P ratios and a high N:P ratio. The forest floor in subalpine and fir plantations had low C:N, C:P, and N:P ratios.

The ratios of C:N, C:P, and N:P of fresh fallen litter are generally in the range of 100 and more, 1000 and more, and between 30 and 50, respectively. They are decreasing gradually by mineralizing C sources [23]. The high C:N and C:P ratios of the forest floor under cedar and cypress plantations suggests that the forest floor consists of relatively young organic materials, i.e., only passing through a short time after senescence. For estimating the decomposition rate of the forest floor organic matter, the residence time of the forest floor can be calculated using annual litterfall data [9,20]. According to a review by Saito [65], total litterfall including branches averaged 5.15 Mg ha^{-1} in cedar plantations and 4.41 Mg ha^{-1} in cypress plantations. Therefore, the mean residence time of the forest floor (forest floor mass/litterfall) of cedar plantations, and cypress plantations are 1.81 and 1.52 years, respectively. When compared with subalpine coniferous forests—having 4.28 Mg ha^{-1} of annual litterfall [65]—the residence time at the forest floor is much longer (8.72 years) than that of cedar, cypress, and broad-leaved forests. The residence time indicates nutrient availability as well. Lang [66] reported that the mean residence time of the forest floor can be used as a P recycling indicator, with turnover rates of the forest floor increasing with accompanying increasing total P stocks in the soil systems of European beech forests. Long residence times in subalpine coniferous forests suggest that the forest floor provides available P and probably N with slow rates.

As for the N:P ratio, fungal biomass and physiology may regulate N and P dynamics, which are both affected by temperature regime. Reich and Oleksyn [67] found that leaf N and P concentrations and N:P ratios decreased with decreasing mean annual temperature. Li et al. [68] also showed that soil microbial N and P concentrations increased, while microbial N:P ratios decreased with decreasing temperature regime. Such effects of temperature on fungal activities would be obvious in moder and mull humus forms in subalpine and cool temperate forests due to high fungal biomass in the F layer [19,61].

Because the land in Japan is widely influenced by volcanic materials, the andic properties of soil often have high P adsorption capacities that act as inhibitors of P availability [69,70]. Higher N:P ratios in the forest floor of cedar and cypress plantations might be affected by the interaction of mineral soils with andic properties because of soil contamination in the thin forest floor.

4.5. Correlation between Dry Weight of the Forest Floor and Nutrient Storage

It was expected that positive correlations between dry weight and nutrient storage in the forest floors existed because the nutrient storage was calculated using the parameter of dry weight. However, the correlations were usually weak for most of the elements, except for N (Table A1). Furthermore, the correlations between dry weight and nutrient concentration in the forest floors were weak (data are not shown; R^2 are 0.000 to 0.268). N dynamics in the forest floor would be closely reflected by C mineralization processes by decomposers, but the nutrient concentration in the forest floor seems to be affected by factors other than forest type and leaf litter decomposition. Geochemical factors, such as mineral associations of P as is discussed above, might influence the P concentration in the forest floors [71,72]. Vogt et al. [20] reported that input of organic matter from fine-roots developed in the forest floor is a crucial factor, decreasing the mean residence time and nutrient turnover rate of forest floor in fir forests of USA. Such mechanisms may occur in our dataset; the subalpine coniferous forests showed relatively weak correlation with low R^2 for all elements. Since deciduous and evergreen broad-leaved forest groups are composed of various tree species, lower R^2 values are unavoidable. Organic acid production in the thick organic layer and nitrate formation through nitrification may militate decrease in cations in the forest floor [73,74]. Other factors were also pointed out for forest floor nutrient conditions: soil fertility [75], forest age [76], and understory vegetation [77]. Furthermore, external environmental factors affect nutrient condition of the forest floor. Acid deposition accelerated the leaching of exchangeable cations [10,78,79]. Thus, accurate estimation of nutrient storage in the forest floor from the dry weight of forest floor mass is not recommended except for N, due to the effects of variation of nutrient concentration by various reasons.

5. Conclusions

A meta-analysis was applied to data for obtaining information on nutrient storage and stoichiometric ratios of the forest floors of predominant forest types in Japan. In cedar and cypress plantations established in fertile sites, the forest floor stored low N and P with high C:N and C:P ratios, suggesting that the forest floor plays only a minor role as a nutrient reservoir. Subalpine coniferous forests and fir plantations in cool climates had large N and P storage with low C:N and C:P ratios in the forest floor, which indicates that nutrient resource and availability are largely dependent on the forest floor despite of slow organic matter decomposition. Japanese cedar plantations are characterized by having large Ca storage in a relatively thin forest floor layer, suggesting a high chemical buffering capacity for acid depositions. Stoichiometry is a useful tool for estimating the quality of the forest floor. The difference in the stoichiometry seems to be reflected by the processes of litter decomposition and the composition of decomposing organic materials in the forest floor. Nitrogen storage was closely related to the dry weight of the forest floor, but P and mineral element storage were not. Because the forest floor is a labile pool of nutrients against forest management and climate conditions, the knowledge of the storage size and stoichiometry of nutrients in the forest floor can provide a better perspective for tree species selection in plantation forestry, ecosystem management, and climate change impact. Lastly, the statistical values in this study may be biased due to the maldistribution of sample positions and imbalances in sample size. Systematic sampling can improve the grasp of the whole aspect of forest floor quality and quantity of Japan.

Supplementary Materials: The following are available online at https://www.mdpi.com/article/10.3390/soilsystems5030051/s1. Table S1. Sample size of forest floor mass and nutrients in the prefecture of Japan; Table S2. Statistics of dry weight (Mg ha^{-1}) of forest floor of different forest types; Table S3. Statistics of nitrogen storage (kg ha^{-1}) in the forest floor of different forest types; Table S4. Statistics of phosphorus storage (kg ha^{-1}) in the forest floor of different forest types; Table S5. Statistics of potassium storage (kg ha^{-1}) in the forest floor of different forest types; Table S6. Statistics of calcium storage (kg ha^{-1}) in the forest floor of different forest types; Table S7. Statistics of magnesium storage (kg ha^{-1}) in the forest floor of different forest types; Table S8. Statistics of

C:N ratio in the forest floor of different forest types; Table S9. Statistics of C:P ratio in the forest floor of different forest types; Table S10. Statistics of N:P ratio in the forest floor of different forest types; Table S11. Statistics of C:Ca ratio in the forest floor of different forest types; Table S12. Statistics of C:Mg ratio in the forest floor of different forest types; Table S13. Statistics of C:K ratio in the forest floor of different forest types.

Funding: This research received no external funding.

Acknowledgments: I thank K. Ishizuka, S. Ohta, and M. Kato for establishing the monitoring protocols for measuring forest floor mass in a project funded by the Forestry Agency of Japan which triggered this study. I also thank two anonymous reviewers and the academic editor for their valuable comments on the manuscript.

Conflicts of Interest: The author declares no conflict of interest.

Appendix A

Table A1. Parameters and the coefficient of determination (R^2) for regression lines between elements/nutrients within and dry weight of the forest floor (FF) using the following equation: Log_{10} (element, kg ha^{-1}) = a + b Log_{10} (FF, Mg ha^{-1}).

Element	Forest Type	a	b	R^2	n
Nitrogen	Cedar	0.916	1.063	0.803	98
	Cypress	0.899	1.083	0.871	52
	Larch	1.053	1.031	0.893	16
	Pine	0.941	1.089	0.924	21
	Todo fir	1.038	1.060	0.803	31
	Subalpine coniferous	1.050	1.045	0.755	25
	Deciduous	1.110	1.048	0.864	49
	Evergreen	1.275	0.839	0.697	22
Phosphorus	Cedar	−0.371	1.054	0.443	43
	Cypress	−0.359	1.003	0.598	26
	Pine	−0.047	0.920	0.903	10
	Todo fir	−0.397	1.337	0.709	30
	Subalpine coniferous	0.089	0.923	0.535	18
	Deciduous	0.233	0.678	0.547	30
	Evergreen	−0.454	1.202	0.725	14
Potassium	Cedar	−0.093	1.107	0.547	96
	Cypress	0.124	1.004	0.609	46
	Larch	−0.087	1.198	0.682	16
	Pine	0.343	0.810	0.753	15
	Todo fir	0.277	0.965	0.700	29
	Subalpine coniferous	0.633	0.680	0.672	13
	Deciduous	−0.372	1.200	0.348	33
	Evergreen	0.223	0.890	0.581	16
Calcium	Cedar	1.292	0.840	0.646	96
	Cypress	0.949	0.836	0.676	44
	Larch	0.878	0.958	0.613	16
	Pine	1.134	0.645	0.631	15
	Todo fir	0.063	1.603	0.574	27
	Subalpine coniferous	1.328	0.668	0.393	12
	Deciduous	1.248	0.522	0.128	33
	Evergreen	1.705	0.397	0.102	16
Magnesium	Cedar	−0.003	1.148	0.612	92
	Cypress	0.166	0.382	0.567	41
	Larch	0.353	0.853	0.576	15
	Pine	0.634	0.567	0.538	15
	Deciduous	−0.117	1.083	0.381	32
	Evergreen	0.694	0.595	0.209	16

References

1. IPCC. *2006 IPCC Guidelines for National Greenhouse Gas Inventories*; Prepared by National Greenhouse Gas Inventories Programme; Eggleston, H.S., Buendia, L., Miwa, K., Ngara, T., Tanabe, K., Eds.; IGES: Kanagawa, Japan, 2006; ISBN 978-4-88788-032-0.
2. Takahashi, M. Comparison of nutrient concentrations in organic layers between broad-leaved and coniferous forests. *Soil Sci. Plant Nutr.* **1997**, *43*, 541–550. [CrossRef]
3. Lee, S.J.; Yim, J.S.; Son, Y.M.; Son, Y.; Kim, R. Estimation of Forest Carbon Stocks for National Greenhouse Gas Inventory Reporting in South Korea. *Forests* **2018**, *9*, 625. [CrossRef]
4. Domke, G.M.; Perry, C.; Walters, B.; Woodall, C.W.; Russell, M.B.; Smith, J.E. Estimating litter carbon stocks on forest land in the United States. *Sci. Total Environ.* **2016**, *557–558*, 469–478. [CrossRef] [PubMed]
5. Joo, S.J.; Yim, M.H.; Nakane, K. Contribution of microarthropods to the decomposition of needle litter in a Japanese cedar (*Cryptomeria japonica* D. Don) plantation. *For. Ecol. Manag.* **2006**, *234*, 192–198. [CrossRef]
6. Petersen, H.; Luxton, M. A Comparative Analysis of Soil Fauna Populations and Their Role in Decomposition Processes. *Oikos* **1982**, *39*, 288. [CrossRef]
7. Saitoh, S.; Fujii, S.; Takeda, H. Evaluation of the bottom-up force of accumulated organic matter on microarthropods in a temperate forest floor. *Eur. J. Soil Biol.* **2011**, *47*, 409–413. [CrossRef]
8. Li, X.; Yin, X.; Wang, Z.; Fan, W. Interaction between decomposing litter and soil fauna of the *Betula ermanii* forest floor of the Changbai Mountains, China. *Can. J. For. Res.* **2014**, *44*, 1507–1514. [CrossRef]
9. Gosz, J.R.; Likens, G.E.; Bormann, F.H. Organic matter and nutrient dynamics of the forest and forest floor in the Hubbard Brook Forest. *Oecologia* **1976**, *22*, 305–320. [CrossRef] [PubMed]
10. Hayashi, S.; Sudo, R. Role of litter in acid buffer capacity in forest area. *Environ. Sci.* **1997**, *10*, 11–19. [CrossRef]
11. Jiang, J.; Wang, Y.; Yu, M.; Cao, N.; Yan, J. Soil organic matter is important for acid buffering and reducing aluminum leaching from acidic forest soils. *Chem. Geol.* **2018**, *501*, 86–94. [CrossRef]
12. Berg, B.; McClaugherty, C. *Plant Litter: Decomposition, Humus Formation, Carbon Sequestration*, 4th ed.; Springer: Berlin/Heidelberg, Germany, 2020; ISBN 978-3-030-59630-9.
13. Osono, T.; Takeda, H. Decomposition of organic chemical components in relation to nitrogen dynamics in leaf litter of 14 tree species in a cool temperate forest. *Ecol. Res.* **2004**, *20*, 41–49. [CrossRef]
14. Aber, J.D.; Melillo, J.M.; McClaugherty, C.A. Predicting long-term patterns of mass loss, nitrogen dynamics, and soil organic matter formation from initial fine litter chemistry in temperate forest ecosystems. *Can. J. Bot.* **1990**, *68*, 2201–2208. [CrossRef]
15. Yue, K.; Ni, X.; Fornara, D.A.; Peng, Y.; Liao, S.; Tan, S.; Wang, D.; Wu, F.; Yang, Y. Dynamics of Calcium, Magnesium, and Manganese During Litter Decomposition in Alpine Forest Aquatic and Terrestrial Ecosystems. *Ecosystems* **2020**, *24*, 516–529. [CrossRef]
16. Kaneko, N.; Katagiri, N.; Miyake, N. Decomposition process of needle litter of Japanese Red Cedar (*Cryptomeria japonica*) by oribatid mites. *J. Jpn. For. Soc.* **1990**, *72*, 158–162. [CrossRef]
17. Ge, X.; Zeng, L.; Xiao, W.; Huang, Z.; Geng, X.; Tan, B. Effect of litter substrate quality and soil nutrients on forest litter decomposition: A review. *Acta Ecol. Sin.* **2013**, *33*, 102–108. [CrossRef]
18. Takahashi, M.; Matoba, S.; Sato, T. Size distribution and concentration of carbon and nitrogen in each size class of organic matter in the organic layers under a Pinus pumila stand. *Jpn. J. For. Environ.* **1996**, *38*, 109–114. [CrossRef]
19. Kanazawa, S.; Wada, H.; Takesima, S.; Takai, Y. The decomposition processes and existence forms of organic matter in subalpine forest soil (Part 1): Microscopic observation and carbon-nitrogen content of fractionated organic layer of Pwh soil type of Mt. Shigayama. *Jpn. J. Soil Sci. Plant Nutr.* **1977**, *48*, 181–186. [CrossRef]
20. Vogt, K.A.; Grier, C.C.; Meier, C.E.; Keyes, M.R. Organic Matter and Nutrient Dynamics in Forest Floors of Young and Mature Abies amabilis Stands in Western Washington, as Affected by Fine-Root Input. *Ecol. Monogr.* **1983**, *53*, 139–157. [CrossRef]
21. Carnol, M.; Bazgir, M. Nutrient return to the forest floor through litter and throughfall under 7 forest species after conversion from Norway spruce. *For. Ecol. Manag.* **2013**, *309*, 66–75. [CrossRef]
22. Tobón, C.; Sevink, J.; Verstraten, J.M. Litterflow chemistry and nutrient uptake from the forest floor in northwest Amazonian Forest ecosystems. *Biogeochemistry* **2004**, *69*, 315–339. [CrossRef]
23. Zechmeister-Boltenstern, S.; Keiblinger, K.M.; Mooshammer, M.; Penuelas, J.; Richter, A.; Sardans, J.; Wanek, W. The application of ecological stoichiometry to plant–microbial–soil organic matter transformations. *Ecol. Monogr.* **2015**, *85*, 133–155. [CrossRef]
24. Moore, T.R.; Trofymow, J.A.; Prescott, C.E.; Fyles, J.; Titus, B.D. Patterns of Carbon, Nitrogen and Phosphorus Dynamics in Decomposing Foliar Litter in Canadian Forests. *Ecosystems* **2006**, *9*, 46–62. [CrossRef]
25. Wen-Ji, M.; Yan-Tao, Z.; Qing-Qing, Z.; Arshad, A.; Qing-Ru, S.; En-Rong, A.Y. C:N:P stoichiometry in forest floor litter of evergreen broad-leaved forests at different successional stages in Tiantong, Zhejiang, eastern China. *Chin. J. Plant Ecol.* **2014**, *38*, 833–842. [CrossRef]
26. Tsukada, M. Vegetation and Climate during the Last Glacial Maximum in Japan. *Quat. Res.* **1983**, *19*, 212–235. [CrossRef]
27. Nakashizuka, T.; Iida, S. Composition, dynamics and disturbance regime of temperate deciduous forests in Monsoon Asia. *Vegetatio* **1995**, *121*, 23–30. [CrossRef]
28. Covington, W.W. Changes in Forest Floor Organic Matter and Nutrient Content Following Clear Cutting in Northern Hardwoods. *Ecology* **1981**, *62*, 41–48. [CrossRef]

29. Zajícová, K.; Chuman, T. Spatial variability of forest floor and topsoil thicknesses and their relation to topography and forest stand characteristics in managed forests of Norway spruce and European beech. *Eur. J. For. Res.* **2020**, *140*, 77–90. [CrossRef]
30. Tokuchi, N.; Takeda, H.; Yoshida, K.; Iwatsubo, G. Topographical variations in a plant-soil system along a slope on Mt Ryuoh, Japan. *Ecol. Res.* **1999**, *14*, 361–369. [CrossRef]
31. Takahashi, M. Direct estimation of carbon mass of organic layer from dry weight. *J. For. Res.* **2005**, *10*, 239–241. [CrossRef]
32. Morita, K. Mineral composition of the fresh litter of major tree species in Japan. *Bull. Gov. For. Exp. Sta.* **1972**, *243*, 33–50.
33. Takahashi, M.; Ishizuka, S.; Ugawa, S.; Sakai, Y.; Sakai, H.; Ono, K.; Hashimoto, S.; Matsuura, Y.; Morisada, K. Carbon stock in litter, deadwood and soil in Japan's forest sector and its comparison with carbon stock in agricultural soils. *Soil Sci. Plant Nutr.* **2010**, *56*, 19–30. [CrossRef]
34. Mashimo, Y. Evaluation of forest growth by quantification of environmental factors. *JARQ Jpn. Agric. Res. Q.* **1978**, *12*, 232–237.
35. Katagiri, N.; Miyake, N.; Fujiwara, Y. Distribution and stand structure of natural Japanese Red-Pine (*Pinus Densiflora* S. et Z.) in Sanbe Forest of Shimane University. *Bull. Fac. Agric.* **1987**, *21*, 39–45.
36. Umezu, K. History and transition of the coastal forest in Japan—Focusing on the Shonai coastal forest. *Tree For. Health* **2016**, *20*, 104–111. [CrossRef]
37. Franklin, J.F.; Maeda, T.; Ohsumi, Y.; Matsui, M.; Yagi, H.; Hawk, G.M. Subalpine Coniferous Forests of Central Honshu, Japan. *Ecol. Monogr.* **1979**, *49*, 311–334. [CrossRef]
38. Takehara, H.; Kubo, T.; Hosokawa, K. Forest soils derived from granite and palaeozoic sedimentary rocks in Kiso region. *J. Jpn. For. Soc.* **1959**, *41*, 436–444. [CrossRef]
39. Takata, K.; Kurinobu, S.; Koizumi, A.; Yasue, K.; Tamai, Y.; Kisanuki, M. Bibliography on Japanese larch (*Larix kaempferi* (Lamb.) Carr.). *Eurasian J. For. Res.* **2005**, *8*, 111–126.
40. Haruki, M. Studies on the material biomass of *Abies sachalinensis* artificial forest. *Res. Bull. Hokkai-Do Univ. For.* **1979**, *36*, 147–254.
41. Yamada, H.; Miyaura, T. Geographic variation in nut size of *Castanopsis* species in Japan. *Ecol. Res.* **2004**, *20*, 3–9. [CrossRef]
42. Sakai, M.; Inoue, K. Amount of migrated coarse organic matter into soil (V)—Monthly changes in Japanese cedar fallen litter on the forest floor by photograph. In *Annual Report of the Shikoku Branch*; Forestry and Forest Products Research Institute: Ibaraki, Japan, 1987; Volume 28, pp. 24–27.
43. Miura, S.; Ugawa, S.; Yoshinaga, S.; Hirai, T.Y.K. Floor Cover Percentage Determines Splash Erosion in *Chamaecyparis obtuse* Forests. *Soil Sci. Soc. Am. J.* **2015**, *79*, 1782–1791. [CrossRef]
44. Ichikawa, T.; Takahashi, T.; Asano, Y. Comparison of Changes in Organic Matter Dynamics due to Stand Age between Artificial Japanese Cedar (*Cryptomeria japonica* D. Don) Forests and Japanese Cypress (*Chamaecyparis obtusa* Sieb. et Zucc.) Forests. *J. Jpn. For. Soc.* **2006**, *88*, 525–533. [CrossRef]
45. Kiyono, Y. Analyses of factors affecting Ao-Layer overage in *Chamaecypayis obtusa* plantations. *J. Jpn. For. Soc.* **1988**, *70*, 71–74. [CrossRef]
46. Miura, S.; Yoshinaga, S.; Yamada, T. Protective effect of floor cover against soil erosion on steep slopes forested with *Chamaecyparis obtusa* (hinoki) and other species. *J. For. Res.* **2003**, *8*, 27–35. [CrossRef]
47. Hattori, S.; Abe, T.; Kobayashi, C.; Tamai, K. Effect of forest floor coverage on reduction of soil erosion in Hinoki plantations. *Bull. For. For. Prod. Res. Inst.* **1992**, *362*, 1–34.
48. Ichikawa, T.; Takahashi, T.; Kobayashi, T. The relation between the population of the earthworm and types of vegetation or soil environment. *J. Jpn. Soc. Reveg. Technol.* **2008**, *34*, 15–20. [CrossRef]
49. Ohta, T.; Niwa, S.; Agetsuma, N.; Hiura, T. Calcium concentration in leaf litter alters the community composition of soil invertebrates in warm-temperate forests. *Pedobiologia* **2014**, *57*, 257–262. [CrossRef]
50. Takeda, H. A 5-year study of litter decomposition processes in a *Chamaecyparis obtusa* Endl. forest. *Ecol. Res.* **1995**, *10*, 95–104. [CrossRef]
51. Akai, T.; Asada, S. Studies on natural reproduction (I) The reproduction of *Chamaecyparis* at Wet Podzolic zone in Kiso district. *Bull. Kyoto Univ. For.* **1967**, *39*, 35–63.
52. Watanabe, M.; Yamaguchi, M.; Tabe, C.; Iwasaki, M.; Yamashita, R.; Funada, R.; Fukami, M.; Matsumura, H.; Kohno, Y.; Izuta, T. Influences of nitrogen load on the growth and photosynthetic responses of Quercus serrata seedlings to O3. *Trees* **2007**, *21*, 421–432. [CrossRef]
53. Yamanaka, T.; Hirai, K.; Aizawa, S.; Yoshinaga, S.; Takahashi, M. Nitrogen-fixing activity in decomposing litter of three tree species at a watershed in eastern Japan. *J. For. Res.* **2011**, *16*, 1–7. [CrossRef]
54. Sawata, S.; Kato, H. Effect of Forest on Soil (Part 2): The base accumulation and other soil properties related to age of Cryptomeria and Japanese cypress stands. *Jpn. J. Soil Sci. Plant Nutr.* **1991**, *62*, 49–58. [CrossRef]
55. Baba, M.; Kato, M.; Sugiura, T.; Kobayashi, H. Calcium accumulation alleviates soil acidification in Japanese cedar (*Cryptomeria japonica*) stands. *Soil Sci. Plant Nutr.* **2004**, *50*, 403–411. [CrossRef]
56. Tanikawa, T.; Ito, Y.; Fukushima, S.; Yamashita, M.; Sugiyama, A.; Mizoguchi, T.; Okamoto, T.; Hirano, Y. Calcium is cycled tightly in Cryptomeria japonica stands on soils with low acid buffering capacity. *For. Ecol. Manag.* **2017**, *399*, 64–73. [CrossRef]
57. Harada, H.; Satoo, H.; Hotta, I.; Hatiya, K.; Tadaki, Y. Study on the nutrient contents of mature Cryptomeria forest. *Bull. Gov. For. Exp. Stn.* **1972**, *249*, 17–74.
58. Takahashi, M. Water soluble elements in decomposing Japanese cedar needle litter. *Soil Sci. Plant Nutr.* **1996**, *42*, 395–399. [CrossRef]

59. Velázquez, E.; Silva, L.R.; Ramírez-Bahena, M.H.; Peix, A. Diversity of potassium-solubilizing microorganisms and their interactions with plants. In *Potassium Solubilizing Microorganisms for Sustainable Agriculture*; Springer: New Delhi, India, 2016; pp. 99–110. ISBN 9788132227762.
60. Osono, T.; Takeda, H. Potassium, calcium, and magnesium dynamics during litter decomposition in a cool temperate forest. *J. For. Res.* **2004**, *9*, 23–31. [CrossRef]
61. Osono, T. Hyphal length in the forest floor and soil of subtropical, temperate, and subalpine forests. *J. For. Res.* **2015**, *20*, 69–76. [CrossRef]
62. Vinichuk, M.; Taylor, A.; Rosén, K.; Johanson, K. Accumulation of potassium, rubidium and caesium (133Cs and 137Cs) in various fractions of soil and fungi in a Swedish forest. *Sci. Total Environ.* **2010**, *408*, 2543–2548. [CrossRef] [PubMed]
63. Yoshida, S.; Muramatsu, Y. Concentrations of radiocesium and potassium in Japanese mushrooms. *Environ. Sci.* **1994**, *7*, 63–70. [CrossRef]
64. Cromack, K.; Todd, R.L.; Monk, C.D. Patterns of basidiomycete nutrient accumulation in conifer and deciduous forest litter. *Soil Biol. Biochem.* **1975**, *7*, 265–268. [CrossRef]
65. Saito, H. Materials for the studies of litterfall in forest stands. *Bull. Kyoto Prefect. Univ. For.* **1981**, *25*, 78–89.
66. Lang, F.; Krüger, J.; Amelung, W.; Willbold, S.; Frossard, E.; Bünemann, E.K.; Bauhus, J.; Nitschke, R.; Kandeler, E.; Marhan, S.; et al. Soil phosphorus supply controls P nutrition strategies of beech forest ecosystems in Central Europe. *Biogeochemistry* **2017**, *136*, 5–29. [CrossRef]
67. Reich, P.B.; Oleksyn, J. Global patterns of plant leaf N and P in relation to temperature and latitude. *Proc. Natl. Acad. Sci. USA* **2004**, *101*, 11001–11006. [CrossRef]
68. Li, P.; Yang, Y.; Han, W.; Fang, J. Global patterns of soil microbial nitrogen and phosphorus stoichiometry in forest ecosystems. *Glob. Ecol. Biogeogr.* **2014**, *23*, 979–987. [CrossRef]
69. Kunito, T.; Tsunekawa, M.; Yoshida, S.; Park, H.-D.; Toda, H.; Nagaoka, K.; Saeki, K. Soil Properties Affecting Phosphorus Forms and Phosphatase Activities in Japanese Forest Soils. *Soil Sci.* **2012**, *177*, 39–46. [CrossRef]
70. Nanzyo, M.; Dahlgren, R.; Shoji, S. Chapter 6 Chemical Characteristics of Volcanic Ash Soils. *Dev. Soil Sci.* **1993**, 145–187. [CrossRef]
71. Stahr, S.; Graf-Rosenfellner, M.; Klysubun, W.; Mikutta, R.; Prietzel, J.; Lang, F. Phosphorus speciation and C:N:P stoichiometry of functional organic matter fractions in temperate forest soils. *Plant Soil* **2017**, *427*, 53–69. [CrossRef]
72. Walker, T.; Syers, J. The fate of phosphorus during pedogenesis. *Geoderma* **1976**, *15*, 1–19. [CrossRef]
73. Johnson, D.W.; Richter, D.D.; Van Miegroet, H.; Cole, D.W. Contributions of Acid Deposition and Natural Processes to Cation Leaching from Forest Soils: A Review. *J. Air Pollut. Control. Assoc.* **1983**, *33*, 1036–1041. [CrossRef]
74. Dijkstra, F.A.; Geibe, C.; Holmström, S.; Lundström, U.S.; Van Breemen, N. The effect of organic acids on base cation leaching from the forest floor under six North American tree species. *Eur. J. Soil Sci.* **2001**, *52*, 205–214. [CrossRef]
75. Vesterdal, L.; Raulund-Rasmussen, K. Forest floor chemistry under seven tree species along a soil fertility gradient. *Can. J. For. Res.* **1998**, *28*, 1636–1647. [CrossRef]
76. Hamburg, S.P.; Yanai, R.D.; Arthur, M.A.; Blum, J.D.; Siccama, T.G. Biotic Control of Calcium Cycling in Northern Hardwood Forests: Acid Rain and Aging Forests. *Ecosystems* **2003**, *6*, 399–406. [CrossRef]
77. MacLean, D.A.; Wein, R.W. Litter production and forest floor nutrient dynamics in pine and hardwood stands of New Brunswick, Canada. *Ecography* **1978**, *1*, 1–15. [CrossRef]
78. Currie, W.S.; Aber, J.D.; Driscoll, C.T. Leaching of nutrient cations from the forest floor: Effects of nitrogen saturation in two long-term manipulations. *Can. J. For. Res.* **1999**, *29*, 609–620. [CrossRef]
79. Wessel, W.W.; Tietema, A. Metal distribution across different pools in the organic layer of a forest under acid deposition and its consequences for the metal dynamics. *Plant Soil* **1995**, *171*, 341–350. [CrossRef]

soil systems

MDPI

Article

Long-Term Effects of Forest Plantation Species on Chemical Soil Properties in Southern Rwanda

Peter Rwibasira [1,2,*], Francois Xavier Naramabuye [3], Donat Nsabimana [3] and Monique Carnol [1,*]

[1] Laboratory of Plant and Microbial Ecology, InBioS, University of Liège, Botany Bât. 22, Chemin de la Vallée 4, 4000 Liège, Belgium
[2] College of Science and Technology (CST), University of Rwanda, KN 73 St., Nyarugenge, Kigali P.O. Box 3900, Rwanda
[3] College of Agriculture, Animal Sciences and Veterinary Medicine (CAVM), University of Rwanda, Huye Campus, Butare P.O. Box 56, Rwanda; naramabuyefrancois@gmail.com or f.x.naramabuye@ur.ac.rw (F.X.N.); donatus_1999@yahoo.fr or d.nsabimana@ur.ac.rw (D.N.)
* Correspondence: p.rwibasira@ur.ac.rw (P.R.); m.carnol@uliege.be (M.C.)

Abstract: Understanding the long-term effects of tree species on soil properties is crucial for the development of forest restoration policies in relation to the choice of species that meet both environmental and local livelihood needs. This study was performed in the Arboretum of Ruhande, Southern Rwanda, where monocultures of 148 deciduous and 56 conifer species have been established in 0.25 ha replicated plots from 1933 onwards. We investigated the effects of six exotic and two native tree species planted in monoculture plots and native species mixed within one self-regenerated plot on soil properties in two layers (0–5 cm and 5–10 cm depth). We measured general soil properties (pH, SOM, exchangeable base cations) and water-soluble C and N as a proxy for soil functioning. Changes in soil properties were observed in the upper soil layer for all tree species. Planting Eucalyptus species caused soil acidification, whereas soil exchangeable cations and pH were higher under native species (*Entandrophragma excelsum* and *Polyschias fulva*) and mixed native species. The effects of tree species were more pronounced for hot water-extractable C and N than for other soil properties. Their analyses could be used for detecting changes in soil functioning linked to vegetation types.

Keywords: soil quality; soil functions; Eucalyptus species; soil acidity; exchangeable cations; water-extractable C and N; Ruhande Arboretum; Rwanda

Citation: Rwibasira, P.; Naramabuye, F.X.; Nsabimana, D.; Carnol, M. Long-Term Effects of Forest Plantation Species on Chemical Soil Properties in Southern Rwanda. *Soil Syst.* **2021**, *5*, 59. https://doi.org/10.3390/soilsystems5040059

Academic Editor: Klaus von Wilpert

Received: 31 July 2021
Accepted: 20 September 2021
Published: 26 September 2021

1. Introduction

Plants and soils are key components of terrestrial ecosystems, and changes in vegetation cover may lead to changes in soil properties, especially in the forest topsoil [1,2]. Soils provide important ecosystem functions, such as nutrient cycling, carbon sequestration in soil organic matter [3], and provision of fiber and food through the supply of water and nutrients to the vegetation [4]. In turn, trees are an important soil-forming factor, and tree species can affect soils through various mechanisms, including nutrient uptake and return to the soil, soil organic matter dynamics, changes in soil acidity via root–soil exchange, and protection from erosion [2,5,6]. As a result, physical, chemical, and biological properties as well as the related processes may be affected by tree species [7] and thus influence the nutrient supply capacity of the soils to the trees. In tropical forests, soil fertility relies heavily on the internal cycling of nutrients through the rapid decomposition of above- and belowground litter from vegetation, taking place in the thin upper soil horizon [8,9]. Understanding the effect of tree species is particularly important in tropical forest ecosystems for the long-term preservation of soil quality and for promoting soil functioning.

Recently, there has been much interest and debate about the delimitation of the concepts of soil quality, health, fertility, and ecosystem services [10–13], with sometimes overlapping or contradicting views, leading to confusion across disciplines. Karlen et al. [10]

defined soil quality as "the capacity of a specific kind of soil to function, within natural or managed ecosystem boundaries, to sustain plant and animal productivity, maintain or enhance water and air quality, and support human health and habitation". They also recommended that soil quality should be evaluated based on soil function without, however, providing a specific definition of soil functions. Greiner et al. [14] indicated that soil functions result from the interaction of soil properties and processes and that they are related to ecosystem services and human benefits, as illustrated in the "Cascading framework" [15]. Soil functions can be measured through physical, chemical, and biological soil properties and processes, which are used as the basic tools to evaluate soil quality under different land-use systems [16,17].

Among many soil properties, soil organic matter and/or carbon (SOM, SOC), pH, and base cations are frequently used as primary indicators of forest soil quality [13,18]. Given the importance of soil organic matter for soil functioning [19], several studies investigated tree species-induced changes on total soil organic carbon (SOC) after afforestation [20]. The findings differed, with some studies showing no change [21,22], increased SOC [23,24], and decreased SOC [20,25]. Numerous factors may govern these contradictory results, and, in a review of 43 afforestation studies, Paul et al. [26] identified, in order of importance, previous land use, climate, and forest tree species as key factors influencing forest soil organic matter dynamics. While SOM is recognized as an important global indicator of soil quality, its slow dynamics does not allow for early detection of changes [27]. Further, most SOM might not be available for microbial breakdown; therefore, total SOM might not be a relevant indicator of soil functioning [28]. For example, in a grassland, 60% of SOM was shown to be a recalcitrant pool [29]. SOM undergoes continuous changes that generate distinct chemical and physical organic matter fractions with different turnover rates, from readily available labile to recalcitrant carbon and nitrogen fractions [30,31]. Labile SOM fractions have recently gained interest as indicators of soil quality because they are more sensitive to changes in vegetation cover and land use than the total organic matter [32,33]. Additionally, being the main substrate and energy source for soil microorganisms, labile carbon and nitrogen fractions such as water-extractable C and N are linked to soil nutrient cycling and thus to soil functioning [28,31,34].

Rwanda experienced the loss of its natural forest cover from 30% in 1920 to 8% in 1998 [35]. This deforestation in a country whose topography is dominated by steep sloping hills with heavy precipitation has led to accelerated soil erosion and to the decline of soil fertility [36]. A tree plantation program was initiated in 2010 to promote "in situ soil conservation through agroforestry and forest landscape restoration" [37] and halt the decline of forest cover, counter soil erosion, and land degradation as well as to meet increasing demands for wood. Within this program, the government of Rwanda has mobilized its entire population and non-governmental organizations to plant trees, mainly fast-growing exotic species, and to maintain remaining forests, whereby a target was set in 2010 to restore the country's forest cover from 19.6% to 30% by 2030 [38]. This target was reached in 2020 with 724,695 ha (30.4%) forest cover in the country [39]. This forest cover is composed of the following: 387,425 ha (53.5%) forest plantations, wooded savannahs in the east cover 161,843 ha (22.3%), natural montane forests occupy 130,850 ha (18.1%), shrublands cover 43,963 ha (6.1%), and 613 ha are occupied by bamboo [40]. Of the forest plantations, eucalyptus species are dominant with 89%, followed by 6.5% pines, 3.1% mixed exotic forests, and 1.4% being plantations of native species [41]. While the effects of tree species on soils were extensively studied for temperate ecosystems, data on tropical soils are scarce [42]. The results of most studies may therefore have limited relevance within the context of tropical soils [43]. Additionally, numerous studies were performed in relatively short-term common garden experiments [42]. We need an in-depth understanding of the effects of the planted species on soil quality in tropical ecosystems. Such expertise for local conditions is important for selecting suitable species promoting soil functioning in these tropical forest ecosystems.

The general aim of this study was to assess the long-term effects of tree species planted in Rwanda on chemical soil quality, including water-soluble labile C and N fractions, as a proxy for soil functioning. Specific aims were to (i) determine the differences in soil chemical properties between tree species in two soil layers (0–5 cm and 5–10 cm depth); (ii) characterize hot and cold water-extractable mineral N and organic C and N in soils under different plantation species, and (iii) investigate the relationships between labile C and N fractions and other soil properties in response to tree species. We hypothesized that the exotic eucalyptus species would reduce the chemical quality of the soils in comparison to native species and that labile C and N fractions would be more sensitive to a change in tree species than SOM.

2. Materials and Methods

2.1. Study Site

Soils were sampled in the Arboretum of Ruhande (Southern Rwanda, 2°36′ S, 29°44′ E, Figure 1) located at 1638–1737 m elevation on a flat plateau of the Ruhande hill [44]. This site is characterized by a mean annual rainfall of 1230 mm and a temperature between 17.5 °C and 19 °C. The rainfall has a bimodal regime with irregular short rains from September to December and a short dry season (January to February), followed by a heavy rainy season from March to May and a long dry season from June to August [45]. The soil is classified as ferralsol (also known as oxisols in USDA soil taxonomy), a red-brown colored soil with a sandy loam texture and diffuse horizons [46]. It is developed from weathered Precambrian phyllite and granitic batholith parental rocks coated with a mixture of quartzites and mica schists [47,48].

The site was established in 1933 on cultivated land under the request of the colonial leaders of Rwanda-Urundi territory for forestry research, wood, and seed provision to the rest of the country [49]. The size of the arboretum was progressively increased to reach currently 200 ha with 143 hardwood tree species, including 126 introduced exotic species of which 69 are eucalyptus species and 17 are native species. It also contains 57 deciduous tree species and 3 bamboo species, of which two are native to Rwanda [44]. Trees are planted in replicated monoculture stands of 0.25 ha (50 m × 50 m), resulting in 504 numbered plots (with 454 plots of exotic species) separated by inter-plot paths 6–10 m wide (Figure 2). Thinning and removal of shrubs and other invading vegetation is performed annually on all plots, except on an undisturbed plot (4 ha) of self-regenerated mixed native species (Mns). Plots are managed to maintain a constant density of the main tree species by planting in replacement of dead plants. From the 24 selected plots (see below), six were completely re-established, but they were aged minimum 30 years at the time of this study (Table A1). Neighboring local households are allowed to collect dry wood each Friday for cooking. Given that trees were planted on the same site with similar (agricultural) land-use history and climatic conditions, we expect current differences in soil characteristics to reflect the influence of the planted tree species.

The uniqueness of Arboretum of Ruhande in terms of design, landscape, tree species composition, and presence of other living organisms lies in its multiple roles as a global site for forestry conservation, research, educational activities, and a gene bank of forestry germplasm in addition to being the country's main source of forest planting materials [50]. This botanical garden was recently (May 2018) awarded international recognition through its enrollment into the "Queen's Commonwealth Canopy" projects. This is a network of forest conservation initiatives within Commonwealth countries aiming at forest and biodiversity conservation for future generations [51].

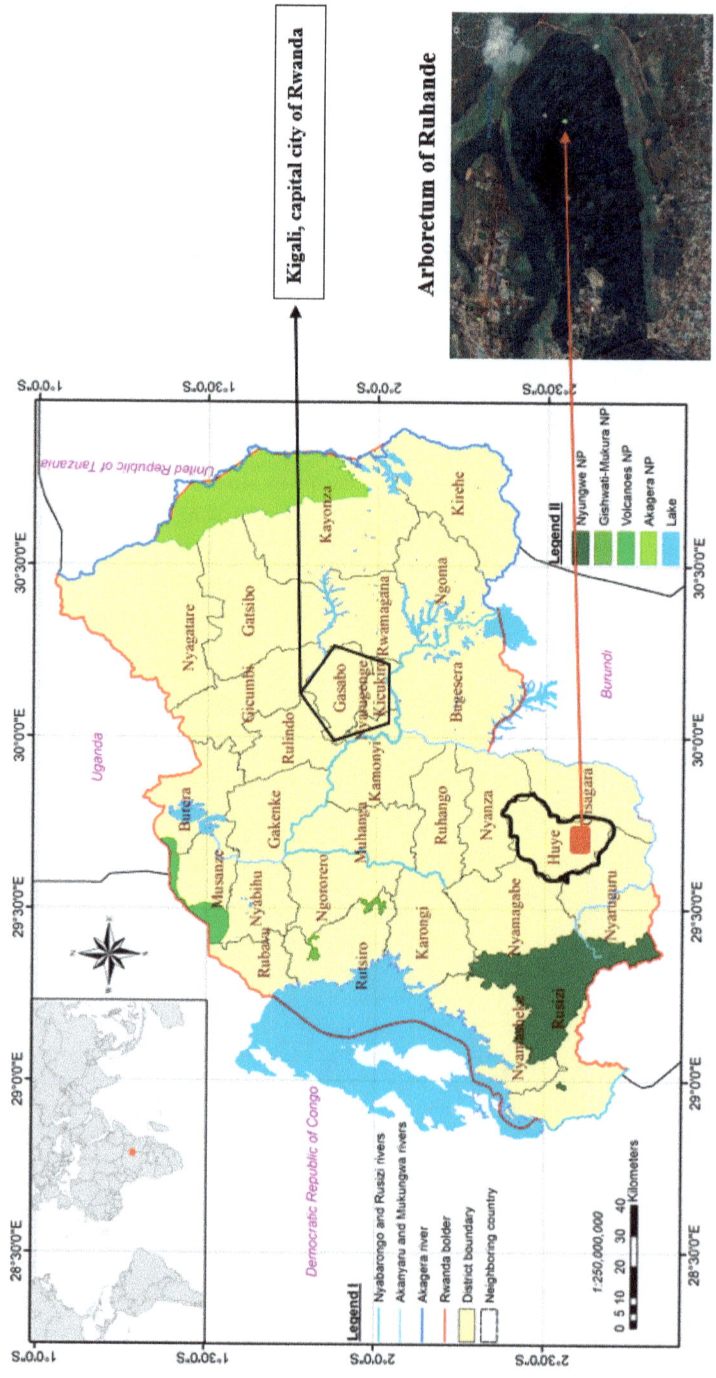

Figure 1. Geographical map of the study site.

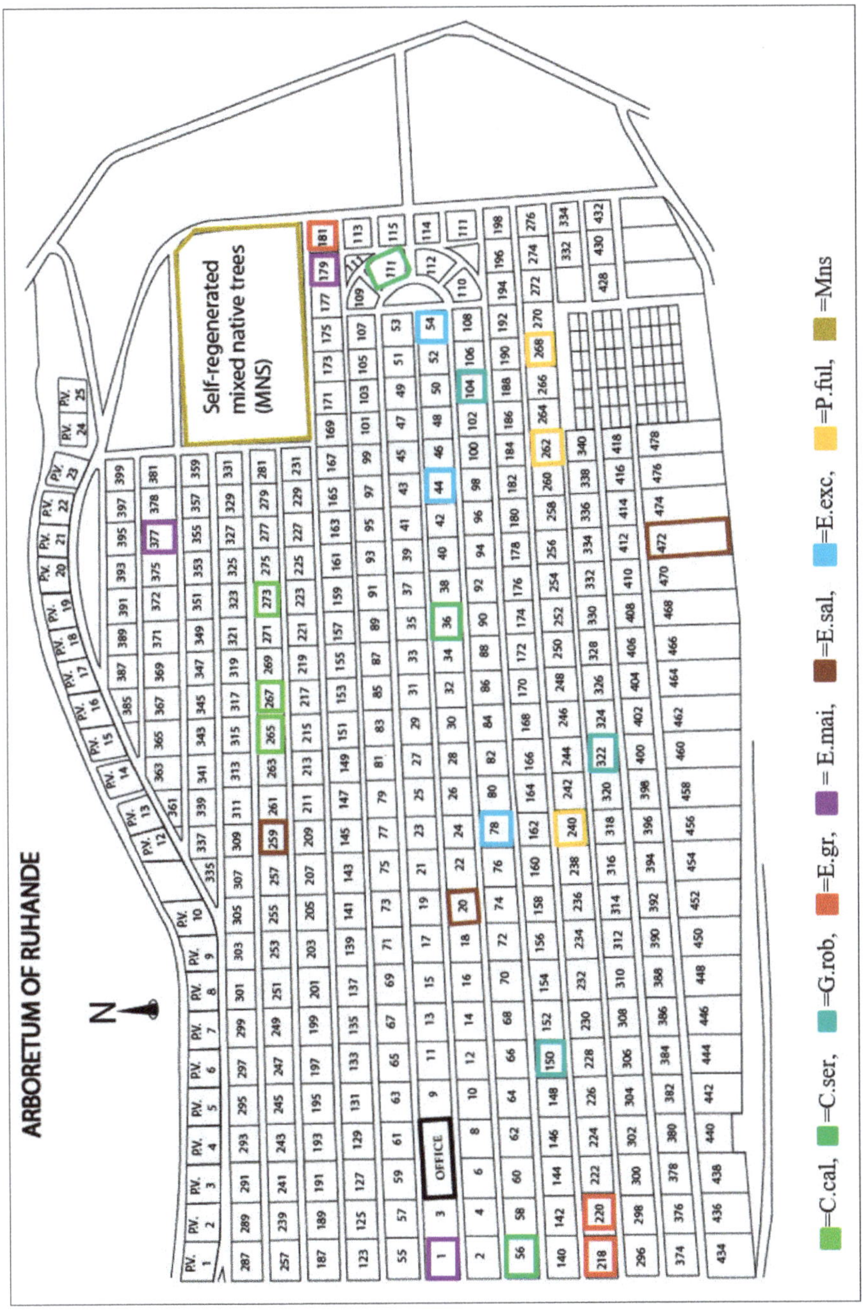

Figure 2. Map of the Arboretum of Ruhande, Rwanda. Studied plots are indicated in colors. More details can be found in Table A1. Adapted with permission from [52]. Copyright 1987 ISAR Foresterie.

2.2. Soil Sampling and Chemical Analyses

Based on the records of forestry seed demands and species adaptability in different regions of the country [53,54], eight species were selected considering three plots per species (Figure 2, Table A1). These included three eucalyptus species (*Eucalyptus grandis*, *Eucalyptus maideni*, and *Eucalyptus saligna*), three agroforestry species (*Calliandra calothyrsus*, *Cedrela serrata*, and *Grevillea robusta*), two native species (*Entandrophragma excelsum* and *Polyscias fulva*), and a self-regenerated plot of native forest (mixed native species = Mns).

Each plot was divided into two sub-plots (25 × 50 m), where soil samples were collected under the trees' canopy at a distance of 1 to 1.5 m from the tree base [55]. One composite sample was taken in each sub-plot by mixing five soil cores (X-shaped sampling) collected using a 30 × 30 cm frame and a shovel. Samples were taken at two soil depths—0–5 cm and 5–10 cm—the most active layers in tropical forest soils with a high rate of organic matter decomposition and nutrient cycling [56]. Thus, we took two composite soil samples per plot at two soil depths. Soils were sieved fresh (4 mm) and stored at 4 °C until analyses.

Gravimetric water content, soil organic matter (SOM), and pH were determined as described by Allen et al. [57]. Briefly, moisture was calculated as the difference between fresh and oven-dried soil at 105 °C for 3 h; SOM was calculated as a weight loss from oven-dry soil after overnight ignition at 550 °C in a muffle furnace. Soil organic carbon (SOC) was estimated by dividing SOM by 1.724 (Van Bemmelen factor), assuming that organic matter contains 58% of organic carbon [58]. The pH_{KCL} was determined in a soil solution (1:2.5 v/v) with 1 M KCl and measured using a pH meter (HI2550 Multiparameter pH Benchtop meter, HANNA® Instruments, Woonsocket, RI, USA). Soil water holding capacity (WHC) was determined using Shaw's method according to Jenkinson and Powlson [59] as the difference between the volume of water (50 mL) added to 25 g of fresh soil and the volume drained after 30 min of saturation in addition to the initial soil moisture content.

Exchangeable cations (Al^{3+}, Ca^{2+}, Fe^{2+}, K^+, Mg^{2+}, Mn^{2+}, Na^+, and Zn^{2+}) were extracted from fresh soil with 0.1 M $BaCl_2$ (1:5 w/v) by agitation for 30 min, followed by centrifugation at 180 rpm [60]. Chemical analysis of the filtered (Macherey Nagel MN 6151/4. Ø 150 mm, Germany) and the acidified (1% HNO_3 Suprapur) $BaCl_2$ extracts was performed using ICP-AESS (Varian, Australia). The sum of exchangeable cations (∑cations) was calculated as the sum of all measured cations, and exchangeable base cations (EBC) were calculated as the sum of Ca^{2+}, K^+, Mg^{2+}, and Na^+; expressed in c mol. kg^{-1}.

Water-extractable C and N were determined using the method of Ghani et al. [61]. Fresh soil was extracted with distilled water (1:6, w/v), shaken (120 rpm, 30 min), centrifuged (4000 rpm, 10 min), and filtered (Whatman #42), representing water-soluble C and N (WSC, WSN) fractions. Hot water-extractable C and N (HWC, HWN) were subsequently extracted from the remaining wet soil, mixed with distilled water (30 mL), and placed in the oven for 16 h at 80 °C. Organic C in the cold (WSC) and the hot water (HWC) extracts was measured using a Total Organic Carbon analyzer (LabToc, Pollution and Process Monitoring, UK). Cold ("WS ... ") and hot water-extractable ("HW ... ") nitrogen forms (N-NH_4: $WSNH_4$, $HWNH_4$; N-NO_3: $WSNO_3$, $HWNO_3$) and total nitrogen (WSNtot, HWNtot) were measured colorimetrically using a continuous flow autoanalyzer equipped with a UV digestor (Autoanalyser3, BranLuebbe, Germany). Organic nitrogen in the extracts (WSN_{org}, HWN_{org}) was calculated as the difference between total nitrogen and mineral nitrogen. Given that most of the mineral N is extracted with cold water, and as ammonium N in hot water extracts comes from hydrolysis of organic N [62], we assumed that HWNtot was entirely deriving from organic N and thus included WSN_{org}, WSNtot, HWN_{org}, HWNtot, WSC:WSN_{org}, and HWC:HWNtot in our analyses.

2.3. Statistical Analysis

We used linear mixed-effects models (LMM) to investigate the differences in soil chemical properties between tree species and soil layers, using *lme4* package and *lmer* function [63] in R, version 3.5.1 [64]. The model used "Species" (9 levels: *C. calothyrsus*,

C. serrata, *G. robusta*, *E. grandis*, *E. maideni*, *E. saligna*, *E. excelsum*, *P. fulva*, and Mixnatives, with three replicates per species), "Layer" (with two levels: upper and lower soil layers), and the interaction between tree species and soil layer (species*layer), which were included in models as fixed effects. "Plot" was included as a random effect to account for the non-independence of the two samples collected within the same plot and the tree age differences between plots. Normality was tested using the Shapiro–Wilk test and/or visual inspection of plotted residuals. Homoscedasticity of random errors was tested using *Levene* test function, part of the Car package in R. Where necessary, response variables were transformed to improve normality and homoscedasticity of errors. Significance of tree species and soil layer effects were analyzed using the model's estimated marginal means (*EMMeans*) function, part of the *multcompView* package in R, using Tukey–Kramer honestly significant difference range post-hoc test to compare all measured parameters across levels at a significant probability of $\alpha = 0.05$. The prediction of response variables explained by the model was determined using a multi-model inference (MuMIn-v1.42.1) package and *r.squaredGLMM* function in R [65].

Pearson's coefficient of correlation was used to determine the correlation between measured variables. Principal component analysis (PCA: using *FactoMineR* and *ggplot2* packages) was used to describe the patterns of variation explained by soil parameters of interest (pH, SOM, EBC, WSC, WSN_{org}, WSNtot, WSC:WSN_{org}, HWC, HWN_{org}, HWNtot, and HWC:HWNtot) between tree species. All statistical analyses and tests were carried out using R software, version 3.5.1 [64].

3. Results

3.1. Chemical Soil Properties in Two Topsoil Layers

Values for all soil parameters (Figure 3, Tables 1 and A2) were significantly higher in the upper (0–5 cm) soil layer compared to the lower (5–10 cm) layer under all tree species (except for Al^{3+}, Fe^{2+}, and Na^{+}). pH, SOM, and EBC were 14%, 57%, and 78% higher in the upper compared to the lower soil layer (4.9, 22%, and 36.3 $cmol_c\ kg^{-1}$ versus, 4.2, 9.6%, and 7.8 $cmol_c\ kg^{-1}$, respectively). Base cations dominated the sum of exchangeable cations, representing 78% (Ca^{2+}), 19.4% (Mg^{2+}), and 2.3% (K^{+}) in the upper soil layer and 65% (Ca^{2+}), 16.5% (Mg^{2+}), and 1.7% (K^{+}) in the lower soil layer (Table A3). In contrast to the other soil parameters, the contribution of Al^{3+}, Fe^{2+}, and Na^{+} to the sum of exchangeable cations was less in the upper soil layer (0.002% Fe^{2+}, 0.2% Al^{3+}, 0.4% Na^{+}) compared to the lower soil layer (0.1% Fe^{2+}, 13.5% Al^{3+}, 2% Na^{+}).

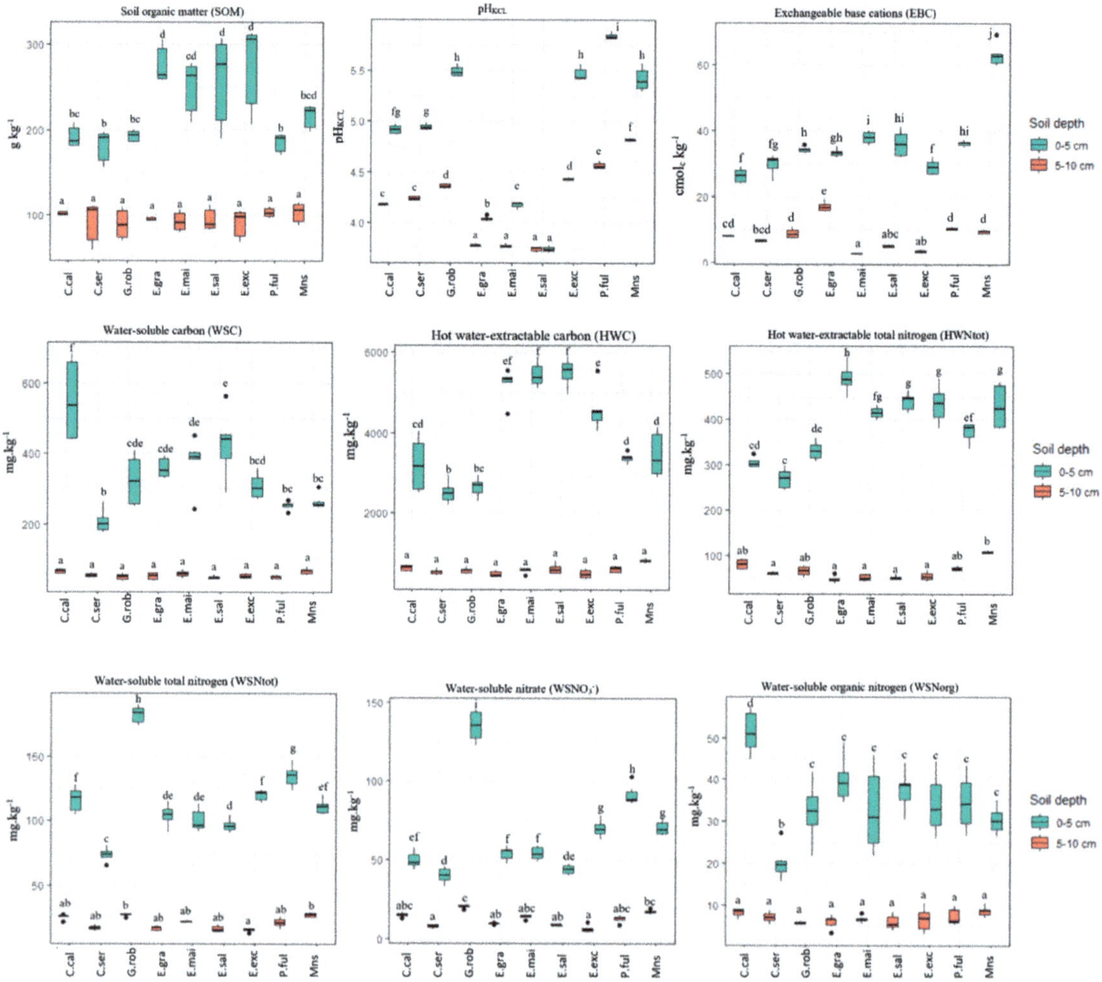

Figure 3. Predicted (LMM) soil properties under eight tree species (C. cal: *Calliandra calothyrsus;* C. ser: *Cedrela serrata;* G. rob: *Grevillea robusta;* E. gra: *Eucalyptus grandis;* E. mai: *Eucalyptus maideni;* E. sal: *Eucalyptus saligna;* E. exc: *Entandrophragma excelsum;* P. ful: *Polyscias fulva*) and in the plot with mixed native species (Mns) at two soil depths (0–5 cm and 5–10 cm). The horizontal black line in the box shows the estimated sample median, while the lower and the upper box boundaries show the first and the third percentiles, respectively. The dots outside the whisker boundaries show observations outside the 5th–95th percentile range. Different letters denote significant differences between tree species and soil layer (mixed linear models, Tukey's HSD, $p < 0.05$).

Table 1. Measured waters' C and N under different tree species in the Arboretum of Ruhande (means ± SEM). Different letters within one parameter denote significant differences between tree species and soil depths (mixed linear models, Tukey's HSD, $p < 0.05$.

Labile C and N Fractions	Soil Layer (cm)	*Calliandra calothyrsus*	*Cedrela serrata*	*Grevillea robusta*	*Eucalyptus grandis*	*Eucalyptus maideni*	*Eucalyptus saligna*	*Entandrophragma excelsum*	*Polyscias fulva*	*Mixed natives*
Water-soluble C and N fractions										
WSC ($mgkg^{-1}$)	0–5	550 ± 49 ^{f}	210 ± 13 ^{b}	320 ± 30 ^{cde}	360 ± 12 ^{cde}	380 ± 29 ^{de}	430 ± 38 ^{e}	310 ± 15 ^{bcd}	250 ± 5 ^{bc}	260 ± 8.9 ^{bc}
	5–10	67 ± 2.5 ^{a}	56± 2.3 ^{a}	50 ± 3.4 ^{a}	52 ± 3.9 ^{a}	60 ± 3.4 ^{a}	50 ± 1.7 ^{a}	53 ± 2.6 ^{a}	50 ± 1.9 ^{a}	68± 3.3 ^{a}
WSNtot ($mgkg^{-1}$)	0–5	120 ± 3.8 ^{f}	73 ± 2.2 ^{c}	180 ± 2.8 ^{h}	100 ± 3.4 ^{de}	100 ± 3.8 ^{de}	96 ± 2 ^{d}	120 ± 1.7 ^{f}	130 ± 3.5 ^{g}	110 ± 2.3 ^{ef}
	5–10	25 ± 0.81 ^{ab}	17 ± 0.73 ^{ab}	26 ± 0.36 ^{b}	16 ± 0.73 ^{ab}	22 ± 0.3 ^{ab}	16 ± 0.99 ^{ab}	15 ± 0.46 ^{a}	21 ± 1.4 ^{ab}	27 ± 0.69 ^{b}
WSN_{org} ($mgkg^{-1}$)	0–5	51 ± 2 ^{d}	20 ± 1.6 ^{b}	32 ± 2.8 ^{c}	40 ± 2.2 ^{c}	33 ± 4.1 ^{c}	37 ± 1.9 ^{c}	34 ± 2.9 ^{c}	35 ± ±2.7 ^{c}	30 ± 1.3 ^{c}
	5–10	8.1 ± 0.44 ^{a}	7.1 ± 0.5 ^{a}	5.8 ± 0.12 ^{a}	5.9 ± 0.64 ^{a}	6.6 ± 0.37 ^{a}	5.9 ± 0.69 ^{a}	6.6 ± 1.2 ^{a}	7.2 ± 0.83 ^{a}	8.6 ± 0.49 ^{a}
$WSNH_4$ ($mgkg^{-1}$)	0–5	15 ± 0.66 ^{d}	13 ± 0.45 ^{cd}	15 ± 0.75 ^{d}	9.9 ± 0.92 ^{b}	13 ± 1.1 ^{cd}	15 ± 0.53 ^{d}	15 ± 1.2 ^{d}	8.7 ± 0.3 ^{b}	10 ± 0.24 ^{bc}
	5–10	1.8 ± 0.14 ^{a}	2 ± 0.27 ^{a}	0.52 ± 0.01 ^{a}	0.76 ± 0.06 ^{a}	0.9 ±0.07 ^{a}	1.2 ± 0.2 ^{a}	2.2 ± 0.2 ^{a}	0.82 ± 0.03 ^{a}	0.8 ± 0.05 ^{a}
$WSNO_3$ ($mgkg^{-1}$)	0–5	50 ± 2.1 ^{ef}	40 ± 2 ^{d}	130 ± 4.1 ^{i}	54 ± 1.8 ^{f}	54 ± 1.7 ^{f}	44 ± 1.2 ^{de}	70 ± 2.2 ^{g}	91 ± 2.6 ^{h}	70 ± 1.8 ^{g}
	5–10	15 ± 0.49 ^{abc}	7.6 ± 0.55 ^{a}	20 ± 0.39 ^{c}	9.3 ± 0.17 ^{ab}	14 ± 0.54 ^{abc}	8.7 ± 0.21 ^{ab}	6.1 ± 0.93 ^{a}	13 ± 0.89 ^{abc}	17 ± 0.53 ^{bc}
WSC/WSNtot	0–5	4.8 ± 0.5 ^{f}	2.9 ± 0.2 ^{abcd}	1.9 ± 0.2 ^{a}	3.5 ± 0.1 ^{bcde}	3.8 ± 0.3 ^{def}	4.5 ± 0.4 ^{ef}	2.9 ± 0.1 ^{abc}	1.9 ± 0.1 ^{a}	2.4 ± 0.0 ^{ab}
	5–10	2.7 ± 0.2 ^{abcd}	3.4 ± 0.1 ^{bcde}	1.9 ± 0.1 ^{a}	3.3 ± 0.3 ^{bcde}	2.8 ± 0.1 ^{abc}	3.2 ± 0.1 ^{bcd}	3.6 ± 0.2 ^{cdef}	2.5 ± 0.2 ^{abc}	2.6 ± 0.1 ^{abc}
WSC/WSN_{org}	0–5	11 ± 1 ^{a}	11 ± 0.9 ^{a}	11 ± 1.8 ^{a}	9.2 ± 0.5 ^{a}	12 ± 1.5 ^{a}	12 ± 1.3 ^{a}	9.3 ± 0.9 ^{a}	7.5 ± 0.6 ^{a}	8.7 ± 0.2 ^{a}
	5–10	8.6 ± 0.83 ^{a}	8 ± 0.34 ^{a}	8.8 ± 0.58 ^{a}	9.6 ± 1.5 ^{a}	9 ± 0.2 ^{a}	8.9 ± 0.79 ^{a}	9.6 ± 1.7 ^{a}	7.4 ± 0.81 ^{a}	8 ± 0.55 ^{a}
Hot water-extractable C and N fractions										
HWC ($mgkg^{-1}$)	0–5	3200 ± 280 ^{cd}	2500 ± 110 ^{b}	2700 ± 97 ^{bc}	5200 ± 150 ^{ef}	5400 ± 120 ^{f}	5500 ± 140 ^{f}	4600 ± 210 ^{e}	3400 ± 51 ^{d}	3500 ± 230 ^{d}
	5–10	640 ±32 ^{a}	540 ± 25 ^{a}	590 ± 25 ^{a}	500 ± 29 ^{a}	580 ± 25 ^{a}	630 ± 51 ^{a}	500 ± 46 ^{a}	620 ± 31 ^{a}	830 ± 16 ^{a}
HWNtot ($mgkg^{-1}$)	0–5	300 ± 4.5 ^{cd}	270 ± 9.4 ^{c}	330 ± 8 ^{de}	490 ± 13 ^{h}	420 ± 5.2 ^{fg}	440 ± 8 ^{g}	430 ± 17 ^{g}	370 ± 8.9 ^{ef}	430 ± 20 ^{g}
	5–10	80 ± 4.6 ^{ab}	61 ± 1.2 ^{a}	66 ± 4.4 ^{ab}	48± 3.1 ^{a}	52 ± 3.2 ^{a}	51 ± 1.7 ^{a}	55 ± 3.8 ^{a}	72 ± 1.5 ^{ab}	110± 1.3 ^{b}
HWN_{org} ($mgkg^{-1}$)	0–5	240 ± 1.9 ^{cd}	220 ± 7.5 ^{c}	240 ± 5.5 ^{cd}	400 ± 11 ^{g}	330 ± 4.8 ^{ef}	360 ± 6.2 ^{f}	340 ± 13 ^{ef}	270 ± 6 ^{d}	320 ± 18 ^{e}
	5–10	65 ± 4.3 ^{ab}	50 ± 1.7 ^{a}	57 ± 3.8 ^{ab}	41 ± 2.9 ^{a}	44 ± 2.9 ^{a}	43 ± 1.6 ^{a}	45 ± 3.9 ^{a}	62 ± 1.7 ^{ab}	91 ± 1.2 ^{b}
$WSNH_4$ ($mgkg^{-1}$)	0–5	50 ± 2.4 ^{b}	43 ± 2.1 ^{b}	73 ± 3.1 ^{cde}	77 ± 2.8 ^{de}	63 ± 1.8 ^{c}	68 ± 2 ^{cd}	82 ± 3.7 ^{e}	94 ± 4.9 ^{f}	97 ± 2.1 ^{f}
	5–10	12 ± 0.35 ^{a}	9.1 ± 0.75 ^{a}	6.4 ± 0.42 ^{a}	5.2 ± 0.15 ^{a}	5.1 ± 0.35 ^{a}	5.9 ± 0.81 ^{a}	9.3 ± 0.59 ^{a}	8.2 ± 0.43 ^{a}	14 ± 0.42 ^{a}
$WSNO_3$ ($mgkg^{-1}$)	0–5	12 ± 0.98 ^{cd}	8 ± 0.19 ^{bc}	17 ± 0.7 ^{de}	17 ± 1.4 ^{de}	21 ± 0.97 ^{e}	14 ± 1.3 ^{d}	14 ± 1.2 ^{d}	12 ± 0.92 ^{cd}	12 ± 3.2 ^{cd}
	5–10	3 ± 0.35 ^{ab}	1.6 ± 0.11 ^{a}	2.8 ± 0.25 ^{ab}	2 ± 0.26 ^{a}	2.2 ± 0.29 ^{a}	1.9 ± 0.03 ^{a}	0.99 ± 0.03 ^{a}	2.4 ± 0.29 ^{a}	3.1 ± 0.21 ^{ab}
HWC/HWNtot	0–5	11 ± 0.9 ^{bcd}	9.4 ± 0.3 ^{abc}	8 ± 0.1 ^{a}	11 ± 0.1 ^{bcd}	13.9 ± 0.1 ^{e}	13 ± 0.2 ^{de}	11 ± 0.3 ^{bcd}	9.1 ± 0.2 ^{ab}	8 ± 0.1 ^{a}
	5–10	8 ± 0.1 ^{a}	8.9 ± 0.5 ^{ab}	9 ± 0.4 ^{ab}	10 ± 0.3 ^{bcd}	11 ± 0.8 ^{cde}	12 ± 0.6 ^{de}	9.1 ± 0.5 ^{ab}	8.5 ± 0.4 ^{ab}	7.7 ± 0.1 ^{a}
HWC/HWN_{org}	0–5	13 ± 1.2 ^{defg}	12 ± 0.4 ^{abcdef}	11 ± 0.2 ^{abcde}	13 ± 0.2 ^{defg}	16 ± 0.2 ^{h}	15 ± 0.3 ^{gh}	14 ± 0.4 ^{efgh}	13 ± 0.3 ^{cdefg}	11 ± 0.2 ^{abcde}
	5–10	9.8 ± 0.17 ^{abc}	11 ± 0.85 ^{abcde}	10 ± 0.54 ^{abcd}	12 ± 0.38 ^{bcdefg}	13 ± 0.95 ^{defg}	14 ± 0.77 ^{fgh}	11 ± 0.6 ^{abcde}	10 ±0.5 ^{ab}	9.1 ± 0.16 ^{a}

Labile (water-soluble and hot water-extractable) carbon and nitrogen also differed between soil layers (Figure 3, Tables 1 and A4). The amounts of water-soluble C (WSC) and hot water-extractable C (HWC) were about seven times higher in the upper than in the lower soil layer. Different components of water-extractable N also varied significantly with soil depth (Table 1). In the upper soil layer, across tree species, cold water extractable N comprised nitrate ($WSNO_3$, 52%), ammonium ($WSNH_4$, 13.4%), and organic nitrogen (WSN_{org}, 34.6%) (Table A4). In the lower soil layer, these proportions accounted for 58.4% nitrate ($WSNO_3$), 6.7% ammonium (WSNH4), and 34.9% organic nitrogen (WSN_{org}). The proportions extracted by hot water also differed with soil depth where nitrate, ammonium, and organic nitrogen accounted for 3.5%, 18.7%, and 77.8%, respectively, in the upper soil layer against 3.4%, 12.6%, and 84%, respectively, in the lower soil layer (Table A4).

3.2. *Effects of Tree Species on Water-Extractable C and N and Other Soil Properties*

Most differences in soil properties between tree species were found in the upper 0–5 cm soil layer (Table A3; Figure 3). In this layer, pH was highest under *Polyscias fulva* (pH_{KCL} = 5.8), followed by the two native species stands (Mns and *Entandrophragma excelsum*) and *Grevillea robusta* (Figure 2). Soils under *Calliandra calothyrsus* and *Cedrela serrata* had an intermediate pH (pH_{KCL} = 4.9), while all eucalyptus species showed the lowest soil pH values (*Eucalyptus saligna* < *Eucalyptus grandis* < *Eucalyptus maideni*). The SOM content was significantly higher under most eucalyptus species and *Entandrophragma excelsum*, while it was not different between the other species.

Water-extractable labile C and N (Table 1, Figure 3) also differed under tree species. Water-soluble organic carbon (WSC) was significantly higher under *Calliandra calothyrsus*, followed by some eucalyptus species. Hot water-extractable carbon (HWC) showed the highest values under eucalyptus species and *Entandrophragma excelsum*, while values were not significantly different under the other tree species. Water-soluble total nitrogen (WSN_{tot}) was highest under *Grevillea robusta* followed by native species (*Entandrophragma excelsum* and *Polyscias fulva*) and *Calliandra calothyrsus* with intermediate values under eucalyptus species and lowest concentration under *Cedrela serrata* (Figure 3).

Unlike WSNtot, hot water-extractable total nitrogen (HWNtot) showed similar differences between tree species as HWC, with the highest values under eucalyptus species and *Entandrophragma excelsum* and similar values under the other tree species. The highest percentage of water-soluble mineral nitrogen relative to total water-soluble nitrogen was measured under *Grevillea robusta* (WSNmin = 82%; $WSNO_3$ = 73% + $WSNH_4$ = 9%), while the lowest percentage was measured under *Entandrophragma excelsum* (WSNmin = 39%; $WSNO_3$ = 12% + $WSNH_4$ = 27%). The proportion of water-soluble organic nitrogen (WSN_{org}) was highest under *Entandrophragma excelsum* (WSN_{org} = 61%) and lowest under *Grevillea robusta* (WSN_{org} = 18%). The proportions of WSN_{org} under the other tree species ranged between 26% and 44%. In the hot water N extracts, organic nitrogen dominated fractions for all species. The highest proportion of mineral nitrogen was measured under *Polyscias fulva* (HWNmin, 28%; $HWNO_3$ = 3% + $HWNH_4$ = 25%), while the lowest proportion was measured under *Eucalyptus saligna* (HWNmin, 18%; $HWNO_3$ = 3% + $HWNH_4$ = 15%). Consequently, WSN_{org} was higher under *Eucalyptus saligna* (WSN_{org} = 82%) followed by *Polyscias fulva* (HWN_{org} = 72%). WSC/WSN_{org} ranged from 7.5 to 12, with no significant difference between tree species (Table 1).

The sum of exchangeable base cations (EBC: Ca^{2+}, Mg^{2+}, K^+, and Na^+) was significantly higher under mixed native species, followed by eucalyptus species, *Polyscias fulva*, and *Grevillea robusta* compared to *Calliandra calothyrsus*, *Cedrela serrata*, and *Entandrophragma excelsum*. Soil base cations such as Ca^{2+}, Mg^{2+}, and K^+ dominated with nearly 97% of the total exchangeable cations, and they generally showed the higher concentrations under native and Eucalyptus species.

In the lower soil layer (5–10 cm), there were no significant differences between tree species for SOM, WSC, WSN_{org}, $WSNH_4$, HWC, $HWNH_4$, and $HWNO_3$. Significant differences between species were observed for pH, EBC, ∑cations, and individual cations such as Ca^{2+}, Mg^{2+}, K^{+}, and Al^{3+} (Table A3). There was also a significant effect of tree species for WSNtot, $WSNO_3$, WSC/WSN, HWNtot, HWN_{org}, and HWC/HWNtot (Table 1). The highest pH (pH_{KCL} = 4.8) was measured under the Mns, followed by the plot of monospecific native species and agroforestry species (*Polyscias fulva* > *Entandrophragma excelsum* = *Grevillea robusta* > *Calliandra calothyrsus* = *Cedrela serrata*), while the lowest pH (pH_{KCL} = 3.7) was measured under eucalyptus species. EBC ranged from 2.8 ± 0.02 $cmol_c kg^{-1}$ (*Eucalyptus maideni*) to 17 ± 0.6 $cmol_c kg^{-1}$ (*Eucalyptus grandis*); this trend was similar to $\sum$cations, which ranged from 5.4 ± 0.07 $cmol_c kg^{-1}$ under *Eucalyptus maideni* to 17 ± 0.2 $cmol_c kg^{-1}$ under *Eucalyptus grandis*. Exchangeable Ca^{2+} was significantly higher under *Eucalyptus grandis*, intermediate under Mns, *Polyscias fulva*, *Grevillea robusta*, and *Calliandra calothyrsus*, and lower values were measured under *Cedrela serrata*, *Eucalyptus maideni*, *Eucalyptus saligna*, and *Entandrophragma excelsum*. Mg^{2+} was higher under *Eucalyptus grandis*, *Polyscias fulva*, and Mns, whereas the values of Mg^{2+} were lower under *Eucalyptus maideni*, *Eucalyptus saligna*, and *Entandrophragma excelsum*, with intermediate values under agroforestry species (*Calliandra calothyrsus*, *Cedrela serrata*, and *Grevillea robusta*). Similar to Ca^{2+} and Mg^{2+}, the concentration of K^{+} was also significantly higher under *Eucalyptus grandis* and Mns but not different for the remaining tree species. There was high variability in the exchangeable Al^{3+} concentration between tree species in the lower soil layer. The concentration of Al^{3+} in the soil classified tree species in the following order: *Eucalyptus maideni* > *Entandrophragma excelsum* > *Eucalyptus saligna* > *Cedrela serrata* > *Grevillea robusta* = *Calliandra calothyrsus* > Mns > *Polyscias fulva* > *Eucalyptus grandis*.

The proportions of water-soluble nitrogen fractions in the lower soil layer (Table A4) showed that the mineral nitrogen was dominant with the highest percentage under *Grevillea robusta* (WSNmin = 78%; $WSNO_3$ = 76% + $WSNH_4$ = 2%) and the lowest percentage under *Entandrophragma excelsum* (WSNmin = 56%; $WSNO_3$ = 41% + $WSNH_4$ = 15%). The other species had WSNmin percentages ranging between 57% and 69%. The water-soluble organic nitrogen ranged between 22% (*Grevillea robusta*) and 44% (*Entandrophragma excelsum*). Hot water-extractable fractions contained mostly organic N ranging from 72% to 82% of the HWNtot and 81% to 86% in the 0–5 cm and the 5–10 cm soil layers, respectively. The hot water-extractable mineral N forms were dominated by N-NH_4^+ (15% to 25%) in the 0–5 cm soil layer and 10% to 17% in the 5–10 cm soil layer. The less abundant hot water-extractable mineral N fraction was N-NO_3- that ranged from 2% to 5% in both 0–5 cm and 5–10 cm soil layers.

3.3. Relationships between Water-Extractable Elements (C, N) and Other Soil Properties

The correlation between soil properties (pH, SOM, and EBC) and water-extractable C and N fractions (WSC, WSN_{org}, WSNmin, $WSNO_3$, WSC/N_{org}, HWC, HWNtot, and HWC/HWNtot) showed significant correlations within each of the two soil layers (Figure 4).

In the upper soil layer (0–5 cm), soil pH was negatively correlated with SOM, all water-soluble and hot water-extractable C and N fractions, and HWC/HWNtot, except WSNmin and $WSNO_3$, which were positively correlated with pH (Figure 4). There was a significant positive correlation between SOM and all the above-mentioned water-extractable C and N fractions, except WSNmin and $WSNO_3$ (r = -0.2). The strongest positive correlation was found between SOM and HWC (r = 0.8), HWNtot (r = 0.7), and HWC/HWNtot (r = 0.5). EBC showed a weak positive correlation with HWNtot and a weak negative correlation with WSC, WSN_{org}, and HWC/HWN; no significant correlation was found with the other water-extractable C and N fractions (Figure 4). In the lower soil layer (5–10 cm), soil pH was positively correlated with HWC, HWNtot, WSN_{org}, WSNmin, and WSNO3, while it was negatively correlated with HWC/HWN. The relationship patterns between SOM and water-extractable C and N fractions showed a positive correlation with HWC, HWNtot, and WSN_{org}, while it was negatively correlated with WSC/WSN_{org}. The strength of the correlation between SOM and water-extractable C and N fractions was comparatively lower compared to the upper soil layer, and there was no significant correlation between EBC and water-extractable C and N fractions (Figure 4).

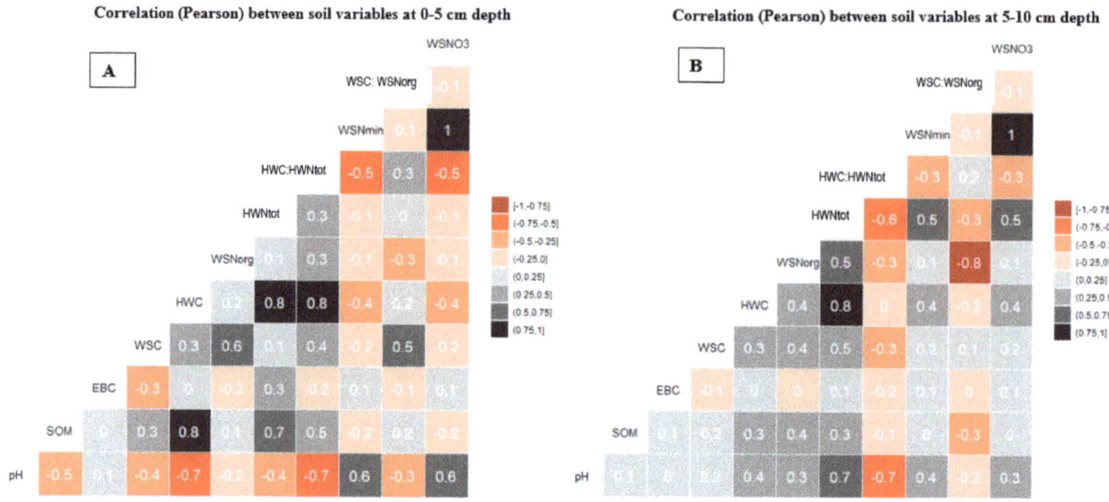

Figure 4. Pearson correlation matrices showing the relationship between soil properties and water-extractable C and N fractions in the upper ((**A**) 0–5 cm) and lower ((**B**) 5–10 cm) soil layers. Relationships between parameters are indicated by the values at the intersection of parameters and interpreted within color contrast as shown in the legends.

Principal component analysis (PCA) of soil properties (pH, SOM, and EBC) and water-extractable C and N fractions (WSC, WSN_{org}, WSNtot, WSC:WSN_{org}, HWC, HWN_{org}, HWNtot, and HWC/HWNtot) for the upper and the lower soil layers showed differences in the patterns of the tree species clustering based on these soil properties (Figure 5). In the upper soil layer (0–5 cm), the total variance explained by the first two principal components was 62%. SOM, HWC, HWN_{org}, HWNtot, and C/N ratio of hot water extracts (HWC/HWNtot) had the highest positive loadings on PC1 (43%), while pH and WSNtot showed the highest loading to the negative side of PC1 (Figure 5A). Eucalyptus species and *Entendrophragma excelsum* clustered separately along with the positive side of PC1, while species such as *Polyscias fulva* and *Grevillea robusta* clustered along its negative side. EBC positively loaded highest on the second PC (19%), while WSC, WSN_{org} C/N ratio of water-soluble C, and organic N (WSC/WSN_{org}) had their negative loading to PC2. Mns plot clustered separately from the other plots along the positive side of the second axis, and *Calliandra calothyrsus*, *Cedrela serrata*, and *Grevillea robusta* overlapped on its negative side (Figure 5A).

In the lower soil layer (5–10 cm), the first two principal components explained 59% of the combined variation in PCA input variables between tree species (Figure 5B, Table 2). The positive loadings on PC1 (46%) were observed for pH, WSN_{org}, WSNtot, HWNorg, HWC, HWNtot, and EBC, while the HWC/HWNtot was highly loaded on its negative side. On the PC2 (13%), WSC/WSN_{org} showed a positive loading, while SOM showed a negative loading (Figure 5B, Table 2). In this soil layer, most of the tree species clustered around the center of biplot quadrants with a tendency for the plots of Mns, *Polyscias fulva*, *Grevillea robusta*, and *Calliandra calothyrsus* to overlap on the positive side of the PC1. Eucalyptus species overlapped with both the negative side of the PC1 and the positive side of the PC2. The clustering patterns of species such as *Entandrophragma excelsum* and *Cedrela serrata* showed a stretching of statistical ellipses across the intersection of PCA axes towards both sides of PC2.

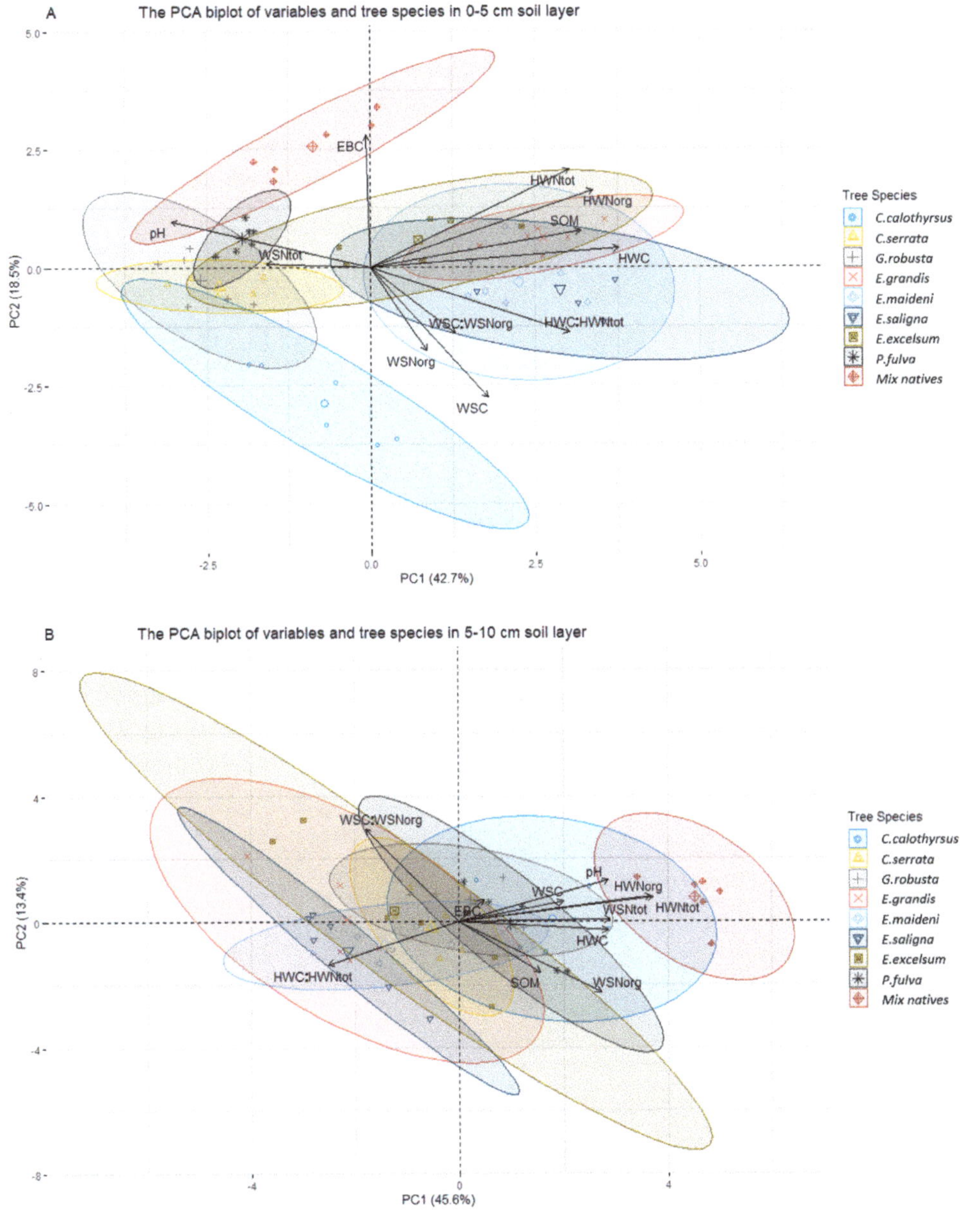

Figure 5. PCA biplot of soil chemical properties and tree species for the upper (**A**) and the lower (**B**) soil layers. The first two principal components explained 62.5% of the combined variation in soil parameters at 0–5 cm soil depth and 54.9% at 5–10 cm soil depth between tree species. Statistical ellipses at 95% confidence level group tree species (represented by different symbols and colors) based soil variables depicted by vectors (pH; SOM = soil organic matter; EBC = exchangeable basic cations; WSC = water-soluble C, WSN_{org} = water soluble organic N; WSNtot = water-soluble total N, WSC:WSN_{org} = water soluble organic C/N ratio; HWC = hot water-extractable C, HWN_{org} = hot water-extractable organic N, HWNtot = hot water-extractable total N, and HWC:HWNtot= hot water-extractable C/N ratio).

Table 2. Principal component analysis (PCA) of 11 selected soil chemical properties measured in 108 samples under nine treatments (tree species) at two soil layers. Variable loadings higher than 0.6 are in bold, expressing a significant weight of variables on PC, and the first five principal components explaining 87%–91% of the cumulative total variance are presented. The sign on variable loadings indicates the direction of the variable on PC axes. Analyzed soil variables (pH; SOM = soil organic matter; EBC = exchangeable basic cations; WSC = water-soluble C, WSN_{org} = water soluble organic N; WSNtot = water-soluble total N, WSC:WSN_{org} = water soluble organic C/N ratio; HWC = hot water-extractable C, HWN_{org} = hot water-extractable organic N, HWNtot = hot water-extractable total N, and HWC:HWNtot= hot water-extractable C/N ratio).

	Upper Soil Layer (0–5 cm)					**Lower Soil Layer (5–10 cm)**				
Principal Components	**PC1**	**PC2**	**PC3**	**PC4**	**PC5**	**PC1**	**PC2**	**PC3**	**PC4**	**PC5**
Eigenvalues	4.69	2.03	1.57	1.03	0.71	5.02	1.472	1.19	1.02	0.89
% variance	42.66	18.52	14.27	9.43	6.48	45.64	13.38	10.85	9.34	8.16
Cumulative % of the total variance	42.66	61.19	75.46	84.89	91.38	45.64	59.03	69.88	79.22	87.39
Loadings (weight) of variables on PCs (%)										
pH_{KCL}	**−0.77**	0.24	0.22	0.25	−0.15	**0.73**	0.33	−0.21	−0.22	−0.09
SOM	**0.81**	0.19	0.07	0.17	−0.25	0.40	−0.41	0.04	0.50	0.38
WSC	0.45	**−0.70**	0.24	0.34	0.29	0.51	0.16	0.50	−0.01	0.59
WSN_{org}	0.22	−0.45	**0.80**	−0.20	0.21	**0.70**	−0.58	−0.04	−0.19	0.18
WSNtot	−0.40	0.017	**0.63**	0.48	−0.23	**0.74**	0.01	0.08	−0.07	−0.20
WSC:WSN_{org}	0.33	−0.35	−0.55	**0.66**	0.06	−0.46	**0.76**	0.35	0.19	0.13
HWC	**0.96**	0.10	0.05	−0.05	−0.08	**0.73**	−0.07	0.41	0.30	−0.39
HWN_{org}	**0.86**	0.41	0.14	0.04	−0.05	**0.94**	0.20	0.04	0.08	−0.14
HWNtot	**0.77**	0.53	0.23	0.14	−0.05	**0.95**	0.19	0.04	0.05	−0.09
HWC:HWNtot	**0.77**	−0.35	−0.11	−0.24	−0.03	**−0.64**	−0.35	0.46	0.29	−0.32
EBC	−0.01	**0.71**	0.02	0.14	0.64	0.12	0.18	−0.59	**0.67**	0.04
Contribution of variables to PCs (%)										
pH_{KCL}	12.96	2.98	3.34	6.25	3.21	10.82	7.78	3.90	4.79	0.94
SOM	14.30	1.85	0.35	3.09	9.37	3.20	11.63	0.18	24.70	16.20
WSC	4.43	24.50	3.79	11.33	12.24	5.33	1.81	21.35	0.03	38.75
WSN_{org}	1.03	10.07	40.84	4.15	6.21	9.79	22.89	0.15	3.78	3.92
WSNtot	3.54	0.01	26.03	22.46	7.58	11.08	0.01	0.55	0.50	4.67
WSC:WSN_{org}	2.35	6.33	19.55	42.61	0.58	4.25	39.38	10.45	3.68	1.98
HWC	19.85	0.49	0.19	0.24	0.89	10.72	0.34	14.47	9.28	17.71
HWN_{org}	15.96	8.55	1.34	0.22	0.42	17.77	2.88	0.15	0.64	2.44
HWNtot	12.76	13.79	3.63	2.01	0.37	18.31	2.62	0.16	0.32	1.06
HWC:HWNtot	12.77	6.21	0.84	5.56	0.19	8.36	8.37	18.14	8.36	12.02
EBC	0.01	25.16	0.05	2.02	58.90	0.30	2.23	30.45	43.87	0.25

4. Discussion

Given that trees species were planted on the same site with similar land-use history and climatic conditions, the Arboretum of Ruhande provided a unique set-up for investigating the effects of tree species used for forest plantations in Rwanda on soil chemical properties. We thus base the interpretation of the results on the assumption that the current differences in soil characteristics reflect the influence of the planted trees.

4.1. Importance of the Thin Upper Soil Layer (0–5 cm Depth)

The present study showed higher values for all analyzed soil properties in 0–5 than 5–10 cm soil layers (except for Al^{3+} and Fe^{2+}), regardless of tree species, although the two soil layers were visibly indistinguishable under most species. SOM, EBC, water-soluble, and hot-water-extractable C and N were two to nine-fold higher compared to the 5–10 cm layer. This vertical distribution was particularly marked for parameters related to soil organic matter content and water-extractable C and N. The water soluble fractions represent the amount of the readily mineralizable C and N in soil [66] and have been linked to soil functions which provide nutrients for the trees. Physical protection and the preservation of soil properties and processes of this layer are therefore of utmost importance [67].

In a previous study conducted at the same site, Nsabimana et al. [44] showed that planting trees increased the levels of soil carbon, nitrogen, base saturation, and exchangeable cation pools in the upper 10 cm of the soil compared to agricultural lands in the same agroecological zone. In the present study, we observed that planted trees influenced soil fertility only in the uppermost soil layer (0–5 cm), with higher values of SOM and exchangeable base cations than the values reported by Nsabimana et al. [44] a decade before at this site and compared to those reported for other tropical forest soils [42,68].

In contrast to high Al saturation and low amounts of exchangeable cations generally characterizing highly weathered and acidic tropical soils dominated by kaolinitic clays [69], we observed that the sum of exchangeable cations was relatively high and dominated by calcium (75%), whereas aluminum represented only 3% of the sum of exchangeable cations. Similarly, high base saturation (87%) with a dominance of Ca^{2+} was reported at this site [44] and for other sites in the same agro-ecological zone with base saturations between 45% and 85% [70,71]. The high proportion of Ca^{2+} could be related to plant litter Ca content, soil pH, and the nature of clay minerals at this site. In tropical nutrient-poor soils, organic acidity is promoted by plants (and soil microorganisms) through the production and the release of organic acids into the soil solution as a "nutrient acquisition strategy" [72]. This may lead to an exchange acidity dominated by protons, allowing for high base saturation events at certain pH values [73]. Further, the presence of interstratified kaolinite-smectite, as reported for soils from some subtropical and tropical climates [74,75], may explain the relatively high exchange capacity measured in this study.

4.2. Effects of Tree Species on Chemical Soil Properties

Tree species effects were mostly observed in the upper soil layer (except for Al and Fe). This may indicate that the changes in aboveground litter quality and quantity, rather than mineral weathering and root exudation, most likely influenced soil chemical properties. In contrast to Bauters et al. [42], who found a significant effect of tree species on soil pH and carbon content until about 30 cm deep in tropical forest plantations, our results highlighted the importance of this thin uppermost 0–5 cm layer in these highly weathered tropical forest soils.

Planting trees is one of the key strategies for restoring degraded forests and soils, especially in tropical soils with inherently poor chemical properties [76]. In our study, the pH under eucalyptus species was 0.6 pH units lower than under exotic agroforestry species (*Calliandra calothyrsus*, *Cedrela serrata*, and *Grevillea robusta*) and 1.7 pH units lower than under native species (*Entandrophragma excelsum*, *Polyscias fulva*, and self-regenerated mixed natives) in the upper layer. Soil acidification under eucalyptus species was reported in previous studies conducted at this site [44], in forest plantations near this site [77], and in other tropical [78,79] and non-tropical regions [80]. The relatively higher concentrations of exchangeable Al^{3+} and Fe^{2+} measured in soils under *Eucalyptus saligna* and *Eucalyptus grandis* compared to other species in this study could be related to the acidifying effect of these species, leading to Al^{3+} and Fe^{2+} release [42] with potential toxic effects for plant roots [72]. Two main mechanisms were suggested for the effects of tree species on soil pH: (1) input of organic acids from litter decomposition and root exudates, (2) increased proton release in the soil to compensate for the high plant uptake and storage of base cations [18,81]. We measured higher pH and exchangeable base cations under mixed native species (Mns) plots compared to other plots. The Mns plots were characterized by high tree density and vegetation diversity dominated by mature native trees accompanied by shrubs and grasses. All species together might have contributed to high quality and quantity of litter as a natural regeneration setup [82] compared to other adjacent monoculture plots. Therefore, we suggest that soil pH, SOM content, water-extractable C and N, and exchangeable cations were likely influenced by the species-specific litter chemical quality.

In the upper soil layer, the clear grouping by tree species and high loadings of variables such as SOM, pH, and hot water-extractable C and N fractions (HWC, HWNorg, HWNtot, and HWC:HWNtot) on the first principal component (PC1 = 43%) may indicate that these properties were the most influential set of variables in explaining the variation between species. A previous study [44] associated eucalyptus plantations with soil organic matter accumulation and decreased pH. This is in line with our PCA results, where the first PC representing soil organic matter-related properties and pH were associated with a cluster of eucalyptus species (*E. grandis, E. maideni,* and *E. saligna*). The high loadings of pH and WSNtot associated with *Grevillea robusta, Polyscias fulva,* and mixed native species indicate increased soil pH and N availability under these species. The second set of influential variables included EBC, WSC, and WSN_{org} loading high on the second principal component (PC2, 18.5%). As described by Ahmed et al. [83], these variables represent the quality and the bioavailability of mineralizable organic matter and related nutrient cycling processes. The high positive loading of EBC associated with Mns plots may be due to the capacity of this undisturbed self-regenerated native forest containing highly dense and diverse vegetation (trees, shrubs, and grasses) for improving soil chemical quality in terms of nutrients cycling. The observed relationship of water-soluble C and organic N (WSC and WSN_{org}) with *Calliandra calothyrsus* may be due to the characteristics of this plant used in agroforestry as an N-fixing tree [84].

In the 5–10 cm soil layer, the two axes of the PCA explained 59% of the variation between tree species. *Calliandra calothyrsus* and Mns plots grouping was explained by pH, WSN_{org}, WSNtot, HWC, HWNorg, HWNtot, and HWC:HWNtot (PC1, 46%). The remaining species overlapped around the center of the biplot, indicating the lack of species influence on selected soil variables. The multivariate analysis of covariation between chemical properties and tree species in this study suggests that the influence of tree species is mainly limited to the upper soil layer (0–5 cm). This first principal component could be interpreted as a measurement of soil acidity and bioavailability of hot water-extractable C and N fractions, reflecting the quality of SOM and its mineralization process in this soil layer. The results from the present study allowed us to consider this upper layer as a highly sensitive layer to vegetation changes in this tropical forest ecosystem.

4.3. *Differences in Water-Extractable C and N between Tree Species*

Soil organic matter has been used for many years as one of the major indicators of soil quality, given its important role in controlling soil chemistry as well as physical and biological processes [85]. However, it may take many decades to detect a change in the total soil organic C pool, given its slow rate of change [86]. Water-soluble and hot water-extractable C and N analyzed in this study are labile components of soil organic matter that could reflect early changes in soil–plant interactions [87]. Water-soluble fractions contain dissolved organic components almost similar to those measured directly in the soil solution using lysimeters and suction devices [88]. Hot water-extractable fractions consist of an easily decomposable pool of SOM, including microbial biomass, that serves as the source of energy and substrate to soil microorganisms, and its decomposition provides nutrients to plants [61,86]. This implies that labile fractions of SOM, especially those extracted with hot water, might be used as a proxy for soil microbial biomass and activity [28,61]. The influence of tree species on soil function, as represented by water-soluble and hot water-extractable C and N, was observable through the discrimination of tree species and also through the correlation of these fractions with other soil properties. These fractions are closely related to the decomposability of the plant's detritus, which is influenced by the litter chemistry [89] and might thus be used as a proxy for soil functioning [28]. Labile C and N fractions were significantly correlated to SOM, EBC, and pH in both upper and lower soil layers, and correlation between HWC and soil organic matter was greater than that for WSC, as also observed by Ghani et al. [61]. While the mineralizable organic N decreased with soil depth, nitrates increased with soil depth. This is likely due to water solubility and leaching of nitrates towards the lower soil horizons [90] and the fact that water-soluble and hot water-extractable C and N fractions originate mainly from above-ground litter rather than root exudates [91]. The dominance of organic N compared to other nitrogen forms may be explained by the fact that most of the mineral N was already extracted by the previous cold water extraction. Hot water (80 °C) extracts the organic matter not only from decomposing plant litter but also from soil microorganisms [92].

5. Conclusions

The present study was conducted to evaluate the effects of forest tree species on chemical soil quality in Rwanda. The most important changes in soil pH, SOM, water-extractable labile C and N fractions, and base cations were observed in the thin upper soil layer (0–5 cm) across tree species, which made it possible to recognize the importance of this thin upper soil layer for soil fertility. Eucalyptus species led to soil acidification while soil pH and nutrients increased under native species (*Entandrophragma excelsum* and *Polyschias fulva*) and Mns plots. Hot water-extractable C and N fractions strongly correlated with most of the analyzed soil parameters and were more sensitive in discriminating tree species effects than other soil properties analyzed. This reflects the suitability of this methodological approach for detecting subtle changes that might be linked to forest trees and its potential to be used as a proxy to SOM analysis. In selecting forest trees, priority should be given to the species which do not negatively alter chemical soil quality.

Author Contributions: Conceptualization, P.R., F.X.N., D.N. and M.C.; Data curation, P.R.; Formal analysis, P.R.; Funding acquisition, D.N. and M.C.; Investigation, P.R.; Methodology, P.R. and M.C.; Project administration, M.C.; Resources, M.C.; Software, P.R.; Supervision, F.X.N., D.N. and M.C.; Validation, M.C.; Visualization, P.R.; Writing—original draft, P.R., M.C.; Writing—review & editing, P.R., F.X.N., D.N. and M.C. All authors have read and agreed to the published version of the manuscript.

Funding: This research was funded by UR-ARES Project, a partnership project between the University of Rwanda and the Académie de Recherche et Enseignement Supérieur (ARES, Belgium Fédération Wallonie-Bruxelles).

Institutional Review Board Statement: Not applicable.

Informed Consent Statement: Not applicable.

Data Availability Statement: The data that support the findings of this study are available on request from the corresponding authors (P.R. and M.C.)

Acknowledgments: The authors are very thankful to the staff of the Laboratory of Plant and Microbial Ecology (ULiège) for scientific and technical support: B. Bosman, A. Degueldre, M.C. Requier, and A. Tahiri. We would like also to thank the Rwanda Forestry Authority (RFA) and the staff of the Arboretum of Ruhande for their support during soil sampling.

Conflicts of Interest: The authors declare no conflict of interest.

Appendix A

Table A1. Selected study plots in the Arboretum of Ruhande, Rwanda. (Rwibasira et al., Long-term effect of forest plantation species on soil chemical properties in Southern Rwanda).

Plot ID	Species	Native/Exotic	Latitude	Longitude	Elevation	Age in 2016 (Years)
Plot273	*Calliandra calothyrsus (C. cal)*	Exotic	02°36′69″ S	29°45′30″ E	1722 m	31
Plot265	*Calliandra calothyrsus (C. cal)*	Exotic	02°36′71″ S	29°45′18″ E	1713 m	31
Plot267	*Calliandra calothyrsus (C. cal)*	Exotic	02°36′72″ S	29°45′21″ E	1714 m	31
Plot56	*Cedrela serrata (C. ser)*	Exotic	02°36′94″ S	29°44′79″ E	1713 m	70
Plot111	*Cedrela serrata (C. ser)*	Exotic	02°36′75″ S	29°45′60″ E	1709 m	79
Plot36	*Cedrela serrata (C. ser)*	Exotic	02°36′83″ S	29°45′30″ E	1730 m	73
Plot150	*Grevillea robusta (G. rob)*	Exotic	02°36′97″ S	29°44′96″ E	1713 m	75
Plot322	*Grevillea robusta (G. rob)*	Exotic	02°36′94″ S	29°45′19″ E	1709 m	69
Plot104	*Grevillea robusta (G. rob)*	Exotic	02°36.84 S	29°45.51 E	1720 m	35
Plot218	*Eucalyptus grandis (E. gr)*	Exotic	02°37′03″ S	29°44′83″ E	1707 m	70
Plot220	*Eucalyptus grandis (E. gr)*	Exotic	02°37′05″ S	29°44′86″ E	1706 m	65
Plot181	*Eucalyptus grandis (E. gr)*	Exotic	02°36′65″ S	29°45′64″ E	1680 m	65
Plot179	*Eucalyptus maideni (E. mai)*	Exotic	02°36′66″ S	29°45′61″ E	1685 m	70
Plot377	*Eucalyptus maideni (E. mai)*	Exotic	02°36′59″ S	29°45′32″ E	1695 m	82
Plot1	*Eucalyptus maideni (E. mai)*	Exotic	02°36′89″ S	29°44′78″ E	1732 m	67
Plot472	*Eucalyptus saligna (E. sal)*	Exotic	02°37′01″ S	29°45′12″ E	1710 m	82
Plot259	*Eucalyptus saligna (E. sal)*	Exotic	02°36′93″ S	29°45′38″ E	1709 m	36
Plot20	*Eucalyptus saligna (E. sal)*	Exotic	02°36′89″ S	29°45′06″ E	1729 m	59
Plot78	*Entandrophragma excelsum (E. exc)*	Native	02°36′90″ S	29°45′12″ E	1727 m	67
Plot44	*Entandrophragma excelsum (E. exc)*	Native	02°36′81″ S	29°45′42″ E	1727 m	64
Plot54	*Entandrophragma excelsum (E. exc)*	Native	02°36′78″ S	29°45′57″ E	1718 m	45
Plot240	*Polyscias fulva (P. ful)*	Native	02°36′96″ S	29°45′15″ E	1714 m	80
Plot262	*Polyscias fulva (P. ful)*	Native	02°36′91″ S	29°45′46″ E	1695 m	80
Plot268	*Polyscias fulva (P. ful)*	Native	02°36′88″ S	29°45′54″ E	1693 m	80
MNS1	Mix natives species (Mns)	Native	02°36′65″ S	29°44′65″ E	1700 m	83
MNS1	Mix natives species (Mns)	Native	02°36′68″ S	29°45′51″ E	1692 m	83
MNS3	Mix natives species (Mns)	Native	02°36′59″ S	29°45′63″ E	1680 m	83

Table A2. Descriptive statistics of soil parameters for two soil layers (0–5 cm and 5–10 cm) across all samples (two samples per plot, eight tree species, one mixed plot) (Rwibasira et al., Long-term effect of forest plantation species on soil chemical properties in Southern Rwanda).

Soil Parameters	Layer	N	Mean	SD	Median	Min	Max	Skew	Kurtosis	SE
pH_{KCL}	0–5 cm	54	4.89	0.71	4.96	3.71	5.89	−0.33	−1.38	0.1
	5–10 cm	54	4.21	0.36	4.24	3.71	4.83	0.04	−1.25	0.05
SOM (%)	0–5 cm	54	22.49	4.71	20.55	15.53	31.32	0.63	−1.03	0.64
	5–10 cm	54	9.6	1.4	9.87	5.93	11.56	−0.95	0.21	0.19
SOC (gkg^{-1})	0–5 cm	54	130.4	27.31	119.2	90.1	181.66	0.63	−1.04	3.72
	5–10 cm	54	55.67	8.13	57.22	34.41	67.06	−0.94	0.2	1.11
WSC ($mgkg^{-1}$)	0–5 cm	54	340.8	115.1	323	175.2	683.5	1.1	1.05	15.6
	5–10 cm	54	56.31	9.41	55.9	39.77	78.74	0.36	−0.61	1.28
$WSNO_3$ ($mgkg^{-1}$)	0–5 cm	54	67.7	28.82	58.13	32.77	145.64	1.34	0.93	3.92
	5–10 cm	54	12.31	4.68	12.3	3.93	20.82	0.18	−1.07	0.64
$WSNH_4$ ($mgkg^{-1}$)	0–5 cm	54	12.85	2.97	12.97	7.62	18.34	−0.12	−1.28	0.4
	5–10 cm	54	1.23	0.67	0.95	0.47	2.9	0.96	−0.22	0.09
WSN_{org} ($mgkg^{-1}$)	0–5 cm	54	34.73	9.74	34.41	15.84	57.02	0.22	−0.51	1.32
	5–10 cm	54	6.85	1.74	6.56	2.99	10.42	0.03	−0.48	0.24
WSNmin ($mgkg^{-1}$)	0–5 cm	54	80.54	29.07	70.65	45.01	162.93	1.47	1.38	3.96
	5–10 cm	54	13.54	4.34	13.4	5.53	21.33	0.22	−1.2	0.59
WSNtot ($mgkg^{-1}$)	0–5 cm	54	115.2	29.68	111.2	65.1	190.1	0.97	0.63	4.04
	5–10 cm	54	20.39	4.86	20.38	12.86	28.78	0.1	−1.48	0.66

Table A2. *Cont.*

Soil Parameters	Layer	N	Mean	SD	Median	Min	Max	Skew	Kurtosis	SE
WSC/WSN_{org}	0–5 cm	54	10.07	2.86	9.28	5.89	18.79	1.13	1.33	0.39
	5–10 cm	54	8.66	2.21	8.43	4.61	15.5	0.96	1.55	0.3
WSC/WSNtot	0–5 cm	54	3.11	1.2	2.71	1.34	6.39	0.86	0.19	0.16
	5–10 cm	54	2.88	0.66	2.87	1.5	4.43	0.02	−0.63	0.09
HWC ($mgkg^{-1}$)	0–5 cm	54	3994.6	1201.2	3904.7	2203.4	5893.4	0.11	−1.51	163.4
	5–10 cm	54	603.43	119.8	601.8	382.16	888.07	0.48	−0.33	16.31
$HWNO_3$ ($mgkg^{-1}$)	0–5 cm	54	14.13	4.86	14.11	4.57	23.36	-0.03	−0.79	0.66
	5–10 cm	54	2.22	0.85	2.01	0.83	3.92	0.32	−0.94	0.12
$HWNH_4$ ($mgkg^{-1}$)	0–5 cm	54	71.7	18.6	71.05	37.07	104.15	-0.01	−0.95	2.53
	5–10 cm	54	8.35	3.15	7.97	4.18	14.97	0.5	−0.91	0.43
HWN_{org} ($mgkg^{-1}$)	0–5 cm	54	301.64	62.12	294.93	194.96	440.04	0.18	−1.2	8.45
	5–10 cm	54	55.44	16.45	53.03	31.94	95.43	0.9	−0.04	2.24
HWNtot ($mgkg^{-1}$)	0–5 cm	54	387.47	73.9	388.71	242.65	536.49	−0.1	−1	10.06
	5–10 cm	54	66.01	19.36	61.47	40.02	111.71	0.9	−0.01	2.63
HWC/HWN_{org}	0–5 cm	54	13.08	2.12	12.64	10.08	17.6	0.37	−1.14	0.29
	5–10 cm	54	11.3	2.15	10.93	8.52	17.22	0.86	0.03	0.29
HWC/HWNtot	0–5 cm	54	10.2	1.91	9.85	7.45	13.77	0.34	−1.26	0.26
	5–10 cm	54	9.48	1.84	9.16	7.01	14.21	0.89	0.09	0.25
Al^{3+} ($cmol_c$ kg^{-1})	0–5 cm	54	0.06	0.05	0.04	0.00	0.16	0.93	−0.31	0.01
	5–10 cm	54	1.22	0.9	0.94	0.01	2.82	0.38	−1.37	0.12
Ca^{2+} ($cmol_c$ kg^{-1})	0–5 cm	54	28.5	7.89	27.27	17.39	52.9	1.42	1.74	1.07
	5–10 cm	54	5.97	3.55	5.97	1.54	16.6	1.02	0.81	0.48
Fe^{2+} ($cmol_c$ kg^{-1})	0–5 cm	54	0.00	0.00	0.00	0.00	0.01	1.22	0.89	0.00
	5–10 cm	54	0.01	0.03	0.00	0.00	0.25	6.93	46.94	0.00
K^+ ($cmol_c$ kg^{-1})	0–5 cm	54	0.61	0.17	0.57	0.31	0.96	0.29	−0.94	0.02
	5–10 cm	54	0.21	0.14	0.15	0.08	0.56	1.47	0.81	0.02
Mg^{2+} ($cmol_c$ kg^{-1})	0–5 cm	54	6.9	2.87	5.67	4.45	15.17	1.74	1.94	0.39
	5–10 cm	54	1.51	0.55	1.49	0.67	2.34	0.05	−1.42	0.07
Mn^{2+} ($cmol_c$ kg^{-1})	0–5 cm	54	0.09	0.07	0.06	0.02	0.29	1.22	0.68	0.01
	5–10 cm	54	0.01	0.01	0.01	0.00	0.03	0.35	−0.88	0.00
Na^+ ($cmol_c$ kg^{-1})	0–5 cm	54	0.15	0.06	0.14	0.07	0.28	0.32	−1.16	0.01
	5–10 cm	54	0.18	0.03	0.19	0.09	0.24	-0.74	0.86	0.00
Zn^{2+} ($cmol_c$ kg^{-1})	0–5 cm	54	0.00	0.00	0.00	0.00	0.00	1.08	0.57	0.00
	5–10 cm	54	0.00	0.00	0.00	0.00	0.00	0.96	−0.28	0.00
EBC ($cmol_c$ kg^{-1})	0–5 cm	54	36.16	10.46	33.96	23.98	69.18	1.75	2.39	1.42
	5–10 cm	54	7.87	4.1	7.75	2.71	19.27	0.88	0.39	0.56
$\sum$cations ($cmol_c$ kg^{-1})	0–5 cm	54	36.31	10.41	34.07	24.32	69.24	1.77	2.42	1.42
	5–10 cm	54	9.11	3.31	8.62	5.08	19.31	1.29	1.33	0.45

Table A3. Measured soil characteristics under different tree species in the Arboretum of Ruhande (means ± SEM). Different letters within one parameter denote significant differences between tree species and soil depths (mixed linear models, Tukey's HSD, $p < 0.05$ (Rwibasira et al., Long-term effect of forest plantation species on soil chemical properties in Southern Rwanda).

Soil Parameters	Layer (cm)	*Calliandra calothyrsus*	*Cedrela serrata*	*Grevillea robusta*	*Eucalyptus grandis*	*Eucalyptus maideni*	*Eucalyptus saligna*	*Entandrophragma excelsum*	*Polyscias fulva*	*Mixed natives*
pH_{KCl}	0–5	4.9 ± 0.01 ^{fg}	4.9 ± 0.01 ^{fg}	5.5 ± 0.02 ^{h}	4.0 ± 0.01 ^{b}	4.2 ± 0.01 ^{c}	3.7 ± 0.01 ^{a}	5.5 ± 0.02 ^{h}	5.8 ± 0.01 ^{i}	5.4 ± 0.04 ^{h}
	5–10	4.2 ± 0.0 ^{c}	4.2 ± 0.0 ^{c}	4.4 ± 0.0 ^{d}	3.8 ± 0.0 ^{a}	3.8 ± 0.0 ^{a}	3.7 ± 0.0 ^{a}	4.4 ± 0.0 ^{d}	4.6 ± 0.0 ^{e}	4.8 ± 0.0 ^{f}
SOM (%)	0–5	19.1 ± 0.52 ^{bc}	18.1 ± 0.81 ^{b}	19.2 ± 0.26 ^{bc}	21.6 ± 0.59 ^{bcd}	27.6 ± 0.92 ^{d}	25 ± 1.33 ^{cd}	25.8 ± 2.24 ^{d}	18.5 ± 0.46 ^{b}	27.5 ± 2.2 ^{d}
	5–10	10.2 ± 0.1 ^{a}	9.19 ± 1.0 ^{a}	8.93 0.7 ^{a}	10.3 ± 0.5 ^{a}	9.58 ± 0.08 ^{a}	9.27 ± 0.4 ^{a}	9.5 ± 0.5 ^{a}	10.3 ± 0.2 ^{a}	9.06 ± 0.7 ^{a}
SOC ($g\ kg^{-1}$)	0–5	110 ± 3 ^{bc}	100 ± 4.7 ^{b}	110 ± 1.6 ^{bc}	160 ± 5.4 ^{d}	140 ± 7.7 ^{cd}	150 ± 13 ^{d}	160 ± 13 ^{d}	110 ± 2.7 ^{b}	130 ± 3.4 ^{bcd}
	5–10	59 ± 0.57 ^{a}	53 ± 6 ^{a}	52 ± 4.3 ^{a}	56 ± 0.52 ^{a}	54 ± 2.8 ^{a}	55 ± 3.2 ^{a}	53 ± 4.1 ^{a}	60 ± 1.3 ^{a}	60 ± 3 ^{a}
Al^{3+}	0–5	0.079 ± 0.00 ^{a}	0.085 ± 0.01 ^{a}	0.026 ± 0.00 ^{a}	0.006 ± 0.00 ^{a}	0.067 ± 0.01 ^{a}	0.04 ± 0.00 ^{a}	0.15 ± 0.00 ^{a}	0.01 ± 0.00 ^{a}	0.03 ± 0.00 ^{a}
($cmol_c\ kg^{-1}$)	5–10	0.93 ± 0.035 ^{d}	1.2 ± 0.036 ^{e}	0.77 ± 0.04 ^{d}	0.027 ± 0.00 ^{a}	2.6 ± 0.083 ^{h}	2.2 ± 0.05 ^{f}	2.4 ± 0.059 ^{g}	0.34 ± 0.01 ^{b}	0.53 ± 0.01 ^{c}
Ca^{2+}	0–5	19 ± 0.77 ^{e}	24 ± 1.3 ^{fg}	28 ± 0.33 ^{hi}	27 ± 0.45 ^{gh}	29 ± 0.59 ^{hi}	31 ± 1.6 ^{i}	23 ± 0.85 ^{f}	28 ± 0.25 ^{hi}	48 ± 1.2 ^{j}
($cmol_c kg^{-1}$)	5–10	6 ± 0.06 ^{bc}	4.8 ± 0.067 ^{abc}	7 ± 0.47 ^{bc}	14 ± 0.64 ^{d}	1.6 ± 0.026 ^{a}	3.8 ± 0.097 ^{ab}	2.2 ± 0.081 ^{a}	7.8 ± 0.15 ^{c}	6.7 ± 0.18 ^{bc}
Fe^{2+} ($cmol_c kg^{-1}$)	0–5	0.004 ± 0.00 ^{ef}	0.001 ± 0.00 ^{ab}	0.001 ± 0.00 ^{a}	0.0008 ± 0.00 ^{a}	0.002 ± 0.00 ^{abcd}	0.002 ± 0.00 ^{abc}	0.002 ± 0.00 ^{abcd}	0.001 ± 0.00 ^{ab}	0.001 ± 0.00 ^{a}
	5–10	0.003 ± 0.0 ^{bcdef}	0.003 ± 0.00 ^{abcdef}	0.004 ± 0.00 ^{def}	0.0028 ± 0.00 ^{abcdef}	0.005 ± 0.00 ^{f}	0.045 ± 0.04 ^{cdef}	0.004 ± 0.00 ^{def}	0.003 ± 0.00 ^{abcde}	0.003 ± 0.00 ^{abcde}
K^{+} ($cmol_c kg^{-1}$)	0–5	0.63 ± 0.023 ^{efg}	0.52 ± 0.016 ^{cde}	0.7 ± 0.045 ^{gh}	0.81 ± 0.018 ^{hi}	0.41 ± 0.014 ^{bc}	0.47 ± 0.012 ^{cd}	0.43 ± 0.042 ^{bcd}	0.64 ± 0.01 ^{fg}	0.85 ± 0.04 ^{i}
	5–10	0.12 ± 0.01 ^{a}	0.12 ± 0.0091 ^{a}	0.13 ± 0.018 ^{a}	0.53 ± 0.009 ^{def}	0.14 ± 0.0044 ^{a}	0.15 ± 0.0083 ^{a}	0.17 ± 0.016 ^{a}	0.17 ± 0.00 ^{a}	0.35 ± 0.01 ^{b}
Mg^{2+}	0–5	6.3 ± 0.14 ^{h}	4.9 ± 0.08 ^{ef}	5.6 ± 0.036 ^{g}	5.1 ± 0.047 ^{fg}	8.2 ± 0.18 ^{i}	4.6 ± 0.053 ^{e}	5.5 ± 0.099 ^{g}	7.7 ± 0.09 ^{i}	14 ± 0.22 ^{j}
($cmol_c kg^{-1}$)	5–10	1.5 ± 0.042 ^{c}	1.5 ± 0.035 ^{c}	1.4 ± 0.089 ^{bc}	2.2 ± 0.054 ^{d}	0.82 ± 0.015 ^{a}	0.91 ± 0.098 ^{a}	0.92 ± 0.048 ^{ab}	2.1 ± 0.04 ^{d}	2.2 ± 0.03 ^{d}
Mn^{2+}	0–5	0.26 ± 0.01 ^{g}	0.061 ± 0.00 ^{d}	0.056 ± 0.00 ^{cd}	0.019 ± 0.00 ^{ab}	0.1 ± 0.00 ^{e}	0.13 ± 0.00 ^{f}	0.12 ± 0.01 ^{ef}	0.02 ± 0.00 ^{ab}	0.033 ± 0.00 ^{bc}
($cmol_c kg^{-1}$)	5–10	0.024 ± 0.00 ^{ab}	0.02 ± 0.00 ^{ab}	0.009 ± 0.00 ^{ab}	0.004 ± 0.00 ^{a}	0.017 ± 0.00 ^{ab}	0.013 ± 0.00 ^{ab}	0.022 ± 0.00 ^{ab}	0.006 ± 0.00 ^{a}	0.013 ± 0.00 ^{ab}
Na^{+} ($cmol_c kg^{-1}$)	0–5	0.091 ± 0.01 ^{a}	0.1 ± 0.00 ^{a}	0.11 ± 0.00 ^{a}	0.12 ± 0.017 ^{ab}	0.19 ± 0.01 ^{cde}	0.21 ± 0.0045 ^{e}	0.21 ± 0.01 ^{e}	0.13 ± 0.02 ^{abc}	0.18 ± 0.02 ^{bcde}
	5–10	0.19 ± 0.01 ^{cde}	0.18 ± 0.00 ^{bcde}	0.2 ± 0.00 ^{de}	0.17 ± 0.0077 ^{bcde}	0.2 ± 0.00 ^{de}	0.18 ± 0.015 ^{bcde}	0.19 ± 0.00 ^{cde}	0.17 ± 0.01 ^{bcde}	0.14 ± 0.013 ^{abcd}
Zn^{2+} ($cmol_c kg^{-1}$)	0–5	0.002 ± 0.00 ^{e}	0.001 ± 0.00 ^{abc}	0.001 ± 0.00 ^{bc}	0.0009 ± 0.00 ^{abc}	0.0007 ± 0.00 ^{abc}	0.0008 ± 0.00 ^{abc}	0.0012 ± 0.00 ^{cd}	0.0006 ± 0.00 ^{abc}	0.0017 ± 0.00 ^{de}
	5–10	0.001 ± 0.00 ^{bcd}	0.0006 ± 0.00 ^{abc}	0.0005 ± 0.00 ^{a}	0.0007 ± 0.00 ^{abc}	0.0006 ± 0.00 ^{abc}	0.0006 ± 0.00 ^{abc}	0.0006 ± 0.00 ^{ab}	0.0006 ± 0.00 ^{abc}	0.0011 ± 0.00 ^{bc}
EBC	0–5	26 ± 0.89 ^{f}	30 ± 1.2 ^{fg}	34 ± 0.36 ^{h}	33 ± 0.48 ^{gh}	38 ± 0.75 ^{i}	36 ± 1.6 ^{hi}	29 ± 0.9 ^{f}	36 ± 0.31 ^{hi}	63 ± 1.4 ^{j}
($cmol_c kg^{-1}$)	5–10	7.9 ± 0.05 ^{cd}	6.6 ±0.08 ^{bcd}	8.7 ± 0.57 ^{d}	17 ± 0.6 ^{e}	2.8 ± 0.02 ^{a}	5 ± ±0.12 ^{abc}	3.4 ± 0.1 ^{ab}	10 ± 0.19 ^{d}	9.3 ± 0.2 ^{d}
Cations	0–5	27 ± 0.9 ^{e}	30 ± 1.2 ^{ef}	34 ± 0.36 ^{g}	33 ± 0.47 ^{fg}	38 ± 0.74 ^{h}	36 ± 1.6 ^{gh}	29 ± 0.91 ^{e}	36 ± 0.31 ^{gh}	63 ± 1.4 ^{i}
($cmol_c kg^{-1}$)	5–10	8.8 ± 0.06 ^{abc}	7.8 ± 0.11 ^{abc}	9.5 ± 0.55 ^{bc}	17 ± 0.6 ^{d}	5.4 ± 0.07 ^{a}	7.2 ± 0.09 ^{abc}	5.9 ± 0.07 ^{ab}	11 ± 0.19 ^{c}	9.9 ± 0.2 ^{c}

Table A4. Proportions of water-soluble and hot water-extractable nitrogen forms under different tree species at two soil layers (0–5 cm and 5–10 cm) (Rwibasira et al., Long-term effect of forest plantation species on soil chemical properties in Southern Rwanda).

Soil Layer	Water Extract. N Fractions	*Calliandra Calothyrsus*	*Cedrela Serrata*	*Grevillea Robusta*	*Eucalyptus Grandis*	*Eucalyptus Maideni*	*Eucalyptus Saligna*	*Entandrophragma Excelsum*	*Polyscias Fulva*	*Mixed natives*
					Water-soluble fractions					
0–5 cm	WSNO3 (%)	43	55	73	52	54	46	12	68	64
	WSNH4 (%)	13	18	9	10	13	16	27	6	9
	WSNorg (%)	44	27	18	38	33	38	61	26	27
5–10 cm	WSNO3 (%)	60	45	76	58	65	55	41	62	64
	WSNH4 (%)	7	12	2	5	4	8	15	4	3
	WSNorg (%)	33	43	22	37	31	37	44	34	33
					Hot water-extractable fractions					
0–5 cm	HWNO3 (%)	4	3	5	3	5	3	3	3	3
	HWNH4 (%)	17	16	22	16	15	15	19	25	23
	HWNorg (%)	79	81	73	81	80	82	78	72	74
5–10 cm	HWNO3 (%)	4	3	4	4	4	4	2	3	3
	HWNH4 (%)	15	15	10	11	10	11	17	11	13
	HWNorg (%)	81	82	86	85	86	85	81	86	84

References

1. Binkley, D.; Menyailo, O. Gaining insights on the effects of tree spiecies on soils. In *Tree Species Effects on Soils: Implications for Global Change*; Kluwer Academic Publishers: Dordrecht, The Netherlands, 2005; pp. 1–16.
2. Carnol, M.; Bazgir, M. Nutrient return to the forest floor through litter and throughfall under 7 forest species after conversion from Norway spruce. *For. Ecol. Manag.* **2013**, *309*, 66–75. [CrossRef]
3. Bauters, M.; Ampoorter, E.; Huygens, D.; Kearsley, E.; De Haulleville, T.; Sellan, G.; Verbeeck, H.; Boeckx, P.; Verheyen, K. Functional identity explains carbon sequestration in a 77-year-old experimental tropical plantation. *Ecosphere* **2015**, *6*, 1–11. [CrossRef]
4. Lathwell, D.J.; Grove, T.L. Soil-Plant Relationships in the Tropics. *Ecology* **2011**, *17*, 1–16. [CrossRef]
5. Hobbie, S.E.; Reich, P.; Oleksyn, J.; Ogdahl, M.; Zytkowiak, R.; Hale, C.; Karolewski, P. Tree Species Effects on decomposition and forest floor dynamics in a common garden. *Ecology* **2006**, *87*, 2288–2297. [CrossRef]
6. Prescott, C.E.; Vesterdal, L. Tree species effects on soils in temperate and boreal forests: Emerging themes and research needs. *For. Ecol. Manag.* **2013**, *309*, 1–3. [CrossRef]
7. Veldkamp, A. Pedogenesis and Soil Forming Factors. In *Land Use and Land Cover*; Encyclopedia of Life Support Systems (EOLSS): Wageningen, The Netherlands, 2000.
8. Sayer, E.J.; Banin, L.F. Tree Nutrient Status and Nutrient Cycling in Tropical Forest—Lessons from Fertilization Experiments. *Tree Physiol.* **2016**, *6*, 275–297. [CrossRef]
9. Schulte, A.; Ruhiyat, D. *Soils of Tropical Forest Ecosystems*; Springer: Berlin/Heidelberg, Germany, 1998.
10. Karlen, D.L.; Mausbach, M.J.; Doran, J.W.; Cline, R.G.; Harris, R.F.; Schuman, G.E. Soil Quality: A Concept, Definition, and Framework for Evaluation (A Guest Editorial). *Soil Sci. Soc. Am. J.* **1997**, *61*, 4–10. [CrossRef]
11. Salomé, C.; Coll, P.; Lardo, E.; Metay, A.; Villenave, C.; Marsden, C.; Blanchart, E.; Hinsinger, P.; Le Cadre, E. The soil quality concept as a framework to assess management practices in vulnerable agroecosystems: A case study in Mediterranean vineyards. *Ecol. Indic.* **2016**, *61*, 456–465. [CrossRef]
12. Baveye, P.C.; Ebaveye, J.; Egowdy, J. Soil "Ecosystem" Services and Natural Capital: Critical Appraisal of Research on Uncertain Ground. *Front. Environ. Sci.* **2016**, *4*, 41. [CrossRef]
13. Bünemann, E.K.; Bongiorno, G.; Bai, Z.; Creamer, R.E.; De Deyn, G.; de Goede, R.; Fleskens, L.; Geissen, V.; Kuyper, T.W.; Mäder, P.; et al. Soil quality—A critical review. *Soil Biol. Biochem.* **2018**, *120*, 105–125. [CrossRef]
14. Greiner, L.; Keller, A.; Grêt-Regamey, A.; Papritz, A. Soil function assessment: Review of methods for quantifying the contributions of soils to ecosystem services. *Land Use Policy* **2017**, *69*, 224–237. [CrossRef]
15. Haines-Young, R.; Potschin, M. *England's Terrestrial Ecosystem Services and Rationale for an Ecosystem Approach*; Ecosystem Services: London, UK, 2008.
16. Kennedy, A.C. Biological Indicators of Soil Health. *J. Environ. Qual.* **1998**, *27*, 1275–1276. [CrossRef]
17. Vogel, H.-J.; Eberhardt, E.; Franko, U.; Lang, B.; Ließ, M.; Weller, U.; Wiesmeier, M.; Wollschläger, U. Quantitative Evaluation of Soil Functions: Potential and State. *Front. Environ. Sci.* **2019**, *7*, 164. [CrossRef]
18. Augusto, L.; Ranger, J.; Binkley, D.; Rothe, A. Impact of several common tree species of European temperate forests on soil fertility. *Ann. For. Sci.* **2002**, *59*, 233–253. [CrossRef]
19. Franzluebbers, A. Soil organic matter stratification ratio as an indicator of soil quality. *Soil Tillage Res.* **2002**, *66*, 95–106. [CrossRef]
20. Wellock, M.L.; LaPerle, C.M.; Kiely, G. What is the impact of afforestation on the carbon stocks of Irish mineral soils? *For. Ecol. Manag.* **2011**, *262*, 1589–1596. [CrossRef]
21. DeGryze, S.; Six, J.; Paustian, K.; Morris, S.J.; Paul, E.A.; Merckx, R. Soil organic carbon pool changes following land-use conversions. *Glob. Chang. Biol.* **2004**, *10*, 1120–1132. [CrossRef]
22. Peri, P.L.; Gargaglione, V.; Pastur, G.J.M.; Lencinas, M.V. Carbon accumulation along a stand development sequence of Nothofagus antarctica forests across a gradient in site quality in Southern Patagonia. *For. Ecol. Manag.* **2010**, *260*, 229–237. [CrossRef]
23. Del Galdo, I.; Six, J.; Peressotti, A.; Cotrufo, M.F. Assessing the impact of land-use change on soil C sequestration in agricultural soils by means of organic matter fractionation and stable C isotopes. *Glob. Chang. Biol.* **2003**, *9*, 1204–1213. [CrossRef]
24. Mao, R.; Zeng, D.-H.; Hu, Y.-L.; Li, L.-J.; Yang, D. Soil organic carbon and nitrogen stocks in an age-sequence of poplar stands planted on marginal agricultural land in Northeast China. *Plant Soil* **2010**, *332*, 277–287. [CrossRef]
25. Ross, D.; Tate, K.; Scott, N.; Feltham, C. Land-use change: Effects on soil carbon, nitrogen and phosphorus pools and fluxes in three adjacent ecosystems. *Soil Biol. Biochem.* **1999**, *31*, 803–813. [CrossRef]
26. Paul, K.; Polglase, P.; Nyakuengama, J.; Khanna, P. Change in soil carbon following afforestation. *For. Ecol. Manag.* **2002**, *168*, 241–257. [CrossRef]
27. Nyberg, G.; Ekblad, A.; Buresh, R.; Högberg, P. Short-term patterns of carbon and nitrogen mineralisation in a fallow field amended with green manures from agroforestry trees. *Biol. Fertil. Soils* **2002**, *36*, 18–25. [CrossRef]
28. Curtin, D.; Beare, M.H.; Qiu, W. Distinguishing functional pools of soil organic matter based on solubility in hot water. *Soil Res.* **2021**, *59*, 319. [CrossRef]
29. Paul, E.A. The nature and dynamics of soil organic matter: Plant inputs, microbial transformations, and organic matter stabilization. *Soil Biol. Biochem.* **2016**, *98*, 109–126. [CrossRef]
30. Choudhary, O.; Gill, J.; Singh, B. Water-Extractable Carbon Pools and Microbial Biomass Carbon in Sodic Water-Irrigated Soils Amended with Gypsum and Organic Manures. *Pedosphere* **2013**, *23*, 88–97. [CrossRef]

31. Haynes, R. Labile Organic Matter Fractions as Central Components of the Quality of Agricultural Soils: An Overview. *Adv. Agron.* **2005**, *85*, 221–268. [CrossRef]
32. Strosser, E. Methods for determination of labile soil organic matter: An overview. *J. Agrobiol.* **2010**, *27*, 49–60. [CrossRef]
33. Zhang, M.; He, Z.; Zhao, A.; Zhang, H.; Endale, D.M.; Schomberg, H.H. Water-Extractable Soil Organic Carbon and Nitrogen Affected by Tillage and Manure Application. *Soil Sci.* **2011**, *176*, 307–312. [CrossRef]
34. Wang, Q.; Wang, S.; Huang, Y. Comparisons of litterfall, litter decomposition and nutrient return in a monoculture Cunninghamia lanceolata and a mixed stand in southern China. *For. Ecol. Manag.* **2008**, *255*, 1210–1218. [CrossRef]
35. Habiyaremye, G.; Jiwen, G.; de la Paix Mupenzi, J.; Balogun, W.O. Demographic pressure impacts on forests in Rwanda. *Afr. J. Agric. Res.* **2011**, *6*, 4533–4538. [CrossRef]
36. Clay, D.C.; Lewis, L.A. Land use, soil loss, and sustainable agriculture in Rwanda. *Hum. Ecol.* **1990**, *18*, 147–161. [CrossRef]
37. MINILAF. *Forest Sector Strategic Plan 2018–2024*; GoR: Kigali, Rwanda, 2018.
38. Warnest, M.; Sagashya, D.; Nkurunziza, E. *Emerging in a Changing Climate-Sustainable Land Use Management in Rwanda*; International Federation of Surveyors (FIG): Rome, Italy, 2012.
39. RFA. *Rwanda Forestry Status*; GoR: Kigali, Rwanda, 2020; Available online: https://www.rfa.rw (accessed on 6 May 2020).
40. MoE. *Rwanda Forest cover Mapping*; GoR: Kigali, Rwanda, 2019.
41. IUCN. A Thousand Hills Turn Green: How Rwanda Became a Leader in Forest Landscape restoration. *Int. Union Conserv. Nat.* 2020. Available online: https://iucn.maps.arcgis.com/apps/MapJournal/index.html?appid=0c40be7f83c64bb9b7c8f3372358c6f9 (accessed on 4 April 2020).
42. Bauters, M.; Verbeeck, H.; Doetterl, S.; Ampoorter, E.; Baert, G.; Vermeir, P.; Verheyen, K.; Boeckx, P. Functional Composition of Tree Communities Changed Topsoil Properties in an Old Experimental Tropical Plantation. *Ecosystems* **2016**, *20*, 861–871. [CrossRef]
43. Bauters, M.; Verbeeck, H.; Demol, M.; Bruneel, S.; Taveirne, C.; Van der Heyden, D.; Cizungu, L.; Boeckx, P. Parallel functional and stoichiometric trait shifts in South American and African forest communities with elevation. *Biogeosciences* **2017**, *14*, 5313–5321. [CrossRef]
44. Nsabimana, L.; Klemedtson, B.; Kaplin, G. Wallin, Soil carbon and nutrient accumulation under forest plantations in southern Rwanda. *African J. Environ. Sci. Technol.* **2008**, *2*, 142–149.
45. Meteorwanda, Climatology of Rwanda. 2021. Available online: https://www.meteorwanda.gov.rw/index.php?id=30 (accessed on 30 April 2021).
46. Nsabimana, D.; Klemedtson, L.; Kaplin, B.; Wallin, G. Soil CO2 flux in six monospecific forest plantations in Southern Rwanda. *Soil Biol. Biochem.* **2009**, *41*, 396–402. [CrossRef]
47. Steiner, K. Using farmers' knowledge of soils in making research results more relevant to field practice: Experiences from Rwanda. *Agric. Ecosyst. Environ.* **1998**, *69*, 191–200. [CrossRef]
48. Moeyersons, J. The topographic thresholds of hillslope incisions in southwestern Rwanda. *Catena* **2003**, *50*, 381–400. [CrossRef]
49. Kalinganire, A. Performance of Grevillea robusta in plantations and on farms under varying environmental conditions in Rwanda. *For. Ecol. Manage.* **1995**, *80*, 279–285. [CrossRef]
50. Burren, C. *Les Eucalyptus au Rwanda: Analyse de 60 ans D'expérience avec Référence Particulière à l'Arboretum de Ruhande*; Switzerland-ISAR/Intercooperation: Bern, Switzerland, 1995.
51. Commonwealth. Arboretum of Ruhande: A 500 Acre Stretch of Botanical Garden, QCC Proj. 2018. Available online: https://queenscommonwealthcanopy.org/projects/arboretum-of-ruhande (accessed on 30 August 2020).
52. ISAR. *Arboretum de Ruhande 1987: Note Explicative à L'intention des Visiteurs Élaborée par les Collaborateurs du Département de Foresterie*, 3rd ed.; ISAR: Butare, Rwanda, 1987.
53. Ndayambaje, J.D. *State of Rwanda's Biodiversity for Food and Agriculture*; FAO-Rwanda: Kigali, Rwanda, 2016.
54. Iiyama, M.; Mukuralinda, A.; Ndayambaje, J.D.; Musana, B.; Ndoli, A.; Mowo, J.G.; Garrity, D.; Ling, S.; Ruganzu, V. Tree-Based Ecosystem Approaches (TBEAs) as Multi-Functional Land Management Strategies—Evidence from Rwanda. *Sustainability* **2018**, *10*, 1360. [CrossRef]
55. Bini, D.; dos Santos, C.A.; Bouillet, J.-P.; Gonçalves, J.L.M.; Cardoso, E.J.B.N. Eucalyptus grandis and Acacia mangium in monoculture and intercropped plantations: Evolution of soil and litter microbial and chemical attributes during early stages of plant development. *Appl. Soil Ecol.* **2013**, *63*, 57–66. [CrossRef]
56. Zalamea, M.; González, G.; Lodge, D. Physical, Chemical, and Biological Properties of Soil under Decaying Wood in a Tropical Wet Forest in Puerto Rico. *Forests* **2016**, *7*, 168. [CrossRef]
57. Rorison, I.H.; Allen, S.E.; Grimshaw, H.M.; Parkinson, J.A.; Quarmby, C. Chemical Analysis of Ecological Materials. *J. Appl. Ecol.* **1976**, *13*, 650. [CrossRef]
58. Périé, C.; Ouimet, R. Organic carbon, organic matter and bulk density relationships in boreal forest soils. *Can. J. Soil Sci.* **2008**, *88*, 315–325. [CrossRef]
59. Jenkinson, D.S.; Powlson, D.S. The effects of biocidal treatments on metabolism in soil—V: A method for measuring soil biomass. *Soil Biol. Biochem.* **1976**, *9*, 209–213. [CrossRef]
60. Hendershot, W.H.; Duquette, M. A simple barium chloride for determining cation exchange capacity and exchangeable cations. *Soil Sci. Soc. Am. J.* **1986**, *50*, 605–608. [CrossRef]

61. Ghani, A.; Dexter, M.; Perrott, K. Hot-water extractable carbon in soils: A sensitive measurement for determining impacts of fertilisation, grazing and cultivation. *Soil Biol. Biochem.* **2003**, *35*, 1231–1243. [CrossRef]
62. Gregorich, E.; Beare, M.; Stoklas, U.; St-Georges, P. Biodegradability of soluble organic matter in maize-cropped soils. *Geoderma* **2003**, *113*, 237–252. [CrossRef]
63. Bates, D.; Maechler, M.; Bolker, B.M.; Walker, S.C. Fitting linear mixed-effects models using lme4. *J. Stat. Softw.* **2015**, *76*, 1–48.
64. Venables, W.N.; Ripley, B.D. Package MASS. Available online: http://www.r-project.org (accessed on 17 October 2012).
65. Barton, K. MuMIn: Multi-Model Inference. 2018. Available online: https://cran.rproject.org/package=MuMIn (accessed on 15 September 2021).
66. Gelsomino, A.; Azzellino, A. Multivariate analysis of soils: Microbial biomass, metabolic activity, and bacterial-community structure and their relationships with soil depth and type. *J. Plant Nutr. Soil Sci.* **2011**, *174*, 381–394. [CrossRef]
67. FAO. Topsoil characterization for sustainable land management. In *Land and Water Development Division Soil Resources, Management and Conservation Service*; FAO: Rome, Italy, 1998; p. 71.
68. Adugna, A.; Abegaz, A. Effects of soil depth on the dynamics of selected soil properties among the highlands resources of Northeast Wollega, Ethiopia: Are these sign of degradation? *Solid Earth Discuss.* **2015**, *7*, 2011–2035. [CrossRef]
69. Cleveland, C.C.; Townsend, A.R.; Schmidt, S.K.; Constance, B.C. Soil microbial dynamics and biogeochemistry in tropical forests and pastures, southwestern costa rica. *Ecol. Appl.* **2003**, *13*, 314–326. [CrossRef]
70. Yamoah, C.F.; Burleigh, J.R.; Malcolm, M.R. Application of expert systems to study of acid soils in Rwanda. *Agric. Ecosyst. Environ.* **1990**, *30*, 203–218. [CrossRef]
71. Mbonigaba, J.J.M.; Nzeyimana, I.; Bucagu, C.; Culot, M. Caractérisation physique, chimique et microbiologique de trois sols acides tropicaux du Rwanda sous jachères naturelles et contraintes à leur productivité. *Biotechnol. Agron. Soc. Environ.* **2009**, *13*, 545–558.
72. Fujii, K. Soil acidification and adaptations of plants and microorganisms in Bornean tropical forests. *Ecol. Res.* **2014**, *29*, 371–381. [CrossRef]
73. Duchaufour, P. *Pédologie: Sol, Végétation, Environnement*; Elsevier Masson: Paris, France, 1994.
74. Dudek, T.; Cuadros, J.; Fiore, S. Interstratified kaolinite-smectite: Nature of the layers and mechanism of smectite kaolinization. *Am. Miner.* **2006**, *91*, 159–170. [CrossRef]
75. Ryan, P.C.; Huertas, F.J. Reaction Pathways of Clay Minerals in Tropical Soils: Insights from Kaolinite-smectite Synthesis Experiments. *Clays Clay Miner.* **2013**, *61*, 303–318. [CrossRef]
76. Celentano, D.; Zahawi, R.; Finegan, B.; Ostertag, R.; Cole, R.J.; Holl, K.D. Litterfall Dynamics Under Different Tropical Forest Restoration Strategies in Costa Rica. *Biotropica* **2010**, *43*, 279–287. [CrossRef]
77. Mugunga, C.P.; Kool, D.; Van Wijk, M.T.; Mohren, G.M.J.; Giller, K.E. Water use by short rotation Eucalyptus woodlots in southern Rwanda. *Agrofor. Syst.* **2015**, *89*, 1119–1139. [CrossRef]
78. Behera, N.; Sahani, U. Soil microbial biomass and activity in response to Eucalyptus plantation and natural regeneration on tropical soil. *For. Ecol. Manag.* **2003**, *174*, 1–11. [CrossRef]
79. Laclau, J.-P.; Ranger, J.; Gonçalves, J.L.M.; Maquère, V.; Krusche, A.; M'Bou, A.T.; Nouvellon, Y.; Saint-André, L.; Bouillet, J.-P.; Piccolo, M.D.C.; et al. Biogeochemical cycles of nutrients in tropical Eucalyptus plantations: Main features shown by intensive monitoring in Congo and Brazil. *For. Ecol. Manag.* **2010**, *259*, 1771–1785. [CrossRef]
80. Rhoades, C.; Binkley, D. Factors influencing decline in soil pH in Hawaiian Eucalyptus and Albizia plantations. *For. Ecol. Manag.* **1996**, *80*, 47–56. [CrossRef]
81. Jobbágy, E.G.; Jackson, R.B. Patterns and mechanisms of soil acidification in the conversion of grasslands to forests. *Biogeochemistry* **2003**, *64*, 205–229. [CrossRef]
82. de Medeiros, E.V.; Duda, G.P.; dos Santos, L.A.R.; Lima, J.R.D.S.; de Almeida-Cortêz, J.S.; Hammecker, C.; Lardy, L.; Cournac, L. Soil organic carbon, microbial biomass and enzyme activities responses to natural regeneration in a tropical dry region in Northeast Brazil. *Catena* **2017**, *151*, 137–146. [CrossRef]
83. Ahmed, A.G.; Murungi, L.K.; Babin, R.; Peng, Y.; Schmidt, I.K.; Zheng, H.; Heděnec, P.; Bachega, L.R.; Yue, K.; Wu, F.; et al. Six, Tree species effects on soil organic matter dynamics: The role of soil cation composition. *For. Ecol. Manage.* **2015**, *36*, 119–127. [CrossRef]
84. Koutika, L.-S.; Epron, D.; Bouillet, J.-P.; Mareschal, L. Changes in N and C concentrations, soil acidity and P availability in tropical mixed acacia and eucalypt plantations on a nutrient-poor sandy soil. *Plant Soil* **2014**, *379*, 205–216. [CrossRef]
85. Wang, Q.; Wang, S.; Yu, X. Decline of soil fertility during forest conversion of secondary forest to Chinese fir plantations in subtropical China. *Land Degrad. Dev.* **2011**, *22*, 444–452. [CrossRef]
86. Bankó, L.; Tóth, G.; Marton, C.L.; Hoffmann, S. Hot-water extractable C and N as indicators for 4p1000 goals in a temperate-climate long-term field experiment: A case study from Hungary. *Ecol. Indic.* **2021**, *126*, 107364. [CrossRef]
87. Gregorich, E.G.; Carter, M.R.; Angers, D.A.; Monreal, C.M.; Ellert, B.H. Towards a minimum data set to assess soil organic matter quality in agricultural soils. *Can. J. Soil Sci.* **1994**, *74*, 367–385. [CrossRef]
88. Ostrowska, A.; Porębska, G.; Kanafa, M. Carbon accumulation and distribution in profiles of forest soils. *Polish J. Environ. Stud.* **2010**, *19*, 1307–1315.
89. Russell, A.E.; Raich, J.W.; Valverde-Barrantes, O.J.; Fisher, R.F. Tree Species Effects on Soil Properties in Experimental Plantations in Tropical Moist Forest. *Soil Sci. Soc. Am. J.* **2007**, *71*, 1389–1397. [CrossRef]

90. Landgraf, D.; Leinweber, P.; Makeschin, F. Cold and hot water–extractable organic matter as indicators of litter decomposition in forest soils. *J. Plant Nutr. Soil Sci.* **2006**, *169*, 76–82. [CrossRef]
91. Binkley, D.; Menyailo, O. *Tree Species Effects on Soils: Implications for Global Change*; Springer: Krasnoyarsk, Russia, 2004; p. 361. [CrossRef]
92. Ćirić, V.; Belić, M.; Nešić, L.; Šeremešić, S.; Pejić, B.; Bezdan, A.; Manojlović, M. The sensitivity of water extractable soil organic carbon fractions to land use in three soil types. *Arch. Agron. Soil Sci.* **2016**, *62*, 1654–1664. [CrossRef]

soil systems

MDPI

Article

Home-Field Advantage of Litter Decomposition Faded 8 Years after Spruce Forest Clearcutting in Western Germany

Liyan Zhuang [1], Andrea Schnepf [1], Kirsten Unger [2], Ziyi Liang [1,3,*] and Roland Bol [1]

1 Forschungszentrum Jülich GmbH, IBG-3 (Agrosphäre), 52428 Jülich, Germany; l.zhuang@fz-juelich.de (L.Z.); a.schnepf@fz-juelich.de (A.S.); r.bol@fz-juelich.de (R.B.)
2 Institute of Crop Science and Resource Conservation (INRES), Soil Science and Soil Ecology, University of Bonn, Nussallee 13, 53115 Bonn, Germany; kunger@uni-bonn.de
3 Long-Term Research Station of Alpine Forest Ecosystems, Key Laboratory of Ecological Forestry Engineering, Institute of Ecology and Forestry, Sichuan Agricultural University, Chengdu 611130, China
* Correspondence: chn_liangzy@163.com or z.liang@fz-juelich.de

Abstract: Home-field advantage (HFA) encompasses all the processes leading to faster litter decomposition in the 'home' environment compared to that of 'away' environments. To determine the occurrence of HFA in a forest and adjacent clear-cut, we set up a reciprocal litter decomposition experiment within the forest and clear-cut for two soil types (Cambisols and Gleysols) in temperate Germany. The forest was dominated by Norway spruce (*Picea abies*), whereas forest regeneration of European Beech (*Fagus sylvatica*) after clearcutting was encouraged. Our observation that Norway spruce decomposed faster than European beech in 70-yr-old spruce forest was most likely related to specialized litter-soil interaction under existing spruce, leading to an HFA. Elevated soil moisture and temperature, and promoted litter N release, indicated the rapid change of soil-litter affinity of the original spruce forest even after a short-term regeneration following clearcutting, resulting in faster beech decomposition, particularly in moisture- and nutrient-deficient Cambisols. The divergence between forest and clear-cut in the Cambisol of their litter $\delta^{15}N$ values beyond nine months implied litter N decomposition was only initially independent of soil and residual C status. We conclude that clearcutting modifies the litter-field affinity and helps promote the establishment or regeneration of European beech in this and similar forest mountain upland areas.

Keywords: clearcutting; Norway spruce; European beech; litter decomposition; N; Ca; home-field advantage (HFA); carbon and nitrogen stable isotopes

Citation: Zhuang, L.; Schnepf, A.; Unger, K.; Liang, Z.; Bol, R. Home-Field Advantage of Litter Decomposition Faded 8 Years after Spruce Forest Clearcutting in Western Germany. *Soil Syst.* **2022**, *6*, 26. https://doi.org/10.3390/soilsystems6010026

Academic Editor: Klaus von Wilpert

Received: 1 February 2022
Accepted: 12 March 2022
Published: 15 March 2022

1. Introduction

Forest cover change contributes to complex feedbacks on forest ecosystems along chronosequences [1] and results in the disruption of ecological processes, including microclimate and soil nutrients mineralization [2–4]. Clearcutting, changing the dominated species, and forest growth dynamics leave behind a significant shift in ecosystem-scale species communities, influencing the decomposition pattern during regeneration. Such as, nutrient-rich litter or logging residual in early successional stages is associated with faster decomposition and turnover rates, while slower organic matter recycling and infertile soil had usually found under older forests [5,6]. The change in the decomposition process determines organic matter sequestration and hence forest growth [7]. A better understanding of litter decomposition and nutrient cycling is necessary for an effective management strategy to promote forest regeneration, especially after deforestation or decades of regeneration [8].

In the last century, large forest areas in central Europe were converted into monocultures of fast-growing spruce. Spruce monocultures are generally known for their low biodiversity and soil deterioration due to acidification and nitrogen leaching [9,10]. To maintain the ecological, sociological, and cultural functions of the forest, the conversion of existing Norway spruce into more natural broadleaved and mixed forests is the main silvicultural aim in Germany and other European countries [11,12]. Some studies have shown that spruce decomposition was accelerated in its originated coniferous stands relative to away from it [13–15]. It is usually considered that soil decomposer organisms may adapt to break down particular substrate in individual ecosystems, thereby accelerating the decomposition of litter from which it is derived (i.e., home) than away from that plant [16,17], which has been termed the home field advantage (HFA) of litter decomposition [18].

Moreover, the data review analysis from Ayres et al. [19] concluded that HFA is widespread in forest ecosystems and suggested that ~30% of the variability of litter decomposition at a global scale can be explained by HFA. Clearcutting brings about a high plant abundance of pioneer species (i.e., high nutrient concentration and low lignin: N ratios) and modified soil abiotic conditions (including nutrient leaching, soil temperature, moisture, and pH) [20–23], resulting in shifts in the functioning of decomposer communities, such as decreased fungal biomass and change in bacterial community structure [24,25]. The resultant association between individual species and site condition can affect soil properties that enhance the decomposition of its own litter, creating an HFA effect for the species-own litter [24]. At the same time, case studies indicated that warmer and moister conditions after clearcutting drive faster litter breakdown by higher soil decomposer activity irrespective of HFA [26,27]. Soil decomposer communities changes when a forest is clear-cut due to the shift in plant communities and soil physical condition, and then microbial differences in ability might arise through local adaption with its "new" home environment (or a 'home' litter) [28], however, studies rarely investigate HFA after removing the dominant species as in clearcuttings. There is a need for gathering reliable scientific knowledge on the influence of clearcutting on original 'home' and 'away' litter decomposition in the new clear-cut.

There is increasing evidence that the strength of HFA is associated with the interaction between local litter quality and specialized microbes. For example, greater fungal biomass in spruce plantations could partly explain the HFA for spruce in its habitat due to the better degradation of recalcitrant fractions through fungi adaption [29]; that is, conifers should favor soil decomposition dominated by fungi and fungivorous microarthropods, in comparison to broadleaved species [30]. Moreover, across succession, soil communities have gone through a wider range of litter qualities contributing to a broader functional capacity to degrade various litter types [28], so decomposer ability in succession may increase with regeneration. However, recent studies pointed out that litter quality was not an important determinant of HFA [31], while the greater dissimilarity between 'home' and 'away' litter indicated strong HFA [32].

Measurements of plant $\delta^{13}C$ and $\delta^{15}N$ abundance have been shown to be useful indicators of forest organic matter dynamics [33]. The difference between the isotopic signature of residual litter and litter degradation or litter nutrient dynamics are considered as the inherent tracers for understanding the progression of decomposition and nutrient mineralization/immobilization [34]. Labile compounds with faster mineralization rates exhibit higher $\delta^{13}C$ values rather than $\delta^{13}C$-depleted recalcitrant lignin [35]. In addition, microbial processes enrich carbon with $\delta^{13}C$ in relation to bulk litter [36]. The changes in foliar$\delta^{15}N$ values are positively associated with nitrate leaching following forest clearcutting [37,38], that is, the foliar $\delta^{15}N$ often relate to N availability, clearcutting increases nitrification and nitrate loss rate, resulting in much of the $\delta^{15}N$-depleted nitrate leaching out, but $\delta^{15}N$-enriched ammonium retaining These findings have provided us with a meaningful point that the alteration of isotopic C and N signature between litter types during decomposition are useful indicators of nutrient status after disruption of the forest.

In the Eifel National Park (Wüstebach, Germany), clearcutting operations were carried out in spruce monoculture in 2013 as the first step of conversion from planted spruce monoculture to natural forest. This significantly affected soil nutrient leaching [39,40], moisture [41] as well as soil respiration [42]. To test the validity of the HFA change long with clearcutting management, we carried out a reciprocal transplant litter decomposition on a 70-yr spruce forest and a clear-cut after short-term (8-yr) regeneration. In addition, we tested the importance of litter quality on the strength/occurrence of HFA. The difference of litter mass loss and nutrient release, as well as isotopic $\delta^{13}C$ and $\delta^{15}N$ discrimination between spruce and beech, were determined to figure out this question.

2. Materials and Methods

2.1. Site Description

The study area is located in Wüstebach (50°30′15.3″ N, 6°20′03.0″ E), situated within the Eifel National Park of western Germany. The climate is mild and humid, with the mean annual air temperature of 7 °C and the mean annual precipitation of about 1200 mm [43]. Winter is moderately cold with periods of snow. Norway spruce replaced European beech as the dominant canopy species for timber production since the 1940s. In the last decades, the Park authority has started accelerating the 'natural' regeneration towards a beech forest by clear-cuts of a significant proportion of the Norway spruce monoculture (~90%) [44]. The ground cover vegetation in these clear-cut stands is formed mainly by young samplings of alder [*Alnus glutinosa* (L.) Gaertn], European beech (*Fagus sylvatica*) with an admixture of early pioneer species, i.e., scrubs, bushes after 8-year regeneration. Norway spruce (*Picea abies*) is the dominant tree species in the remaining uncut forest. Five subplots were selected for this study ranging from 595 m in the northern part to 628 m in the south in forest and clear-cut, respectively. Soils at the stands are classified as Cambisols and Gleysols, and Gleysols nearby stream is moister than Cambisol. For more information about soil properties, refer to Siebers and Kruse. [40] and Wiekenkamp et al. [42].

2.2. Litter Decomposition Experiments

Between 2019 and 2020, a reciprocal litter transplant experiment was established in the forest and clear-cut. In September 2019, freshly senesced spruce needles and beech leaves were collected from 6 sampling sites at the forest and clear-cut ecosystems, respectively. Within each collection, each substrate was collected from a minimum of 6 different plant individuals to ensure the representativeness of the pool collected. According to the purpose of forest management, we assumed that spruce is the home environment for the forest, while the home environment for beech is clear-cut.

All samples were air-dried to constant mass. 2.5 g of Spruce needles or Beech leaves were filled into each polyethylene litterbag (10 × 8 cm; 0.25 mm mesh size), respectively. The mesh size permits the entry of bacteria, fungi, and micro-fauna [45]. In October 2019, five sampling locations were selected for clear-cuts and adjacent forests on both Cambisols and Gleysols, respectively. At each subplot, 4 litterbags of each species were placed on the soil surface after getting rid of the humus layer or grass. Litterbags were retrieved after 1, 3, 9, 12 months. Altogether, we prepared 160 litterbags (4 sampling times × 2 stands × 2 soil types × 2 species × 5 replicates) in total. Harvested litterbags were transported to the laboratory. Oven-dried and weighed after removing soil particles and other extraneous materials.

C and N contents of each sampling were measured by a CNS analyzer. The natural abundance of $\delta^{13}C$ and $\delta^{15}N$ was measured by stable isotope ratios mass spectrometry. The total phosphorus (P), Calcium (Ca) were determined after microwave digestion with H_2O_2-HNO_3 using inductively coupled plasma mass spectrometry (ICP-MS). The soil temperature and moisture in Wüstebach were measured with the wireless sensor network with 600 ECH2O EC-5 and 300 ECH2O 5TE sensors.

2.3. Data Statistics

To determine the strength and direction of home-field effects on litter decay rate, the home-field advantage index (HFAi) for mass loss and the release of C and N was calculated following Ayres et al. [19] and adapted from Veen et al. [31] as

$$\text{HFAi}\ (\%) = \left(\frac{A_{RL_a} + B_{RL_b}}{2} / \frac{A_{RL_b} + B_{RL_a}}{2}\right) \times 100 - 100 \quad (1)$$

where i_{RL_j} represents the relative mass or nutrient loss of species i in environment j. Single sample t-tests were used to test whether the HFAi differed from 0.

HFAi stands for the additional decomposition or mineralization at home versus away environment and is a net value for both species (A and B) in the reciprocal experiment.

The mean HFA (% increase in k value at home versus away environment) for each litter type was calculated according to [46]:

$$\text{The mean HFA} = (k_{\text{home}} - k_{\text{away}})/k_{\text{away}} \times 100 \tag{2}$$

where k_{home} and k_{away} are the decomposition constants of a given species at home and in away environments, respectively.

The $\delta^{13}C$ and $\delta^{15}N$ values are expressed as

$$\delta\ (‰) = \left(\frac{R_{sample}}{R_{standard}} - 1\right) \times 1000 \tag{3}$$

where R_{sample} and $R_{standard}$ represent either ^{13}C: ^{12}C or ^{15}N: ^{14}N ratios of sample and standard material, respectively. The stable isotope ratio values are expressed in parts per million (‰) relative to international standards. Vienna Pee Dee Belemnite (VPDB) for carbon isotope and atmospheric nitrogen for nitrogen isotopes. The analytical precisions for carbon isotopes ± 0.1‰ and ± 0.3‰ for nitrogen isotopes.

Mass remaining (%) was calculated from dry mass at sampling date divided by initial dry mass. The decomposition rate (k value, yr^{-1}) was estimated according to the exponential regression $y = e^{-kt}$, y (%) is mass remaining over time t, k is the decomposition rate by Olson. [47]. Nutrients remaining (%) of each sample were estimated as nutrient content at each sampling time divided by initial nutrient content and expressed by % of the initial amount. We performed t-tests: (1) to test the significance of initial quality and residuals after one year of decomposition between beech and spruce; (2) to test k values of beech and spruce in clear-cuts and forest on Cambisols and Gleysols; (3) to examine the environmental differences between forest and clear-cut at each sampling point; and finally, (4) to determine if the HFAi was significant between soil types. Repeated measure ANOVAs were used to compare the significance of soil types, stands, and species on various nutrients remaining over time. Three-way ANOVAs were calculated to compare the three factors: soil types, stand, and species on nutrient remaining. A series of stepwise regressions were conducted to detect the variance relationship, like nutrients and stoichiometry on mass loss between soil types, stands, and species. All statistical analysis was performed using SPSS22.0 for the Windows software package.

3. Results

3.1. Environmental Difference between Stands and Soil

Soil types and forest management greatly influence soil environmental conditions (Figure 1). On average, the soil moisture content was significantly higher in clear-cut than in the forest at both soil types, ranging from 36.6–55%, and Cambisols showed a larger difference in soil moisture by 8.7% than Gleysols by 2.7%, comparing between clear-cut and forest. At both soil types, the temperature at the forest floor was approximately 1 °C higher in clear-cut than in the forest. The results of the t-test revealed that the soil moisture and temperature conditions were mostly higher in clear-cut than forest with times, particularly at Cambisols.

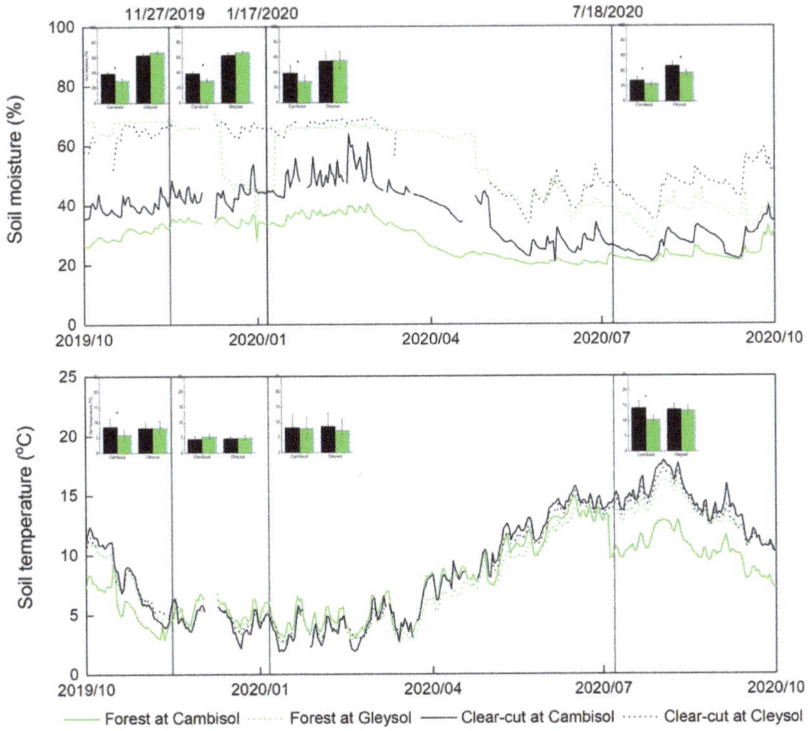

Figure 1. The soil moisture (%) and temperature (°C) dynamics in the top layer during one year of decomposition. Bar charts indicate mean values with error bars at each sampling time. The green bar indicated forest; the black bar indicated clear-cut.

3.2. Initial Litter Quality and Litter Nutrients after One Year of Decomposition

Initial litter quality differed between species. European beech had significantly better initial quality than Norway spruce for C, N, P, and Ca, as well as lower C: N and C:P ratios (Table 1). After one year of decomposition, nutrients concentration and C stoichiometry were significantly different between species and stands. Most nutrient concentrations decreased, except for the N and C:P ratio. Furthermore, C concentration increased in the forest but decreased in clear-cuts for both species during decomposition.

Table 1. Nutrient concentrations and compound ratios of beech and spruce litter before and after one year of decomposition.

	Initial Litter Quality		Residual Quality after 1 Year of Decomposition			
	Beech	Spruce	Forest Beech	Spruce	Clear-Cut Beech	Spruce
C (%)	47.1 ± 0.2a	48.4 ± 0.1b	48.2 ± 0.8a	49.5 ± 0.6b	46.1 ± 0.6a	47.9 ± 0.5b
N (%)	2.1 ± 0.1a	1.2 ± 0.0b	3.0 ± 0.1a	1.9 ± 0.8b	2.8 ± 0.1a	1.6 ± 0.1b
P (mg kg^{-1})	278.4 ± 10.4a	254.9 ± 8.6b	105.2 ± 6.3a	81.0 ± 8.9b	104.5 ± 6.0a	61.5 ± 4.9b
C:N	22.4 ± 0.5a	39.0 ± 1.3b	20.2 ± 0.6a	34.4 ± 0.7b	20.8 ± 0.8a	35.1 ± 0.4b
Ca (mg kg^{-1})	2.18 ± 0.04a	1.67 ± 0.01b	1.02 ± 0.09a	0.74 ± 0.08b	1.20 ± 0.05a	0.79 ± 0.06b

The lower-case letter indicates the significance between species at the same stands. Different lowercase letters indicate significant differences between beech and spruce in each site ($p < 0.05$).

3.3. The Effect of Home-Field Advantage on Litter Decomposition Rates

A significant home-field advantage was shown for the two soil types in this experiment (HFAi = 11 at Cambisols and HFAi = 4 at Gleysols, Table 2). A pattern of the higher decomposition rate of the spruce in the original spruce forest after one year of decomposition followed by *k* values (Figure 2c,d). However, there was no promotion between forest and clear-cut stands in the initial three months. Spruce has a lower 3-month *k* value in forest than in clear-cut (Figure 2a,b), while beech in clear-cut decomposed faster than in forest at Cambisols, but slower when decomposing in Gleysols. Moreover, after one year of decomposition, the *k* value of beech in clear-cut decreased but was higher than the forest stand at Cambisols, while the *k* value of beech in Gleysols did not differ significantly.

Table 2. Home-field advantage index of mass loss and C and N release on Cambisol and Gleysol.

	Cambisol	Gleysol
Mass loss	11.2 ± 0.5a	3.7 ± 1.0b
C release	14.0 ± 2.5a	10.7 ± 0.9b
N release	28.3 ± 0.9a	43.1 ± 5.4b

The lower-case letter indicates significance between stands. Different lowercase letters indicate significant differences between the two soils ($p < 0.05$).

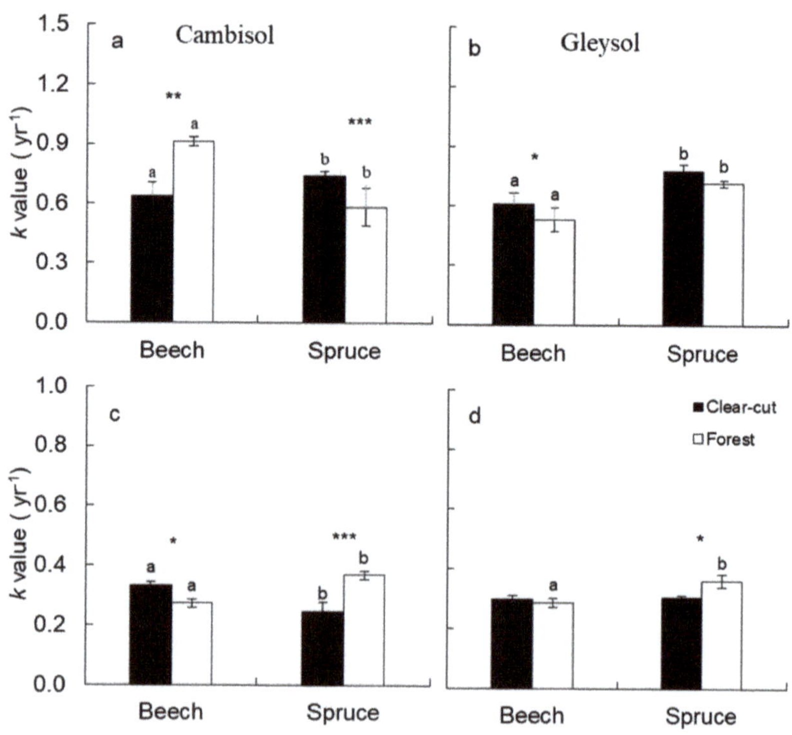

Figure 2. k-values (yr-1) of beech and spruce in clear-cut and forest at Cambisols and Gleysols after 3 months (**a**,**b**) and one year (**c**,**d**) of decomposition. *: $p < 0.05$; **: $p < 0.01$; ***: $p < 0.001$ indicate significance between stands with same species; lower-case letters indicated significance between species at same stands.

The k values for beech and spruce varied among stands and soil types with times. k decreased in time for both spruce and beech. The 3-month k values were on average 2- to 3-fold higher compared to the 1-year values (Figure 2). The decomposition rate of spruce in the first three months was significantly higher than beech in most stands except for forest stand at Cambisols (Figure 2a,b). Significantly or slightly higher k value of spruce showed in forest at all plots except for clear-cut at Cambisols after one year of decomposition, when comparing with beech litter.

3.4. C and N Dynamics and Their HFA

Our results indicate that overall C and N release increased at "home" compared with "away" (Table 2). The difference on C release was stronger in Cambisols (14% vs. 10% in Gleysols) for spruce decomposed in forest, while HFAi of N promoted a higher N release in Gleysols (43%) than in Cambisols (28%).

The significance of litter C dynamics varied through time and different treatments (Figure 3a,b, Table 3). A loss of C could be observed in all substrates within the year-long decomposition. Spruce litter lost most C fraction in this original forest during the study period, while beech C in the forest was released rapidly in the first three months and leveled out by the times, which was 6.6% faster on average for spruce in the forest.

Table 3. Three-way ANOVA analysis of F-value on the effect of soil types, stands, species, and their interactions on nutrient remaining over decomposition.

Effects	Remaining				
	df	C	N	P	Ca
Soil type	1	6.8 *	0.7	8.4 *	0.1
Stand × Soil type	1	2.9	2.1	4.3	15.6 **
Soil type × Species	1	31.6 ***	0.6	0.0	0.0
Stand × Soil type × Species	1	6.3 *	1.3	8.5 *	3.2

*: $p < 0.05$; **: $p < 0.01$; ***: $p < 0.001$.

Both the spruce and beech did not show significance in the absolute amount of N release in one year period (Figure 3c,d, $p > 0.05$). N immobilization for both species appeared in forest after the first 47 days and almost up to its initial N amount. In clear-cuts, the N remaining generally decreased such that, on average, 9% of the total amount of N was released after one year of decomposition (Figure 3c,d). Irrespective of species and soil types, decomposing litter in clear-cut mineralized relatively more N compared to the initial amount than in forest. There was no significant difference in the retention of N between soil types (Table 3).

3.5. The Dynamic of Litter Nutrients Release during Decomposition

Most nutrients indicated significant mineralization over time (Figure 3, Table 3) and observed net mineralization in all substrates following Figure 3. Both leaf litters released P rapidly one year after the start of the decomposition, losing approximately 80% of their initial amount of P (Figure 3e,f, $p < 0.001$). Beech (26%) retained more P than spruce (18%) in one year period ($p < 0.05$). Figure 3e,f shows a similar pattern of P mineralization between forest and clear-cut for both species ($p > 0.05$). On average, the final Ca remaining was significantly higher in forest (70%) than in clear-cut (65%), regardless of species (Figure 3g,h). Furthermore, the interaction between stands and soil types revealed that spruce Ca release was faster than beech (Figure 3g,h, Table 3).

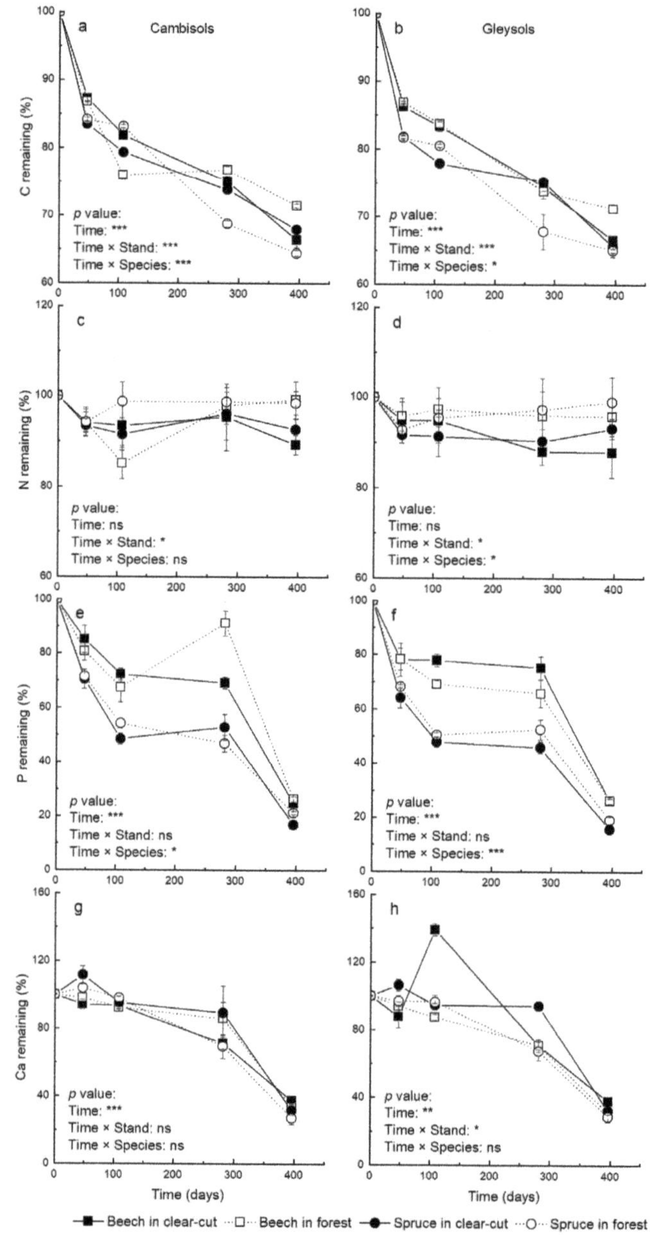

Figure 3. Nutrients (C, N, P, and Ca) remaining of beech and spruce in Cambisols (**a**,**c**,**e**,**g**) and Gleysols (**b**,**d**,**f**,**h**) after one year of decomposition. Error bars represent standard errors. Repeated measure ANOVA indicated significance of stand and species with time periods: ns $p > 0.05$, * $p < 0.05$, ** $p < 0.01$,*** $p < 0.001$.

3.6. Correlation between Litter Mass Loss Rate and Residual Quality

Litter decomposition rate was associated with changing substrate quality in all subplots (Table 4). Decomposition rate of spruce in forest increased with litter N concentration but decreased with litter Ca concentration (R^2 = 0.97). Decomposition of spruce in clear-cut positively changed with N but negatively correlated with P concentration (R^2 = 0.93). Beech decomposition rate in forests was positively related to C: N ratios (R^2 = 0.67) but also decreased with litter Ca concentration when decomposed in clear-cuts (R^2 = 0.85).

Table 4. Stepwise regression of the correlation between litter mass loss rate and nutrient concentrations and stoichiometry of beech and spruce under forest and clear-cut over decomposition. Data indicates significant variables related to decomposition, followed by R^2.

	Variables	Coefficients	R^2
Spruce			
Clear-cut	N, P	0.65, −0.36	0.93
Forest	N, Ca	0.77, −0.23	0.97
Beech			
Clear-cut	Ca, C:N	−0.51, −0.49	0.85
Forest	C:N	−0.82	0.66

3.7. Isotopic Change during Decomposition

The $\delta^{13}C$ values of decomposing litters leveled off over time across litter types. The initial $\delta^{13}C$ values were −32.7 ‰ in beech leaves and increased by 0.13 ‰ on average. For spruce needles, the initial $\delta^{13}C$ value was −30.6 ‰ and decreased by 0.14 after 1 year of decomposition. The initial $\delta^{15}N$ values ranged from −3.3 ‰ in beech leaves and −4.7‰ in spruce needles (Figure 4). The $\delta^{15}N$ values for both species were finally higher in clear-cut than in forest. In the first 9 months, $\delta^{15}N$ became enriched in all subplots but then depleted in forest, while a larger decrease happened in Cambisols. Over the same period, the $\delta^{15}N$ value in clear-cut became higher throughout the experimental period in Cambisols (−3.0 and −3.7 ‰ for beech and spruce, respectively), but it slightly dropped since July in moister Gleysols (−3.2 and −4.1 ‰ for beech and spruce, respectively) (Figure 4). Linear regression plots of N isotopic against C concentration (%) were negatively significant among species over both stands ($p < 0.05$, Figure S2). While the relationship of the $\delta^{13}C$ values and C: N was only linearly significant in forest for spruce (R^2 = 0.83, $p < 0.01$).

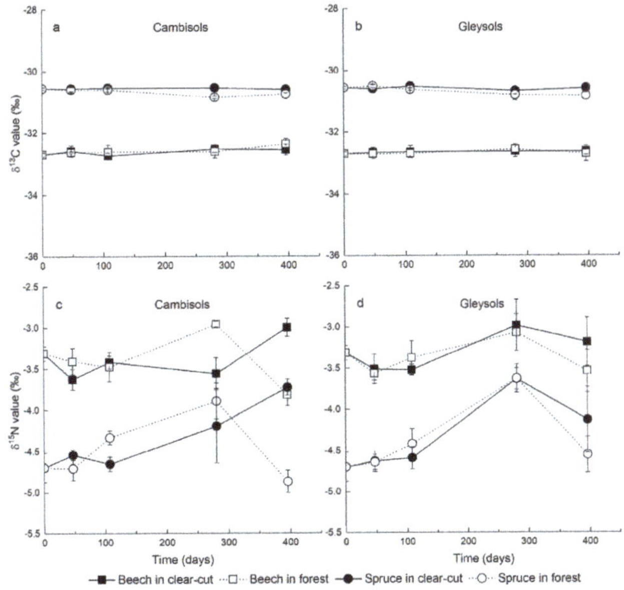

Figure 4. The change of isotopic $\delta^{13}C$ and $\delta^{15}N$ value of beech and spruce over one year of decomposition. Error bars represent standard errors.

4. Discussion

4.1. HFA in Forest and Post-Harvest Decomposition

Our results indicated a positive effect of litter-site interaction on litter decomposition rate at the home of 11% and 4% and thus a net HFA in the spruce forest, which verified our first hypothesis. A meta-analysis has noticed that decomposition HFA is widespread in forest ecosystems, with on average 4.2% promotion in the home habitat [46]. Low-quality spruce with low nutrient content and high C:N decomposed faster in spruce forest, probably due to the presence of more fungal communities well adapted for degrading recalcitrant litter [31]. After 8-year clearcutting, both soil physical (soil moisture and temperature, Figure 1) and chemical conditions [40] had been markedly elevated, which can influence the development and succession of microorganisms that can assimilate substrate [48]. And thus, we found a significant suppression on spruce decomposition and a slight promotion on beech decomposition in clear-cut. On the other hand, post-harvest regeneration of understory species improves soil nutrient availability and forest sites quality, masking the original soil-litter affinity on pre-harvest forest [49]. This could also account for the lower decomposition rate for spruce in the clear-cut and potentially masked the mean HFA for spruce after a short-term regeneration.

4.2. Litter Chemistry Regulated Decomposition of Norway Spruce in Original Forest

Decomposition and mineralization in the initial phase are generally characterized by the leaching of soluble nutrients and by decomposition of soluble and non-lignified cellulose and hemicellulose [50]. Winter snow cover and snowmelt in this initial period physically breakdown litter tissue and accelerated nutrients release and mineralization [51], resulting in a higher k value in clear-cut versus forest, and thus no HFA was detected in the initial 3-month decomposition.

The decomposition difference between litter types was correlated to the concentration of C and N, and C: N ratios. Our results also corroborated this hypothesis that litter N concentration served as the most critical nutrient to regulate the degradation of spruce, and beech was decreased with increased C: N ratios, according to the stepwise regression (Table 4). Slower N release was detected in forest, which decreased litter C:N ratios and promoted the generation of brown and white rot fungi [52], and benefited the degradation of the lignin-rich substrate (i.e., spruce). Faster spruce Ca release strengthen the soil acidification that maintains the soil pH, sustaining the home-field effect. Although litter quality well-regulated the litter mass loss, litter quality independently did not serve as a predictor of mean HFA in this case (supplementary, Figure S1). This result is supported by evidence from Veen et al. [31]. This may be because HFA is not restricted by single litter types, but the heterogeneity of litter quality between the 'home' and 'away' habitats [32]. Alternatively, the occurrence of HFA is likely system-dependent, suggesting that transplants between labile litter from nutrient-rich ecosystems and recalcitrant litter from nutrient-limited ecosystems better induce HFA [53,54]. The results from this work were limited to spruce and beech only; a wider assessment between species and ecosystems is necessary for relevant controlling to determine the magnitude and the direction of HFA for plant traits.

4.3. Clearcutting Promoted Beech Decomposition and Nutrient Release Pattern

Beech leaves decomposed faster during the first year in clear-cut, which is accompanied by an increase in the mineralization rate of C and N in beech leaves and higher in immobilization in spruce needles. In addition, the less home effect of C, N release was observed after clearcutting. Thus, a transfer from spruce to beech would facilitate the potential utilization of nutrients by trees. The shift of dominated trees species by clearcutting treatment would inherently influence the regeneration in this site through litter input quality [40].

Decay rates for beech in both stands were tightly related to C: N ratios. Beech with lower C: N ratios contributed to a faster decomposition rate for beech ($k = 0.31$ on average) than spruce ($k = 0.29$ on average) in clear-cuts. Changing environmental conditions would directly affect litter mass loss after rapid shifts in plant community composition [32], contributing to the relatively elevated mass-loss rate for beech in clear-cuts than in spruce forests and the suppression of mean HFA effect for spruce. Moreover, removing the forest canopy elevated atmosphere C and N deposition with precipitation promotes soil nutrient availability in the short term [40] and restructures the local fungal community in soil [25]. This would further hinder the litter decomposition and nutrient turnover rate in these successional stands.

4.4. Soil Moisture as a Mediator of Litter Decomposition and HFA

Our results indicated that the decay of beech in clear-cuts differed between soil conditions; that is, beneath Cambisols, the decay rates of beech were significantly higher in clear-cuts than in the forest, and interestingly, it was faster than spruce in clear-cuts. However, a minor difference in clear-cut was observed when decomposing on Gleysols, as well as a decline in the HFA of decomposition and C concentration. Gleysols nearby the stream is moister than Cambisols. The microbial breakdown is likely limited with a high soil moisture level [55], probably resulting in insignificant decomposition between species and stands. Additionally, given the importance of the water-driven decomposition determines a weak mass loss in low-quality litter [56,57], contributing to a similar k value (from spruce) between soil types.

Across soil types, the results showed lower HFA on mass and C on Gleysol. A saturated soil environment has been identified to reduce soil microbial decomposition [58]. Soil microbial communities of high soil moisture are generally N limited due to the less nutrient availability [59], resulting in higher N accumulation and N release HFA in Gleysol.

4.5. Dynamics of the Natural Abundance of $\delta^{13}C$ and $\delta^{15}N$ during Decomposition

Isotopic discrimination during litter decomposition has been observed in several studies involving selective consumption of various C compounds. Litter C concentration with $\delta^{15}N$ value, in this case, was negatively significant. This correlation signifies that $\delta^{15}N$ discrimination between litter types is due to the preferential recalcitrant fraction in substrates, which is consistent with several studies [35,60]. Microbial analysis suggests that ^{15}N was transferred actively aboveground by saprotrophic fungi [61] via promotion in the lignin or tannin degradation by fungi-based microbes. Suggesting that decreased ^{15}N values by retaining more litter N from forest floor than from clear-cut do contribute to higher microbial uptake and hence faster spruce litter C degradation in 'home' forest, strengthening the HFA.

In our study, we found a negligible change of $\delta^{13}C$ between stands during decomposition; residual C pools with slightly $\delta^{13}C$ distinct were needed to account for the duration of the experiment. A report from Ngao and Cotrufo. [62] indicated litter $\delta^{13}C$ discrimination appeared particularly in late stages of litter decomposition owing to the increase in the $\delta^{13}C$ of decomposition litter α-cellulose. Future long-term litter decomposition studies on the discrimination of natural abundance of isotope between species types and ecosystems are therefore recommended.

5. Conclusions

Spruce decomposed faster in spruce forest while beech decomposed faster in clear-cut, tightly associating with litter quality, indicating the occurrence of decomposition HFA at forest and clear-cut. Promoted decomposition and C mineralization for spruce in forest could be implied through relatively higher residual N concentration. Since the clear-cut in 2013, plant community and soil environment had shifted the historical resources from the original forest that facilitated faster beech decomposition and nutrients turnover rates due to lower C:N, thereby overriding pre-existing species HFA effects, especially at dryer Cambisols. $\delta^{15}N$ diverged after nine months at Cambisol between forest and clear-cut, suggesting that litter N decomposition correlated to soil and residual C status. This has implications for the management of upland forests that are currently still under conifers:

Their regeneration to more natural forests with European beech can be promoted in short-term by intensive forest management.

Supplementary Materials: The following supporting information can be downloaded at: https://www.mdpi.com/article/10.3390/soilsystems6010026/s1, Figure S1: The relationship between total C, N concentrations, δ13C, δ15N, C:N ratios, Ca and mean HFA; Figure S2: The relationship between the initial litter C, N and C:N ratios on the isotopic 13C and δ15N value.

Author Contributions: Designed study concept: L.Z., R.B. and A.S. Sample preparation, field sampling, and obtained data: L.Z., Z.L. and K.U. Wrote and revised the text: L.Z., R.B. and A.S. All authors have read and agreed to the published version of the manuscript.

Funding: Liyan Zhuang and Ziyi Liang acknowledge Ph.D grant support from China Scholarship Council, overall project support was also via German Research Foundation (DFG) under Germany's Excellence Strategy, EXC-2070—390732324—"PhenoRob" and Helmholtz Association grant 2173 Agro-biogeosystems: controls, feedbacks and impact (POF IV: 2021-2026).

Institutional Review Board Statement: Not applicable.

Informed Consent Statement: Not applicable.

Data Availability Statement: The data are available from the authors upon request.

Conflicts of Interest: The authors declare no conflict of interest.

References

1. Farber, S.; Costanza, R.; Childers, D.L.; Erickson, J.; Gross, K.; Grove, M.; Hopkinson, C.S.; Kahn, J.; Pincetl, S.; Troy, A.; et al. Linking Ecology and Economics for Ecosystem Management. *BioScience* **2006**, *56*, 121–133. [CrossRef]
2. Both, S.; Elias, D.M.; Kritzler, U.H.; Ostle, N.J.; Johnson, D. Land use not litter quality is a stronger driver of decomposition in hyperdiverse tropical forest. *Ecol. Evol.* **2017**, *7*, 9307–9318. [CrossRef] [PubMed]
3. Mayer, M.; Prescott, C.E.; Abaker, W.E.; Augusto, L.; Cécillon, L.; Ferreira, G.W.; James, J.; Jandl, R.; Katzensteiner, K.; Laclau, J.-P. Influence of forest management activities on soil organic carbon stocks: A knowledge synthesis. *For. Ecol. Manag.* **2020**, *466*, 118127. [CrossRef]
4. Achilles, F.; Tischer, A.; Bernhardt-Romermann, M.; Heinze, M.; Reinhardt, F.; Makeschin, F.; Michalzik, B. European beech leads to more bioactive humus forms but stronger mineral soil acidification as Norway spruce and Scots pine—Results of a repeated site assessment after 63 and 82 years of forest conversion in Central Germany. *For. Ecol. Manag.* **2021**, *483*, 15. [CrossRef]
5. Trap, J.; Bureau, F.; Brethes, A.; Jabiol, B.; Ponge, J.-F.; Chauvat, M.; Decaens, T.; Aubert, M. Does moder development along a pure beech (*Fagus sylvatica* L.) chronosequence result from changes in litter production or in decomposition rates? *Soil Biol. Biochem.* **2011**, *43*, 1490–1497. [CrossRef]
6. Xuluc-Tolosa, F.J.; Vester, H.F.M.; Ramı, X.; Rez-Marcial, N.; Castellanos-Albores, J.; Lawrence, D. Leaf litter decomposition of tree species in three successional phases of tropical dry secondary forest in Campeche, Mexico. *For. Ecol. Manag.* **2003**, *174*, 401–412. [CrossRef]
7. Zhang, W.; Gao, D.X.; Chen, Z.X.; Li, H.; Deng, J.; Qiao, W.J.; Han, X.H.; Yang, G.H.; Feng, Y.Z.; Huang, J.Y. Substrate quality and soil environmental conditions predict litter decomposition and drive soil nutrient dynamics following afforestation on the Loess Plateau of China. *Geoderma* **2018**, *325*, 152–161. [CrossRef]
8. Allison, S.D.; Lu, Y.; Weihe, C.; Goulden, M.L.; Martiny, A.C.; Treseder, K.K.; Martiny, J.B.H. Microbial abundance and composition influence litter decomposition response to environmental change. *Ecology* **2013**, *94*, 714–725. [CrossRef]
9. Wilpert, K.V. Forest Soils—What's Their Peculiarity? *Soil Syst.* **2022**, *6*, 5–10. [CrossRef]
10. Spiecker, H. *Norway Spruce Conversion: Options and Consequences*; Brill: Leiden, The Netherland; Boston, MA, USA; Köln, Germany, 2004.
11. Otto, H.J. Die verwirklichung des LÖWE_Regierungsprogramms. *AFZ/Wald* **1995**, *50*, 10269.
12. Wulf, M. Forest policy in the EU and its influence on the plant diversity of woodlands. *J. Environ. Manag.* **2003**, *67*, 15–25. [CrossRef]
13. Asplund, J.; Kauserud, H.; Bokhorst, S.; Lie, M.H.; Ohlson, M.; Nybakken, L. Fungal communities influence decomposition rates of plant litter from two dominant tree species. *Fungal Ecol.* **2018**, *32*, 1–8. [CrossRef]
14. Berger, T.W.; Berger, P. Greater accumulation of litter in spruce (Picea abies) compared to beech (*Fagus sylvatica*) stands is not a consequence of the inherent recalcitrance of needles. *Plant Soil* **2012**, *358*, 349–369. [CrossRef] [PubMed]
15. Vesterdal, L. Influence of soil type on mass loss and nutrient release from decomposing foliage litter of beech and Norway spruce. *Can. J. For. Res.* **1999**, *29*, 95–105. [CrossRef]
16. Liu, Q.; Zhuang, L.; Yin, R.; Ni, X.; You, C.; Yue, K.; Tan, B.; Liu, Y.; Zhang, L.; Xu, Z. Root diameter controls the accumulation of humic substances in decomposing root litter. *Geoderma* **2019**, *348*, 68–75. [CrossRef]

17. Fanin, N.; Bertrand, I. Aboveground litter quality is a better predictor than belowground microbial communities when estimating carbon mineralization along a land-use gradient. *Soil Biol. Biochem.* **2016**, *94*, 48–60. [CrossRef]
18. Gholz, H.L.; Wedin, D.A.; Smitherman, S.M.; Harmon, M.E.; Parton, W.J. Long-term dynamics of pine and hardwood litter in contrasting environments: Toward a global model of decomposition. *Glob. Change Biol.* **2000**, *6*, 751–765. [CrossRef]
19. Ayres, E.; Steltzer, H.; Simmons, B.L.; Simpson, R.T.; Steinweg, J.M.; Wallenstein, M.D.; Mellor, N.; Parton, W.J.; Moore, J.C.; Wall, D.H. Home-field advantage accelerates leaf litter decomposition in forests. *Soil Biol. Biochem.* **2009**, *41*, 606–610. [CrossRef]
20. Kang, H.; Gao, H.; Yu, W.; Yi, Y.; Wang, Y.; Ning, M. Changes in soil microbial community structure and function after afforestation depend on species and age: Case study in a subtropical alluvial island. *Sci. Total Environ.* **2018**, *625*, 1423–1432. [CrossRef]
21. Prescott, C.E.; Blevins, L.L.; Staley, C.L. Effects of clear-cutting on decomposition rates of litter and forest floor in forests of British Columbia. *Can. J. For. Res.-Rev. Can. Rech. For.* **2000**, *30*, 1751–1757. [CrossRef]
22. Ishikawa, H.; Osono, T.; Takeda, H. Effects of clear-cutting on decomposition processes in leaf litter and the nitrogen and lignin dynamics in a temperate secondary forest. *J. For. Res.* **2007**, *12*, 247–254. [CrossRef]
23. Achat, D.L.; Deleuze, C.; Landmann, G.; Pousse, N.; Ranger, J.; Augusto, L. Quantifying consequences of removing harvesting residues on forest soils and tree growth—A meta-analysis. *For. Ecol. Manag.* **2015**, *348*, 124–141. [CrossRef]
24. Pennanen, T.; Liski, J.; Bååth, E.; Kitunen, V.; Uotila, J.; Westman, C.J.; Fritze, H. Structure of the microbial communities in coniferous forest soils in relation to site fertility and stand development stage. *Microb. Ecol.* **1999**, *38*, 168–179. [CrossRef] [PubMed]
25. Kohout, P.; Charvátová, M.; Štursová, M.; Mašínová, T.; Tomšovský, M.; Baldrian, P. Clearcutting alters decomposition processes and initiates complex restructuring of fungal communities in soil and tree roots. *ISME J.* **2018**, *12*, 692. [CrossRef] [PubMed]
26. Finér, L.; Jurgensen, M.; Palviainen, M.; Piirainen, S.; Page-Dumroese, D. Does clear-cut harvesting accelerate initial wood decomposition? A five-year study with standard wood material. *For. Ecol. Manag.* **2016**, *372*, 10–18. [CrossRef]
27. Gliksman, D.; Haenel, S.; Osem, Y.; Yakir, D.; Zangy, E.; Preisler, Y.; Grünzweig, J.M. Litter decomposition in Mediterranean pine forests is enhanced by reduced canopy cover. *Plant Soil* **2018**, *422*, 317–329. [CrossRef]
28. Keiser, A.D.; Keiser, D.A.; Strickland, M.S.; Bradford, M.A. Disentangling the mechanisms underlying functional differences among decomposer communities. *J. Ecol.* **2014**, *102*, 603–609. [CrossRef]
29. Paterson, E.; Osler, G.; Dawson, L.A.; Gebbing, T.; Sim, A.; Ord, B. Labile and recalcitrant plant fractions are utilised by distinct microbial communities in soil: Independent of the presence of root and mycorrhizal fungi. *Soil Biol. Biochem.* **2008**, *40*, 1103–1113. [CrossRef]
30. Chomel, M.; Guittonny-Larchevêque, M.; DesRochers, A.; Baldy, V. Home Field Advantage of Litter Decomposition in Pure and Mixed Plantations under Boreal Climate. *Ecosystems* **2015**, *18*, 1014–1028. [CrossRef]
31. Veen, G.F.; Sundqvist, M.K.; Wardle, D.A. Environmental factors and traits that drive plant litter decomposition do not determine home-field advantage effects. *Funct. Ecol.* **2015**, *29*, 981–991. [CrossRef]
32. Veen, G.F.; Freschet, G.T.; Ordonez, A.; Wardle, D.A. Litter quality and environmental controls of home-field advantage effects on litter decomposition. *Oikos* **2015**, *124*, 187–195. [CrossRef]
33. Michener, R.; Lajtha, K. *Stable Isotopes in Ecology and Environmental Science*; John Wiley & Sons: Hoboken, NJ, USA, 2008.
34. Gautam, M.K.; Lee, K.-S.; Song, B.-Y.; Lee, D.; Bong, Y.-S. Early-stage changes in natural ^{13}C and ^{15}N abundance and nutrient dynamics during different litter decomposition. *J. Plant Res.* **2016**, *129*, 463–476. [CrossRef] [PubMed]
35. Osono, T.; Takeda, H.; Azuma, J.-i. Carbon isotope dynamics during leaf litter decomposition with reference to lignin fractions. *Ecol. Res.* **2008**, *23*, 51–55. [CrossRef]
36. Dijkstra, P.; LaViolette, C.M.; Coyle, J.S.; Doucett, R.R.; Schwartz, E.; Hart, S.C.; Hungate, B.A. 15N enrichment as an integrator of the effects of C and N on microbial metabolism and ecosystem function. *Ecol. Lett.* **2008**, *11*, 389–397. [CrossRef] [PubMed]
37. Pardo, L.H.; Hemond, H.F.; Montoya, J.P.; Fahey, T.J.; Siccama, T.G. Response of the natural abundance of 15N in forest soils and foliage to high nitrate loss following clear-cutting. *Can. J. For. Res.* **2002**, *32*, 1126–1136. [CrossRef]
38. Craine, J.M.; Elmore, A.J.; Aidar, M.P.M.; Bustamante, M.; Dawson, T.E.; Hobbie, E.A.; Kahmen, A.; Mack, M.C.; McLauchlan, K.K.; Michelsen, A.; et al. Global patterns of foliar nitrogen isotopes and their relationships with climate, mycorrhizal fungi, foliar nitrient concentrations, and nitrogen availability. *New Phytol.* **2009**, *183*, 980–992. [CrossRef]
39. Jiang, C.; Séquaris, J.-M.; Vereecken, H.; Klumpp, E. Effects of temperature and associated organic carbon on the fractionation of water-dispersible colloids from three silt loam topsoils under different land use. *Geoderma* **2017**, *299*, 43–53. [CrossRef]
40. Siebers, N.; Kruse, J. Short-term impacts of forest clear-cut on soil structure and consequences for organic matter composition and nutrient speciation: A case study. *PLoS ONE* **2019**, *14*, e0220476. [CrossRef]
41. Rosenbaum, U.; Bogena, H.R.; Herbst, M.; Huisman, J.A.; Peterson, T.J.; Weuthen, A.; Western, A.W.; Vereecken, H. Seasonal and event dynamics of spatial soil moisture patterns at the small catchment scale. *Water Resour. Res.* **2012**, *48*, 1–22. [CrossRef]
42. Wiekenkamp, I.; Huisman, J.A.; Bogena, H.R.; Graf, A.; Lin, H.S.; Druee, C.; Vereecken, H. Changes in measured spatiotemporal patterns of hydrogical response after partial deforestation in a headwater catchment. *J. Hydrol.* **2016**, *542*, 648–661. [CrossRef]
43. Havlik, D. Das Klima von Aachen. *Aachen. Geogr. Arb.* **2002**, *36*, 1–20.
44. Etmann, M. Dendrologische Aufnahmen im Wassereinzugsgebiet oberer Wüstebach anhand verschiedener Mess-und Schätzverfahren. Master's Thesis, Westfälische Wilhelms-University Münster, Münster, Germany, 2009.
45. Bradford, M.A.; Tordoff, G.M.; Eggers, T.; Jones, T.H.; Newington, J.E. Microbiota, fauna, and mesh size interactions in litter decomposition. *Oikos* **2002**, *99*, 317–323. [CrossRef]

46. Wang, Q.K.; Zhong, M.C.; He, T.X. Home-field advantage of litter decomposition and nitrogen release in forest ecosystems. *Biol. Fertil. Soils* **2013**, *49*, 427–434. [CrossRef]
47. Olson, J.S. Energy storage and the balance of producers and decomposers in ecological systems. *Ecology* **1963**, *44*, 322–331. [CrossRef]
48. Aneja, M.K.; Sharma, S.; Fleischmann, F.; Stich, S.; Heller, W.; Bahnweg, G.; Munch, J.C.; Schloter, M. Microbial colonization of beech and spruce litter—Influence of decomopsition site and plant litter species on the diversity of microbial community. *Micobial Ecol.* **2006**, *52*, 127–135. [CrossRef]
49. Zhou, X.; Zhu, H.; Wen, Y.; Goodale, U.M.; Li, X.; You, Y.; Ye, D.; Liang, H. Effects of understory management on trade-offs and synergies between biomass carbon stock, plant diversity and timber production in eucalyptus plantations. *For. Ecol. Manag.* **2018**, *410*, 164–173. [CrossRef]
50. Heim, A.; Frey, B. Early stage litter decomposition rates for Swiss forests. *Biogeochemistry* **2004**, *70*, 299–313. [CrossRef]
51. McDowell, N.G.; Allen, C.D.; Marshall, L. Growth, carbon-isotope discrimination, and drought-associated mortality across a Pinus ponderosa elevational transect. *Glob. Change Biol.* **2010**, *16*, 399–415. [CrossRef]
52. Rajala, T.; Peltoniemi, M.; Hantula, J.; Mäkipää, R.; Pennanen, T. RNA reveals a succession of active fungi during the decay of Norway spruce logs. *Fungal Ecol.* **2011**, *4*, 437–448. [CrossRef]
53. St John, M.G.; Orwin, K.H.; Dickie, I.A. No 'home' versus 'away' effects of decomposition found in a grassland-forest reciprocal litter transplant study. *Soil Biol. Biochem.* **2011**, *43*, 1482–1489. [CrossRef]
54. Freschet, G.T.; Aerts, R.; Cornelissen, J.H.C. Multiple mechanisms for trait effects on litter decomposition: Moving beyond home-field advantage with a new hypothesis. *J. Ecol.* **2012**, *100*, 619–630. [CrossRef]
55. Petraglia, A.; Cacciatori, C.; Chelli, S.; Fenu, G.; Calderisi, G.; Gargano, D.; Abeli, T.; Orsenigo, S.; Carbognani, M. Litter decomposition: Effects of temperature driven by soil moisture and vegetation type. *Plant Soil* **2019**, *435*, 187–200. [CrossRef]
56. Cotrufo, M.F.; Soong, J.L.; Horton, A.J.; Campbell, E.E.; Haddix, M.L.; Wall, D.H.; Parton, W.J. Formation of soil organic matter via biochemical and physical pathways of litter mass loss. *Nat. Geosci.* **2015**, *8*, 776. [CrossRef]
57. Mansfield, S.D.; Bärlocher, F. Determination of Soluble Carbohydrates. In *Methods to Study Litter Decomposition: A Practical Guide*; Graça, M.A.S., Bärlocher, F., Mark, O., Eds.; Springer: Dordrecht, The Netherlands, 2005; pp. 85–90.
58. Ahmed, I.U.; Mengistie, H.K.; Godbold, D.L.; Sandén, H. Soil moisture integrates the influence of land-use and season on soil microbial community composition in the Ethiopian highlands. *Appl. Soil Ecol.* **2019**, *135*, 85–90. [CrossRef]
59. Mueller, P.; Schile-Beers, L.M.; Mozdzer, T.J.; Chmura, G.L.; Dinter, T.; Kuzyakov, Y.; de Groot, A.V.; Esselink, P.; Smit, C.; D'Alpaos, A.; et al. Global-change effects on early-stage decomposition processes in tidal wetlands—Implications from a global survey using standardized litter. *Biogeosciences* **2018**, *15*, 3189–3202. [CrossRef]
60. Preston, C.M.; Nault, J.R.; Trofymow, J.A. Chemical Changes During 6 Years of Decomposition of 11 Litters in Some Canadian Forest Sites. Part 2. 13C Abundance, Solid-State 13C NMR Spectroscopy and the Meaning of "Lignin". *Ecosystems* **2009**, *12*, 1078–1102. [CrossRef]
61. Hobbie, E.A.; Högberg, P. Nitrogen isotopes link mycorrhizal fungi and plants to nitrogen dynamics. *New Phytol.* **2012**, *196*, 367–382. [CrossRef]
62. Ngao, J.; Cotrufo, M.F. Carbon isotope discrimination during litter decomposition can be explained by selective use of substrate with differing $\delta^{13}C$. *Biogeosci. Discuss* **2011**, *2011*, 51–82.

soil systems

MDPI

Article

Do the Invasive Earthworms *Amynthas agrestis* (Oligochaeta: Megascolecidae) and *Lumbricus rubellus* (Oligochaeta: Lumbricidae) Stimulate Oxalate-Based Browser Defenses in Jack-in-the-Pulpit (*Arisaema triphyllum*) by Their Presence or Their Soil Biogeochemical Activity?

Ryan D. S. Melnichuk [1], Hüseyin Barış Tecimen [2] and Josef H. Görres [3,*]

1 Innotech Alberta, Edmonton, AB T6N 1E4, Canada; rdm7@ualberta.ca
2 Soil Science and Ecology Department, Faculty of Forestry, Istanbul University-Cerrahpaşa, 34473 Istanbul, Turkey; hbarist@gmail.com
3 Department of Plant and Soil Science, University of Vermont, Burlington, VT 05405, USA
* Correspondence: jgorres@uvm.edu

Abstract: The introduction of invasive earthworms initiates physical and chemical alterations in previously earthworm-free forest soils, which triggers an ecological cascade. The most apparent step is the shift in the herbaceous plant community composition. However, some species, such as *Arisaema triphyllum* (jack-in-the-pulpit), persist where earthworms are present. It has been hypothesized that *A. triphyllum* produces insoluble oxalate, an herbivory deterrent, in the presence of earthworms. This study aimed to distinguish between the effects of earthworm-induced changes in soils and the physical presence of earthworms on oxalate production. As such, a two-way factorial greenhouse trial was conducted using uninvaded soils to test this hypothesis for two invasive earthworm species (*Amynthas agrestis* and *Lumbricus rubellus*). The sequential extraction of oxalates in *A. triphyllum* corms was performed with absolute ethanol, deionized water, acetic acid and HCl, representing fractions of decreasing solubility. Earthworm presence increased water-soluble ($p = 0.002$) and total oxalate ($p = 0.022$) significantly, but only marginally significantly for HCl-soluble oxalate ($p = 0.065$). The corms of plants grown in soils previously exposed to the two species did not differ in oxalate production when earthworms were not present. However, the data suggest that earthworms affect corm oxalate concentrations and that the sequence of invasion matters for oxalate production by *A. triphyllum*.

Keywords: soil properties; forest soil modifications; oxalate; *A. triphyllum*; earthworm invasions; calcifery

Citation: Melnichuk, R.D.S.; Tecimen, H.B.; Görres, J.H. Do the Invasive Earthworms *Amynthas agrestis* (Oligochaeta: Megascolecidae) and *Lumbricus rubellus* (Oligochaeta: Lumbricidae) Stimulate Oxalate-Based Browser Defenses in Jack-in-the-Pulpit (*Arisaema triphyllum*) by Their Presence or Their Soil Biogeochemical Activity? *Soil Syst.* **2022**, *6*, 11. https://doi.org/10.3390/soilsystems6010011

Academic Editor: Klaus von Wilpert

Received: 29 November 2021
Accepted: 12 January 2022
Published: 18 January 2022

1. Introduction

Earthworm invasions into previously uninvaded forests cause severe disturbances in forested ecosystems where the soils and plant community structure have developed in the absence of these invertebrates [1–4]. Specifically, the invasion of woodlands by earthworms results in the loss of organic soil horizons (through consumption of the O_e and O_i layers) and a mixing of organic and mineral material [5], setting off an ecological cascade of effects that reach well beyond the forest soil and the immediately affected ecosystem [6]. With invasions now reaching into the ecosystems of the arctic circle [7,8], earthworms may affect the vast stocks of carbon stored in the taiga and tundra soils [9,10]. In the temperate regions of North America, the invasion is already well advanced and is affecting the plant–soil system [11–13], as well as whole ecosystems [1]. However, plant communities [14] are altered not only by the actions of earthworms but also by subsequent foraging by large ungulate browsers, such as white-tailed deer (*Odocoileus virginianus*) in North America [15]. Here, we examine a hypothesis that earthworms, or the changes they induce in the soils system, can result in the increased production of insoluble oxalates, an herbivory deterrent, in jack-in-the-pulpit (*Arisaema triphyllum*).

The vast majority of earthworms found in northeastern US forests are introduced species and have been recognized as such for some time [5,16–18]. Anthropogenic activities, such as horticulture, agriculture, non-commercial fishing, and recreation, have increased earthworm dispersal during the last century [1,19,20]. In contrast to agricultural lands, where they are regarded as positive indicators of soil quality (USDA-NRCS, 2001), earthworms have negative impacts on previously earthworm-free forest systems [14]. The mixing of organic and mineral soil by earthworms reduces the seedbank and germination function of the forest soil for many understory plants, often resulting in a decline in species richness and evenness [14,21–23], although abundance increases have also been observed, though often involving different species to the ones endemic to the region [21,24]. Additionally, the change in soil structure and the concomitant loss of the O_i and O_e horizons results in a flush of soluble nutrients early in the growing season, which may shift the phenology of the community to favor invasive plant species that may be able to utilize this unusually early abundance of an ordinarily limited resource [25]. Plants that remain are subject to greater browsing pressure from deer because of the lowered plant densities [22].

Some understory, "vermiphile" plants persist in the face of earthworm invasions [14,22,26]. Among plants that indicate the heavy infestation of forests by earthworms are *Arisaema triphyllum*, *Allium tricoccum*, *Carex pensylvanica* and saplings of *Fraxinus* species. Hypotheses regarding their resilience to earthworm invasions are several, but the one we investigated here relates to a potential chemical defense to browsing. Compounds likely to serve as a deterrent to browsers, such as oxalate, are often multifunctional, with additional roles in metabolism and structure [27]. An intermediate product of saccharide metabolism, compounds of oxalate can serve both as nutrient storage as well as an herbivory deterrent. For example, oxalate compounds with metal nutrients (Ca, Mg) to form sharp, insoluble raphides, intracellular needle-like crystals [28], that act as a sink for excess essential metals and represent a physical/chemical browser defense [27]. The oxalic acid exuded from root tips also immobilizes and prevents the uptake of toxic metals (Al, Hg, Pb, Cd) from soils or binds internally with those in plant tissues, making them metabolically inactive [29]. The focus of this study was to determine whether it is the mere presence of earthworms or the soil conditions caused by earthworm activity that affect plant oxalate production as well as the fractionation of oxalate compounds with different solubility.

The proposed mechanism of accelerated calcium oxalate (CaC_2O_4) raphide formation is shown in Figure 1. It is based on the idea that earthworms accelerate the mineralization of Ca from organic matter by providing an additional mechanism of mineralization. However, many earthworms also have calciferous glands, which provide the calcium that neutralizes carbonic acid (a metabolic by-product) to form calcite ($Ca(CO_3)_2$) in their gut [30]. This process, termed calcifery, results in the excretion of calcite granules by the earthworms. Calcite has low solubility, such that calcifery can sequester C and Ca into recalcitrant soil pools. However, the equilibrium chemistry of production and dissolution will elevate both labile and recalcitrant calcium carbonate pools in the soil, reaching a steady state after approximately 6 years [31]. In addition, calcifery requires that Ca is available, mineralized from organic matter, or released from the soil parent material.

Accelerated mineralization by earthworms promotes the availability of Ca and other base cations in soils. Some plants respond to greater Ca availability by storing it in compounds, such as Ca oxalate. We have observed that soils at earthworm-invaded sites had greater concentrations of available Ca (data not presented) which may trigger additional oxalate production in plants so that calcium in excess of cytosol requirement can be neutralized [32]. This adaptation allows these plants to tolerate the greater calcium concentrations that are purportedly associated with earthworms by storing it as calcium oxalate raphides. However, calcium relations in earthworms differ among species [33]. For example, *Lumbricus rubellus* is strongly calciferous, whereas *Aporrectodea caliginosa* is not. Calcium granule production for these two species is one to ten per day and one to ten per month, respectively [33,34].

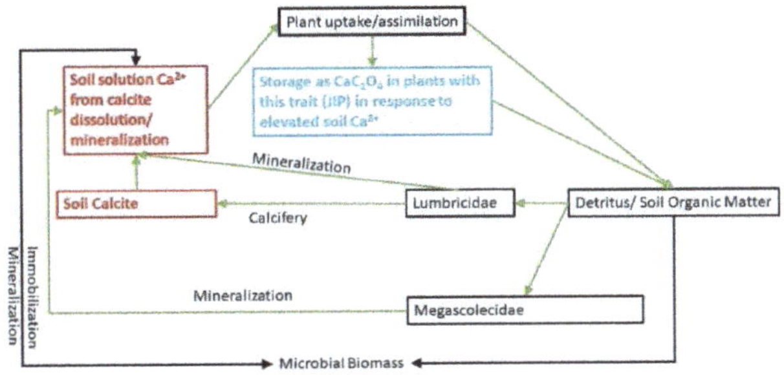

Figure 1. The role of earthworms in the formation of calcium oxalate (CaC_2O_4) formation in vermiphile plants. Both Megascolecidae and Lumbricidae affect the soil solution Ca^{2+} through accelerated mineralization rates that are in addition to microbial mineralization. However, Lumbricidae also affects it through calcifery.

To test the hypotheses that earthworms promote insoluble oxalate production, jack-in-the-pulpit (hereon in JIP, *Arisaema triphyllum*, Figure 2b), a known vermiphile plant species with established naturalized populations common in northern hardwood forest communities, was selected as the plant model [22]. Jack-in-the-pulpit is a calcium oxalate-accumulating plant native to the northeastern US that is anecdotally observed more often in the presence of earthworms. Calcium oxalate raphides in JIP are regarded as defenses against browsers, such as deer [35,36].

Figure 2. (**a**): *Amynthas agrestis* on the forest floor of a sugar maple stand at UVM's Horticultural Research Center in South Burlington, VT, and (**b**) jack-in-the-pulpit with large leaves in a sugar maple stand in the Champlain Valley, VT, USA. Arrow pointing at the inflorescence of the plant.

We selected *Amynthas agrestis* (Figure 2a) and *L. rubellus* as our model earthworms partly because they purportedly differ in soil calcium relations. They are both commonly observed, exotic, epi-endogenic earthworms found in the northeastern USA [17,37]. *A. agrestis* is of the family Megascolecidae and is one of three aggressive and rapidly dispersing earthworm species [20], with origins in far-eastern Asia. This species is part of a second wave of invaders which colonized northern North American soils already invaded by Lumbricidae [38], with potential further effects on soil–plant relations. Megascolecidae generally do not have calciferous glands [39], but it is possible that some species may have this adaptation. *L. rubellus* is of the family Lumbricidae, an almost ubiquitous, strongly calciferous earthworm with origins in Eurasia. Our objective was to separate as much as possible the influence of the edaphic conditions created previously by earthworms from the effect of their presence on plant concentrations of the oxalate compounds differing in solubility. We wanted to see (1) whether there was a difference in oxalate synthesis when calciferous or non-calciferous earthworms were present, (2) whether the effect of earthworms would last when they were removed and (3) whether the invasion sequence mattered, i.e., whether the second wave really does matter. To this end, we conducted an experiment in which the pre-incubation of soils with one or the other earthworm and without earthworms was one factor (SOIL), and the subsequent absence or presence of earthworms (WORM) was another factor (Figure 3A). The combination of pretreatment incubation in the absence of earthworms and the pots without earthworms serves as a control. Sequentially extracting oxalate along a solubility gradient may give insight into the effect of earthworms on oxalate storage as Ca oxalate, the form that deters browsers.

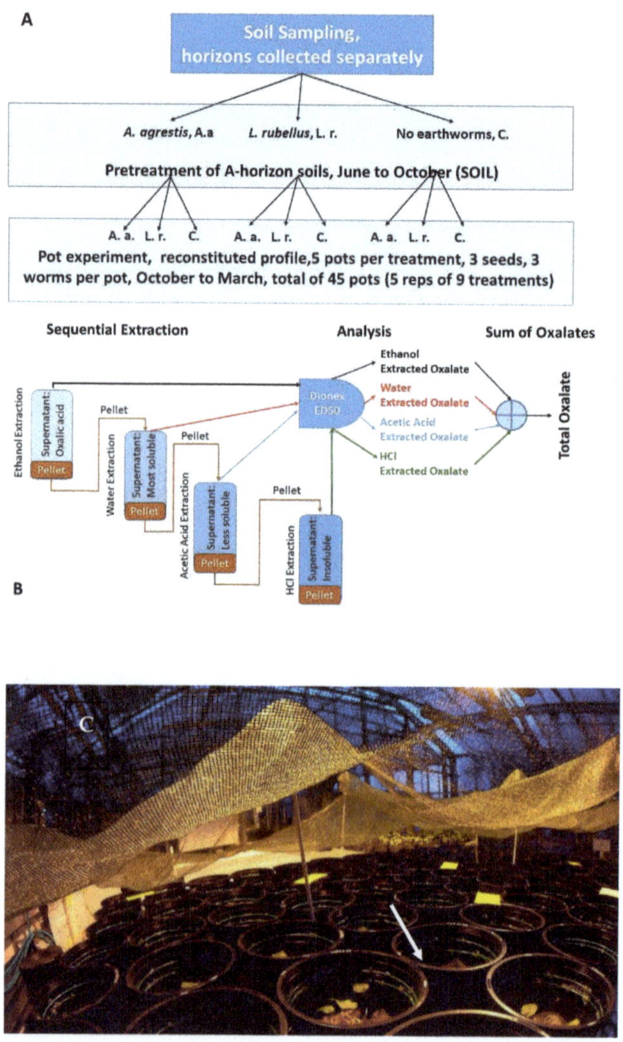

Figure 3. (**A**) Experimental design showing the relationship between soil pretreatments (SOIL) and pot treatments (WORM). A.a.—Amynthas agrestis, L.r.—Lumbricus rubellus, C.—no earthworms added (control). (**B**) Sequential extraction of oxalate fractions, with solid fraction (pellet) from earlier extraction passed to next extraction. Total oxalates represents the sum of all solubility fractions. (**C**) Experimental setting in greenhouse showing shade cloth and pots with hook-up wire (arrow) a month after the initiation of the experiment.

2. Materials and Methods

2.1. Earthworm and Soil Collection

Soil was collected from the University of Vermont Jericho Research Forest (latitude: 44°26′55″ N, longitude: 72°59′48″, elevation 230 m) from an earthworm-free northern hardwood forest stand. The soil is mapped as Duane series, sandy-skeletal, mixed, frigid, ortstein Typic Haplorthods [40]. However, the observed profile better matches a Peru fine sandy loam (coarse-loamy, isotic, frigid Aquic Haplorthods) [40]. Each soil horizon (A, E, B_{hs}, B) was collected from soil pits and stored separately until mesocosm construction. Mineral soil was sieved at 7.5 mm and the homogenized pass fraction was used in the experiment. Part of the A horizon was set aside to acclimate earthworms prior to the start of the experiment. The O_a horizon, which was less than 2.5 cm thick in this soil series, was not collected. The leaf litter collected at the site was found to be predominantly sugar maple (*Acer saccharum*) and American beech (*Fagus grandifolia*). The leaf litter was manually shredded, and all materials were sieved at 7.5 mm and the pass fraction was used in the experiment.

A. agrestis (Figure 2a) were collected from naturalized populations at UVM's Horticulture Research Center (lat.: 44°25′52″, long.: 73°11′57″, elevation 112 m) and *L. rubellus* from a field under a 7-year silage corn–alfalfa rotation at UVM's Miller Dairy Complex (lat.: 44°27′24″, long.: 73°11′20″, elevation 100 m) both in South Burlington, VT. Earthworms were maintained in the aforementioned separate fraction of A horizon soil for 1 week to acclimate and ensure survival under experimental conditions.

In order to minimize variation amongst individuals, seeds of *A. triphyllum*, a dioecious perennial plant (Figure 2b), were purchased from a local sustainable wildflower horticulture center that collects and propagates native varieties (Vermont Wildflower Farm, Vergennes, VT, USA). Seeds were scarified and then stratified for 3 months prior to sowing into pots.

2.2. Greenhouse Pot Trials

The greenhouse trial was designed to examine the effect of soil modification by earthworms (SOIL), representing the "invasion history", and the effect of physical earthworm presence (WORM), representing new introductions, on JIP corm oxalate production. Thus, the experiment had three SOIL pretreatments in which earthworm-free soils were aged in the presence of *L. rubellus* or *A. agrestis*, or no earthworms at all (Figure 3A). Each of these pretreated soils was then subjected to three earthworm (WORM) treatments: No earthworms, *A. agrestis* or *L. rubellus*. In this way, we artificially created physical models of invasion sequences: invasions into soils not recently occupied by earthworms, *A. agrestis* following on from *L. rubellus*, and *L. rubellus* following *A. agrestis*.

To produce pretreated soils (SOIL), A horizon soil was divided into three shallow containers, each with a volume of 59 L (89 cm long, 42 cm wide, 16 cm deep). These containers were constructed to produce the three pre-experimental WORM treatments: (1) Control (no earthworms), (2) *A. agrestis* (140 g fresh biomass) and (3) *L. rubellus* (140 g fresh biomass). Pre-incubation started on 9 June 2012. The A horizon soil was incubated in the presence of the worms for 4.5 months on a 12 h day/night cycle at 15 °C. At the end of the pre-incubation period, earthworms were removed from the pretreated A horizon soil. Then, 45 experimental units were constructed in 4 L, plastic horticultural pots by sequentially layering B, then B_{hs}, then E, then A horizon, and lastly leaf litter into each vessel to simulate the soil profile as observed at the collection site. This was to create soil conditions similar to those found in field soil. Briefly, each pot contained 50 g of leaf litter, 800 mL A horizon, 30 mL E horizon, 400 mL B_{hs} horizon and 2000 mL B horizon. Fifteen pots each received soils pretreated with *A. agrestis*, *L. rubellus* or neither earthworm. To five pots in each SOIL set, three *A. agrestis*, three *L. rubellus* or no earthworms were added on 29 October 2012. To prevent the escape of the earthworms, we modified the pots prior to reconstructing the soil profile. Two rings of single stranded insulated copper electrical wire (40 gage) were glued 1.5 cm apart to the inside of the pots 5 cm below the rim. As earthworms were moving over the two wires their adhesion to the pot walls were broken and they fell back onto the soil surface. In trials, this prevented *A. agrestis* from scaling the pot walls. A fine mesh was added to the bottom of the pots to prevents worms from escaping through the drainage holes.

All pots thus prepared were transferred to a greenhouse on 29 October 2012 and arranged in a complete randomized design under 12-h day/night cycle sodium supplemental lighting and 50% shade cloth. Temperature was maintained between 15 and 21 °C and relative humidity varied at 50–100%. Pots were misted equally with reverse osmosis water for the duration of the experiment at the replacement rate of the controls (WORM control on SOIL control). Each pot received 3 *A. triphyllum* seeds. Germination was high, with 2 or 3 seedlings emerging in each pot. Plants were continuously monitored until harvest, which was carried out one week after senescence when the corms from each treatment were collected on 13 March 2013.

Bulk samples of A horizon were collected from each pot and dried at 50 °C. Standard 2 M KCl (mineral nitrogen) and 1 M NH_4Cl (nutrients) extracts of 5:1 solvent to dry soil ratio were performed. Ammonium chloride extracts were analyzed using ICP-AES (Perkin-Elmer Corp., Norwalk, CT, USA). Mineral nitrogen was quantified colorimetrically on a BioTek 96-well microtiter plate reader (BioTek, Winooski, VT, USA) using a miniaturized method [36]. In addition, relative dissolved organic matter content was analyzed for absorbance at 330 nm with a GENSYS10vis spectrophotometer (Thermo Scientific Spectronics, Rochester, NY, USA) [41].

Upon deconstruction of the soils, the contact between the A and B horizon was still distinct, but earthworm-related bioturbation and burrows were evident with the Bhs and some of the E horizons incorporated into the A and B horizons (Figure 4). The resulting contact between the A and the B horizon was wavy.

Figure 4. Soil after deconstruction of the pots. Dark soil at top is the A horizon, lighter patches below the A horizon are remnants of the E horizon, orange soil is the B horizon. Worm burrows are visible in both A and B horizons.

2.3. Oxalate Quantification

Several forms of variably soluble oxalate are present in plant tissues. We performed a sequential extraction with absolute ethanol, deionized water, 5% acetic acid and 2N HCl, representing the oxalate fractions of decreasing solubility (Figure 3B). This method is operationally defined, but the ethanol fraction should contain only free oxalic acid, the water fraction should contain unbounded, polyhydrate and amorphous fractions of oxalate, the acetic acid fraction should contain the balance of non-crystalline forms of oxalate and the HCl fraction should contain the remaining forms (including Ca oxalate) of oxalate [42–45]. Because of the sequential extraction on the same sample, the total oxalate in a corm could be estimated as the sum of the extractions. Sequential solvent extraction was performed on fresh corms, which are tuber-like, below-ground storage organs, after surficial soil removal. Specifically, each corm was macerated, combined with 25 mL of ethanol and agitated with glass beads for 2 h at ambient conditions before centrifugation (TJ6 centrifuge; Beckman Coulter, Brea, CA, USA) and collection of supernatants. To pellet, 25 mL of water was added and the process was repeated. Acetic acid and HCl extraction were carried out in the same manner, with the exception that the HCl extract was agitated for 4 h. The water content of the corms from each treatment were also analyzed, and no significant differences were observed. Extraction efficiency by mass was compared within treatments and found to not differ significantly.

Oxalate was quantified with a Dionex ED50 Electrochemical Detector (Sunnyvale, CA, USA) equipped with an AS50 Autosampler, an AS11 column and an AG11 guard column. A CSRS suppressor was included in the line to remove cations and contribution of gradient NaOH eluent (Dionex, Sunnyvale, CA, USA). A standard 20 mg/L oxalate solution was mixed at the beginning of the trial; aliquots were frozen and included as QCs in each run. Additionally, within each run, random samples were analyzed in duplicate to ensure replicable results.

2.4. Aggregate-Scale Ca Concentrations

To analyze whether the two earthworm species had different effects on soil microsite water-extractable Ca concentrations, we analyzed earthworm castings and compared them to aggregates formed in soils without earthworms. The experimental set up and analyses methods are given in Tecimen et al. [46]. In brief, aggregate samples were harvested from the following treatments described above: pretreatment control soil with no earthworms, *L. rubellus* or *A. agrestis*. The aggregates samples were extracted with deionized water and then immediately analyzed for Ca with an ion-specific microelectrode (Microelectrodes, Inc., Bedford, NH, USA).

2.5. Statistics

The mesocosm trial was a completely random, fully factorial design; the response variables (ethanol, water, acetic acid and HCl—extractable corm oxalates and total corm oxalates) were tested for homoscedasticity, independence and normality; if the data did not meet the assumptions, a transformation according to the natural log was performed. Subsequently, the data were examined by full two-way factorial ANOVA where the factors were SOIL (soil exposed to no earthworms, *A. agrestis* or *L. rubellus*) and WORM (no added earthworms, and *A. agrestis* or *L. rubellus*). In total, there were 9 treatments, which were replicated 5 times for 45 pots. All analyses were computed using JMP 11 (SAS Institute, Cary, NC, USA). The statistical model for the analysis of variance is given in Equation (1), where all factors are fixed effects:

$$Y = \mu + \alpha_i + \beta_j + \alpha\beta_{ij} + \varepsilon_{ijk} \tag{1}$$

where μ is the treatment mean, α is the SOIL treatment (I = 3), β is the WORM treatment (j = 3) and ε is the error term (k = 5). Significant differences were evaluated post hoc with Tukey's HSD separately for each oxalate extraction. In addition, effect sizes were calculated as η^2 and partial η^2. These are expressions of the fraction of variance explained by each variable. In designs with more than one independent variable, partial η^2 calculates the effect of each variable with the variances of the other variables (and their interaction) removed from the total variance. It should be noted that the maximum value of η^2 and partial η^2 is less than 1. For normally distributed variables, values vary between 0 and 0.64. Additionally, η^2 and partial η^2 values are classified into small ($0.0099 < \eta^2 < 0.0588$), medium ($>0.0588 < \eta^2 < 0.1379$) and large ($\eta^2 > 0.1379$) effects [47].

For the comparison of Ca concentrations in aggregates, results were analyzed with a one-way ANOVA followed by Tukey's HSD test.

3. Results

3.1. Oxalate Fractions

The water-extractable fraction was generally the largest fraction, followed by acetic acid fraction, then the HCl-extractable fraction and finally the ethanol-extractable (free oxalic acid) fraction (Table 1). The ANOVA for ethanol-soluble and acetic acid-soluble oxalate did not detect any significant differences. However, for water-soluble and total oxalate, the ANOVA indicated significant differences ($p = 0.034$ and $p = 0.042$, respectively). The effect tests showed that there was no significant difference in water-soluble oxalate among the SOIL treatments. Yet, for the WORM treatments, there was a difference in both water-soluble oxalate and total oxalates, with both earthworms having 27% greater water-soluble concentrations than the control. Total oxalate concentrations were 22.5% greater in the *L. rubellus* than the Control WORM treatments ($p = 0.027$). There were no significant differences between the *L. rubellus* and *A. agrestis* WORM treatments. The ANOVA showed marginally significant differences in HCl-extractable oxalate among treatments ($p = 0.065$). Here, again, there was no significant differences among the SOIL treatments. However, in the WORM treatments, the HCl extractions were greater in the *L. rubellus* than the *A. agrestis* treatment. The exclusion of a statistical outlier increased the significance of the full ANOVA model and the WORM treatment F values ($F_{8,44} = 2.25$, $p = 0.0465$ and $F_{2,35} = 5.36$, $p = 0.0093$ respectively). Statistical outliers are not necessarily biological outliers, so both values are included in the interest of transparency.

Table 1. Summary of full model and treatment effect statistical values for each extract fraction. The oxalate extracts vary left to right from low to high solubility. The mean and standard error of each treatment factor illustrate the effect where significantly different values are designated by different letters. The effect size (η^2) and the partial effect size (partial η^2%) are given as fractions of variance explained by soil, earthworm and their interaction. Effect size categories S (small), M (medium) and L (large) are also indicated for η^2.

		Oxalate Extract (µg/g)				
Full Model		**Ethanol**	**Water**	**5% Acetic Acid**	**2 N HCl**	**Total**
	df	8,44	8,44	8,44	8,44	8,44
	F-value	1.113	3.954	0.322	2.073	3.150
	p-value	0.378	0.002	0.952	0.065	0.011
Effect tests						
SOIL	df	2,36	2,36	2,36	2,36	2,36
	F-value	0.549	1.391	0.349	0.169	0.935
	p-value	0.582	0.262	0.708	0.845	0.402
	η^2%		0.043 (S)	0.015 (S)	0.003 (S)	0.031 (S)
	Partial η^2%		0.074 (M)	0.016 (S)	0.004 (S)	0.050 (S)
WORM	df	2,36	2,36	2,36	2,36	2,36
	F-value	0.046	6.266	0.205	4.563	4.247
	p-value	0.955	0.005	0.816	0.017	0.027
	η^2%		0.189 (L)	0.087 (M)	0.185 (L)	0.137 (M)
	Partial η^2%		0.255 (L)	0.092 (M)	0.213 (L)	0.187 (L)
SOIL*WORM	df	4,36	4,36	4,36	4,36	4,36
	F-value	1.928	4.079	0.368	1.780	3.709
	p-value	0.127	0.008	0.830	0.154	0.013
	η^2%		0.248 (L)	0.041 (S)	0.129 (M)	0.245 (L)
	Partial η^2%		0.312 (L)	0.042 (S)	0.159 (L)	0.292 (L)
Treatment Mean (SE)						
SOIL	Control	35.6(4.7)	585.2(41.8)	300.7(40.1)	171.6(21.0)	1093.1(80.7)
	A. agrestis	29.6(3.1)	513.6(39.4)	264.8(22.7)	187.9(25.1)	995.9(59.6)
	L. rubellus	36.0(6.4)	533.7(35.2)	273.0(25.1)	177.1(21.5)	1019.7(39.7)
WORM	Control	33.0(5.3)	453.7(39.8) a	269.3(24.5)	161.2(24.6) ab	917.2(52.3) a
	A. agrestis	34.9(4.5)	585.6(40.4) b	296.0(40.1)	147.1(14.0) a	1063.6(69.8) ab
	L. rubellus	33.3(5.0)	593.1(25.5) b	273.1(23.8)	228.2(21.8) b	1127.8(52.2) b

3.2. Bulk Soil Chemistry

Bulk soil analyses showed that available Fe was the only nutrient element with a significant difference among treatments (Table 2) in the overall model ($F_{8,44} = 2.37$, $p = 0.0367$)

with soil pretreatment contributing significantly to the variation ($F_{2,36} = 6.85$, $p = 0.0030$). The average pH of the soil was low at 3.91 [46]. Neither NH_4Cl-extractable Ca nor NO_3 concentrations were significantly different among treatments. To meet parametric statistical requirements, absorbance at 330 nm was logarithmically transformed, resulting in significant difference in the overall model ($F_{8,44} = 5.04$, $p = 0.003$) and within SOIL treatments ($F_{2,36} = 18.8$, $p < 0.0001$) (Table 2). In particular, the soils pretreated by *A. agrestis* had significantly greater dissolved organic matter content, estimated by absorbance, than the control ($p < 0.0001$) and *L. rubellus*-treated soils ($p = 0.0356$), regardless of the WORM treatment.

Table 2. Mean nutrient concentrations (mg/kg soil) of bulk soil after extraction with NH_4Cl. Standard errors (±) are also given. Only Fe and 330 nm absorbance were significantly different among SOIL treatments.

		Ca	K	Na	Al	Fe	Mn	Zn	Mg	NH_4	NO_3	330 nm	
SOIL	**WORM**					(mg/kg Soil)						$\bar{x}$	SE
Control	Control	939 ± 202	126 ± 23.6	46.3 ± 8.2	152 ± 31.9	36.3 ± 5.3	66.3 ± 13.9	7.08 ± 0.39	142 ± 31.6	33.1 ± 9.3	180 ± 40.1	0.267	0.049
	A. agrestis	853 ± 65	129 ± 12.6	42.7 ± 6.4	164 ± 15.1	37.8 ± 2.9	68.3 ± 9.2	7.08 ± 0.20	142 ± 7.8	41.6 ± 10.5	230 ± 60.6	0.260	0.043
	L. rubellus	1002 ± 260	177 ± 38.1	60.7 ± 15.7	170 ± 8.2	35.7 ± 1.3	72.0 ± 17.1	7.03 ± 0.96	178 ± 41.7	64.1 ± 10.9	370 ± 111	0.232	0.035
A. agrestis	Control	1135 ± 127	189 ± 25.8	52.4 ± 6.6	116 ± 20.4	25.4 ± 2.5	96.9 ± 13.5	7.52 ± 0.37	204 ± 18.4	55.6 ± 12.4	301 ± 80	0.493	0.070
	A. agrestis	1102 ± 101	197 ± 31.6	73.4 ± 11.3	124 ± 15.6	26.6 ± 2.2	89.5 ± 11.7	7.95 ± 0.21	200 ± 19.5	67.6 ± 11.0	345 ± 61.3	0.442	0.059
	L. rubellus	1218 ± 109	189 ± 22.2	64.1 ± 13.7	119 ± 16.1	27.9 ± 5.0	102.0 ± 6.6	8.34 ± 0.53	219 ± 26.0	71.6 ± 17.4	366 ± 94.8	0.467	0.038
L. rubellus	Control	959 ± 148	151 ± 30.5	41.1 ± 9.8	126 ± 25.9	29.7 ± 3.9	68.3 ± 11.7	7.06 ± 0.77	165 ± 27.3	39.0 ± 13.2	211 ± 37.7	0.339	0.036
	A. agrestis	1146 ± 121	173 ± 22.5	55.8 ± 6.8	116 ± 14.1	30.7 ± 1.8	85.0 ± 5.3	7.46 ± 0.35	198 ± 21.3	54.5 ± 10.9	303 ± 65.8	0.395	0.051
	L. rubellus	1125 ± 141	191 ± 35.2	84.2 ± 14.0	171 ± 19.2	39.1 ± 2.9	90.0 ± 17.3	8.76 ± 0.66	198 ± 31.0	77.8 ± 13.5	588 ± 205	0.303	0.019

3.3. Aggregate Scale Ca Concentrations

The ANOVA revealed that water-extractable Ca concentrations were significantly different among aggregates formed by the earthworms and aggregates from the control ($F_{2,117} = 14.08$, $p < 0.0001$). The mean Ca concentrations (standard error) were 80.6 (2.63), 92.1 (2.28) and 110.9 (6.13) mg Ca/kg soil, for the control, *A. agrestis* and *L. rubellus* aggregates, respectively. Significant differences occurred between *L. rubellus* and *A. agrestis* ($p < 0.0001$) and *L. rubellus* and the control ($p < 0.0042$).

4. Discussion

4.1. Oxalate Production by JIP

4.1.1. Effect of Earthworms (WORM)

The results support the hypothesis that earthworms influence the oxalate production of JIP when earthworms were physically present (WORM) (Table 1). Only water- and HCl-extractable oxalate fractions and total oxalates varied significantly between treatments. The NH_4Cl-extractable Ca concentrations in the bulk soils were not significantly different among treatments (Table 2), suggesting that there was no effect of soil Ca on oxalate production. However, when looking at water-extractable Ca concentrations in the aggregates, *L. rubellus* SOIL pretreatment had the largest Ca concentration, as one would expect for the calciferous earthworm. The magnitude of HCl-extractable corm oxalate follows the same sequence as the aggregate scale, water-extractable Ca: it is greatest for *L. rubellus* SOIL, and lowest for the control SOIL treatment. The difference here may be both in the sample support (homogenized bulk soil versus frass aggregates) and the type of extraction, i.e., water versus NH_4Cl.

4.1.2. Effect of the Soil Pretreatment (SOIL)

The effect of pretreatment was not as clear cut as that of the WORM treatments because the full linear model showed no significant effect of SOIL, but a significant interaction between the two factors. To resolve the effect of SOIL, we analyzed it separately for the cases where no earthworms were added (WORM control) at the $p = 0.10$ level. We found differences among SOIL treatments for the water-extractable and total oxalates (Figure 5A) with the three SOIL treatments explaining 36% of variation ($\eta^2 = 0.361$) in the water-extractable oxalate and 44% in total oxalate variation ($\eta^2 = 0.441$). Water-extractable oxalate concentrations in the corm grown in the *L. rubellus*-pretreated soil were greater than the concentrations in the corms of the *A. agrestis* pretreatment. Total oxalate concentrations in the corms grown in the calciferous *L. rubellus* soil were greater than in the *A. agrestis* pretreatment and Control pretreatment.

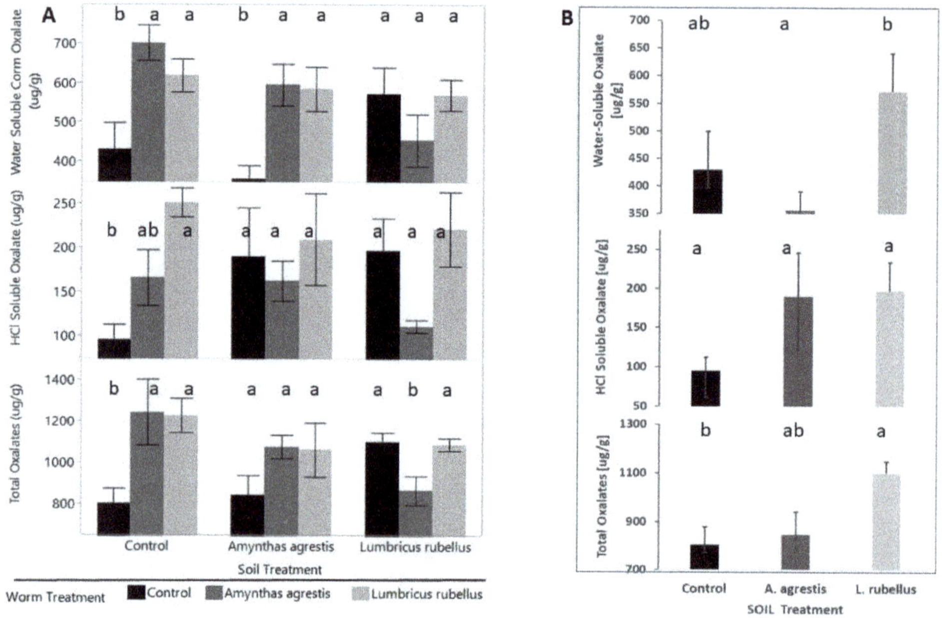

Figure 5. (**A**) Comparisons between SOIL treatment without any worms added, showing some effect of SOIL in the total and water-soluble oxalates. Different letters indicate significant differences at the $p = 0.10$ level. (**B**) Water- and HCl-extractable oxalate and total oxalates as a function of earthworm and soil treatment. Error bars specify one standard error. Different letters indicate significant differences at the $p = 0.05$ level for comparisons of WORM treatments within a SOIL pretreatment.

4.1.3. Effect of Sequence of Earthworm Invasions

In the current second wave of earthworm invasion [20] in the northeastern USA, Megascolecidae are replacing Lumbricidae. The concern is that the new invasion is further changing the ecosystems' response to earthworms. In this experiment, the sequence of invasions is modeled by adding earthworms to soils pretreated by earthworms. The invasion baseline would be the introduction of earthworms to soils previously unaltered by earthworms. In that case, the earthworm effects on water-soluble, HCl-soluble and total oxalates were large and significant, regardless of which species was added (Figure 5B). The effect of earthworm invasions on soil chemistry is well known, although we were looking at the effect of earthworms on oxalates in plant tissue. Hale et al. [5] found large differences in field soil and leaf N concentrations across an invasion front of Lumbricidae in Minnesota. Price-Christenson et al. [48] showed similar effects for *Amynthas agrestis* and *Amynthas tokioensis*.

When examining the effect of earthworm addition to soils previously occupied by earthworms, three trends were interesting. First, when adding either *A. agrestis* or *L. rubellus* to *A. agrestis* SOIL treatment (Figure 5B), the addition of earthworms increased corm water-soluble oxalate concentrations significantly. Secondly, when introducing earthworms to *L. rubellus*-pretreated soils, *L. rubellus* did not affect any of the three oxalate fractions that showed any significance in the linear model: no difference was observed between no earthworm added and *L. rubellus* introductions. We expected that the combination of *L. rubellus*-pretreated soil and the presence of the calciferous *L. rubellus* would give a greater response. It is probable that the effect of an earthworm on JIP oxalate production is plastic, but bound by physiological constraints. Thirdly, water- and HCl-extractable oxalate concentrations were lower when *A. agrestis* were introduced to the soil previously inhabited by *L. rubellus*, although the differences were not significant ($p = 0.32$ and $p = 0.17$ for the water-soluble and HCl-extractable oxalate, respectively). Yet, for total oxalate in the *L. rubellus* SOIL treatment, the addition of *A. agrestis* lowered the oxalate concentration significantly ($p = 0.024$). We did not design the experiment to understand the mechanisms of the plant response. However, the very different response when *A. agrestis* was added to *L. terrestris* soil may suggest that there are different mechanisms that drive oxalate production for the two worms.

It is not clear what the consequence of the reduction of oxalates in JIP would be in the field when *A. agrestis* displaces *L. rubellus* in the current "second wave" of earthworm invasion. Would the effect of oxalates as a grazing defense be lowered? It is important to note that we did not investigate the above-ground tissue, but a below-ground structure that is not accessible to grazers. While such a change could have an impact on forest plant communities, field research and further lab studies need to be conducted to ascertain such an effect.

4.2. Effect of Soil Chemistry

The conventionally held belief that plants can store excess Ca, potentially mineralized from organic matter by earthworms, as Ca oxalate [49,50] is difficult to invoke here because of the lack of significant differences in NH_4Cl-extracted Ca among soil pretreatments (Table 2). Lambkin et al. [31] estimated that for the soils they investigated, a steady state ($Ca(CO_3)_2$) would only be reached after six years. Likely, the short period of time (4.5–8 months) that the soils were exposed to earthworms in this experiment would not have produced discernibly higher Ca accumulations. In addition, the low pH would also have promoted the dissolution of ($Ca(CO_3)_2$).

The soil in the experiment had a very low pH, which promotes high concentrations of aluminum and heavy metals. It has been observed that some plant species exude oxalate in response to aluminum and/or high heavy metal concentrations [50], which may have caused reduced concentrations of oxalate in the corms. We only considered oxalate concentrations at the end of the experimental period, but not the loss of oxalate from the corm in response to potential toxic metals that may be present in the soil at low pH. The pH in the soils of this experiment was 3.91, lower than the pKa of aluminum. We also could not consider the redistribution of oxalate between the different solubility pools, which may also confound differences in oxalate activity among treatments. While high soil aluminum concentrations may have increased the exudation of oxalate, the high availability and uptake of Ca could have increased cytosolic oxalate production. Future studies are encouraged to disentangle what could possibly be opposing factors influencing plant oxalate concentrations, such as plant oxalate responses to soil Al and Ca. If toxic metals result in the greater exudation of oxalate, then metal pollution could lower the resistance of Ca oxalate-accumulating plants to browsing and thus cause further reduction in understory biodiversity.

Downstream biotic influence could have driven the increased production of oxalate in the presence of earthworms. *Pseudomonas oxalaticus* and Actinomycetes both have the ability to decompose oxalate; this is relevant, as each have been found in the guts of earthworms [51]. Where earthworm geophagy takes place in the rhizosphere of the plants exuding oxalate, a decreased soil concentration of oxalate could trigger the plant to increase production. No direct detection of oxalate by plant roots has been reported; however, the indirect detection of shifts in soil Al and Ca concentration could increase the plant exudation of oxalate. Alternatively, experiments in which Al, or other metals, and Ca are manipulated could improve our understanding of their effect on oxalate production.

Soil biochemical processes occur at microsites [52–55] which can be created by earthworms [46,55]. However, nutrient content is measured generally at much larger scales, thus averaging out any differences that may occur at microsites. While estimating nutrient availability at larger scales makes sense for nutrient management, it does not add to our understanding of the soil ecological processes that generally occur at the scale of the organisms involved. When we examined castings (aggregate-scale frass) there was indeed a difference in water-extractable soil Ca concentrations among the treatments. This form of Ca is immediately available to plants, and thus might be more important for signaling to the plant to produce more oxalate. In addition, corms and roots may be intersecting microsites associated with aggregates where chemical stimuli may be concentrated.

We were particularly surprised that there were so few differences in soil nutrient concentrations among treatments, even when comparing soils with earthworms to earthworm-free controls. Others have shown that nitrification is promoted by earthworms [56–58], especially in the drilosphere [56]. Nitrate concentrations in our mesocosms were high compared to other investigators, though our trial was comparatively twice as lengthy [59]. Nitrate concentrations reported by Burtelow et al. [57] were similar to our soils, although they observed differences between earthworm and earthworm-free soils. It is possible that any differences between treatments were masked by the scale of our measurement, which would have averaged out the effect of the drilosphere. Alternatively, the longer experimental period compared to others may have resulted in a depletion of available nitrogen, as no fertilizer was added to the microcosms. Moreover, soil nitrogen may have been immobilized by the plants and soil fauna prior to soil sampling. To note, differences in nitrate concentrations may have not been observed because of our method of analysis; the microplate method employed is best for screening, and so a comparatively low precision may have introduced more error than other methods.

Finally, when comparing the SOIL treatments, *A. agrestis* had greater dissolved organic carbon (measured as light absorbance at 330 nm) than the control or *L. rubellus*-pretreated soils. This either suggests that *A. agrestis* is mobilizing organic carbon for microbial use more so than *L. rubellus* or that the microbiome associated with its castings is not processing it.

4.3. Biological Significance of Results

Statistical and biological significance may not be same in regards to plant oxalate concentration and fitness. Trials on nutrition and herbivory deterrent effectiveness in *Medicago truncata* have demonstrated that oxalate decreases herbivory and nutrition in rats and chewing insects, though these model plants were knock-out genotypes (with no oxalate production) [60,61]. The effects of slight changes in concentration, as seen here, may be more difficult to determine over short experimental durations. A possible future study on the interaction of a castrating pathogen, *Uromyces atriphyllia*, and the reproductive success of JIP in the presence and absence of earthworms would be a natural progression from what has been examined here. In addition, an experiment that examines the natural range of oxalate production in JIP, as well as under various nutrient regimes, would be advised.

The observed effects of earthworms on oxalate concentrations in JIP corms suggest that there might be an effect of previous earthworm occupancy at the time of invasion. When *A. agrestis* invades a soil with *L. rubellus*, the production of oxalate in the plant corm is suppressed. Could this have any effect on the palatability of the plant? The literature suggests that Ca oxalate raphides, needle-like structures, are one defense against herbivores [60,62]. However, a synergistic effect occurs when other defensive factors are present, such as other needle-like structures, such as silica raphides, or chemical defenses, such as cysteine protease [63]. Synergism could amplify the effect of smaller amounts of HCl-extractable oxalate, which is presumably associated with Ca oxalate crystals. When *L. rubellus* invades an *A. agrestis* patch, there may not be an effect on oxalate concentrations. When either worm invades earthworm-free soil, plant oxalate concentrations increase. In our greenhouse experiment, the differences developed quickly over a six-month period.

5. Conclusions

Earthworms increase some oxalate fractions in the corms of JIP, even when the soils had been occupied by earthworms up to the beginning of the assay. In this study the hypothesis that increased concentrations of oxalate in plant tissue were mediated by Ca in the bulk soil was not confirmed. However, focusing on microsites, here represented by earthworm casts, did show greater water-extractable Ca concentrations for *L. rubellus* than for the other treatments.

In this experiment, introducing *A. agrestis* to soils previously inhabited by *L. rubellus* reduced corm oxalate accumulation. In the other two soils (control and habitation by *A. agrestis*), corm accumulations of oxalate were the same for both earthworms at the time scale examined. This might mean that the introduction of *A. agrestis* to soils previously inhabited by *L. rubellus* may affect the plant community by reducing the synthesis of the oxalate compounds potentially involved in deterring herbivores.

Author Contributions: Conceptualization, R.D.S.M. and J.H.G.; methodology, R.D.S.M., H.B.T. and J.H.G.; formal analysis, R.D.S.M. and J.H.G.; investigation, R.D.S.M. and H.B.T.; writing—original draft preparation, R.D.S.M. and J.H.G.; writing—review and editing, H.B.T., R.D.S.M. and J.H.G.; visualization, R.D.S.M.; supervision, J.H.G., project administration, J.H.G.; funding acquisition, J.H.G. All authors have read and agreed to the published version of the manuscript.

Funding: This research was funded by the College of Agriculture and Life Science at the University of Vermont, the Scientific Research Projects Coordination Unit of Istanbul University under grant number 18262.

Conflicts of Interest: The authors declare no conflict of interest.

References

1. Bohlen, P.J.; Scheu, S.; Hale, C.M.; McLean, M.A.; Migge, S.; Groffman, P.M.; Parkinson, D. Non-native invasive earthworms as agents of change in northern temperate forests. *Front. Ecol. Environ.* **2004**, *2*, 427–435. [CrossRef]
2. Hendrix, P.F.; Baker, G.; Callaham, M.A.; Damoff, G.; Fragoso, C.; Gonzalez, G.; James, S.; Lachnicht, S.L.; Winsome, T.; Zou, X. Invasion of exotic earthworms into ecosystems inhabited by native earthworms. In *Biological Invasions Belowground: Earthworms as Invasive Species*; Springer: Dordrecht, The Netherlands, 2006; pp. 87–100.
3. Frelich, L.E.; Hale, C.M.; Reich, P.B.; Holdsworth, A.R.; Scheu, S.; Heneghan, L.; Bohlen, P.J. Earthworm invasion into previously earthworm-free temperate and boreal forests. In *Biological Invasions Belowground: Earthworms as Invasive Species*; Springer: Dordrecht, The Netherlands, 2006; pp. 35–45.
4. Tiunov, A.V.; Hale, C.M.; Holdsworth, A.R.; Vsevolodova-Perel, T.S. Invasion patterns of Lumbricidae into the previously earthworm-free areas of northeastern Europe and the western Great Lakes region of North America. In *Biological Invasions Belowground: Earthworms as Invasive Species*; Springer: Dordrecht, The Netherlands, 2006; pp. 23–34.
5. Hale, C.M.; Frelich, L.E.; Reich, P.B.; Pastor, J. Effects of European earthworm invasion on soil characteristics in northern hardwood forests of Minnesota, USA. *Ecosystems* **2005**, *8*, 911–927. [CrossRef]
6. Frelich, L.E.; Blossey, B.; Cameron, E.K.; Dávalos, A.; Eisenhauer, N.; Fahey, T.; Ferlian, O.; Groffman, P.M.; Larson, E.; Loss, S.R.; et al. Side-swiped: Ecological cascades emanating from earthworm invasions. *Front. Ecol. Environ.* **2019**, *17*, 502–510. [CrossRef]
7. Wackett, A.A.; Yoo, K.; Cameron, E.K.; Olid, C.; Klaminder, J. Global Wo'rming and Darwin Revisited: Quantifying Soil Mixing Rates by Non-native Earthworms in Fennoscandian Boreal and Arctic Ecosystems. In Proceedings of the AGU Fall Meeting Abstracts, New Orleans, LA, USA, 11–15 December 2017; p. B21F-2002.
8. Makarova, O.; Kolesnikova, A. Earthworms (Oligochaeta, Lumbricidae) in the tundra of Eastern Europe. *Biol. Bull.* **2019**, *46*, 438–449. [CrossRef]
9. Arvidsson, E. Invasive Earthworms and Their Effect on Soil Organic Matter: Impact on Soil Carbon 'Quality'in Fennoscandian Tundra. 2021. Available online: https://www.diva-portal.org/smash/get/diva2:1565059/FULLTEXT01.pdf (accessed on 20 October 2021).
10. Lejoly, J.; Quideau, S.; Laganière, J. Invasive earthworms affect soil morphological features and carbon stocks in boreal forests. *Geoderma* **2021**, *404*, 115262. [CrossRef]
11. Lawrence, B.; Fisk, M.C.; Fahey, T.J.; Suarez, E.R. Influence of nonnative earthworms on mycorrhizal colonization of sugar maple (*Acer saccharum*). *New Phytol.* **2003**, *157*, 145–153. [CrossRef] [PubMed]
12. Paudel, S.; Longcore, T.; MacDonald, B.; McCormick, M.K.; Szlavecz, K.; Wilson, G.W.; Loss, S.R. Belowground interactions with aboveground consequences: Invasive earthworms and arbuscular mycorrhizal fungi. *Ecology* **2016**, *97*, 605–614. [CrossRef]
13. Blume-Werry, G.; Krab, E.J.; Olofsson, J.; Sundqvist, M.K.; Väisänen, M.; Klaminder, J. Invasive earthworms unlock arctic plant nitrogen limitation. *Nat. Commun.* **2020**, *11*, 1–10. [CrossRef]
14. Hale, C.M.; Frelich, L.E.; Reich, P.B. Changes in hardwood forest understory plant communities in response to European earthworm invasions. *Ecology* **2006**, *87*, 1637–1649. [CrossRef]
15. Fisichelli, N.; Eisenhauer, N.; Frelich, L.; Reich, P. *Deer and Earthworms Modify Forest Responses to Climate Change*; NPS Climate Change Response Program: Washington, DC, USA, 2013.

16. Reynolds, J.W. The earthworms of Tennessee (Oligochaeta). IV. Megascolecidae, with notes on distribution, biology and a key to the species in the state. *Megadrilogica* **1978**, *3*, 117–129.
17. Reynolds, J.W.; Görres, J.H.; Knowles, M.E. A checklist by counties of earthworms (Oligochaeta: Acanthodrilidae, Lumbricidae and Megascolecidae) in the states of Maine, New Hampshire and Vermont, USA. *Megadrilogica* **2015**, *17*, 125–140.
18. Alban, D.H.; Berry, E.C. Effects of earthworm invasion on morphology, carbon, and nitrogen of a forest soil. *Appl. Soil Ecol.* **1994**, *1*, 243–249. [CrossRef]
19. Cameron, E.K.; Bayne, E.M.; Clapperton, M.J. Human-facilitated invasion of exotic earthworms into northern boreal forests. *Ecoscience* **2007**, *14*, 482–490. [CrossRef]
20. Chang, C.-H.; Johnston, M.R.; Görres, J.H.; Dávalos, A.; McHugh, D.; Szlavecz, K. Co-invasion of three Asian earthworms, Metaphire hilgendorfi, Amynthas agrestis and Amynthas tokioensis in the USA. *Biol. Invasions* **2018**, *20*, 843–848. [CrossRef]
21. Addison, J. Distribution and impacts of invasive earthworms in Canadian forest ecosystems. *Biol. Invasions* **2009**, *11*, 59–79. [CrossRef]
22. Holdsworth, A.R.; Frelich, L.E.; Reich, P.B. Effects of earthworm invasion on plant species richness in northern hardwood forests. *Conserv. Biol.* **2007**, *21*, 997–1008. [CrossRef]
23. Drouin, M.; Bradley, R.; Lapointe, L.; Whalen, J. Non-native anecic earthworms (*Lumbricus terrestris* L.) reduce seed germination and seedling survival of temperate and boreal trees species. *Appl. Soil Ecol.* **2014**, *75*, 145–149. [CrossRef]
24. Nuzzo, V.; Dávalos, A.; Blossey, B. Invasive earthworms shape forest seed bank composition. *Divers. Distrib.* **2015**, *21*, 560–570. [CrossRef]
25. Blouin, M.; Hodson, M.E.; Delgado, E.A.; Baker, G.; Brussaard, L.; Butt, K.R.; Dai, J.; Dendooven, L.; Pérès, G.; Tondoh, J. A review of earthworm impact on soil function and ecosystem services. *Eur. J. Soil Sci.* **2013**, *64*, 161–182. [CrossRef]
26. Corio, K.; Wolf, A.; Draney, M.; Fewless, G. Exotic earthworms of great lakes forests: A search for indicator plant species in maple forests. *For. Ecol. Manag.* **2009**, *258*, 1059–1066. [CrossRef]
27. Franceschi, V.R.; Nakata, P.A. Calcium oxalate in plants: Formation and function. *Annu. Rev. Plant Biol.* **2005**, *56*, 41–71. [CrossRef]
28. Black, O. Calcium oxalate in the Dasheen. *Am. J. Bot.* **1918**, *5*, 447–451. [CrossRef]
29. Mazen, A.; El Maghraby, O. Accumulation of cadmium, lead and strontium, and a role of calcium oxalate in water hyacinth tolerance. *Biol. Plant.* **1997**, *40*, 411–417. [CrossRef]
30. Briones, M.; López, E.; Méndez, J.; Rodríguez, J.; Gago-Duport, L. Biological control over the formation and storage of amorphous calcium carbonate by earthworms. *Mineral. Mag.* **2008**, *72*, 227–231. [CrossRef]
31. Lambkin, D.C.; Gwilliam, K.H.; Layton, C.; Canti, M.G.; Piearce, T.G.; Hodson, M.E. Production and dissolution rates of earthworm-secreted calcium carbonate. *Pedobiologia* **2011**, *54*, S119–S129. [CrossRef]
32. Webb, M.A. Cell-mediated crystallization of calcium oxalate in plants. *Plant Cell* **1999**, *11*, 751–761. [CrossRef] [PubMed]
33. Piearce, T. The calcium relations of selected Lumbricidae. *J. Anim. Ecol.* **1972**, *41*, 167–188. [CrossRef]
34. Canti, M.G.; Piearce, T.G. Morphology and dynamics of calcium carbonate granules produced by different earthworm species: The 7th International Symposium on Earthworm Ecology·Cardiff·Wales·2002. *Pedobiologia* **2003**, *47*, 511–521.
35. Bierzychudek, P. The demography of jack-in-the-pulpit, a forest perennial that changes sex. *Ecol. Monogr.* **1982**, *52*, 335–351. [CrossRef]
36. Sims, G.; Ellsworth, T.; Mulvaney, R. Microscale determination of inorganic nitrogen in water and soil extracts. *Commun. Soil Sci. Plant Anal.* **1995**, *26*, 303–316. [CrossRef]
37. Bellitürk, K.; Görres, J.H.; Kunkle, J.; Melnichuk, R.D.S. Can commercial mulches be reservoirs of invasive earthworms? Promotion of ligninolytic enzyme activity and survival of Amynthas agrestis (Goto and Hatai, 1899). *Appl. Soil Ecol.* **2015**, *87*, 27–31. [CrossRef]
38. Chang, C.; Szlavecz, K.; Bernard, M.; Pitz, S. The second wave of earthworm invasion: Soil organic matter dynamics from the stable isotope perspective. In Proceedings of the 2013 AGU Fall Meeting, San Francisco, CA, USA, 9–13 December 2013. Abstracts B31C-0423.
39. Plisko, J.D.; Nxele, T.C. An annotated key separating foreign earthworm species from the indigenous South African taxa (Oligochaeta: Acanthodrilidae, Eudrilidae, Glossoscolecidae, Lumbricidae, Megascolecidae, Microchaetidae, Ocnerodrilidae and Tritogeniidae). *Afr. Invertebr.* **2015**, *56*, 663–708. [CrossRef]
40. Soil Survey Staff, Natural Resources Conservation Service, United States Department of Agriculture. Web Soil Survey. Available online: http://websoilsurvey.sc.egov.usda.gov/ (accessed on 10 October 2021).
41. Moore, T. Patterns of dissolved organic matter in subarctic peatlands. *Earth Surf. Processes Landf.* **1987**, *12*, 387–397. [CrossRef]
42. Ferguson, I.B.; Turner, N.A.; Bollard, E.G. Problems in fractionating calcium in plant tissue. *J. Sci. Food Agric.* **1980**, *31*, 7–14. [CrossRef]
43. Libert, B.; Franceschi, V.R. Oxalate in crop plants. *J. Agric. Food Chem.* **1987**, *35*, 926–938. [CrossRef]
44. Zindler-Frank, E.; Hönow, R.; Hesse, A. Calcium and oxalate content of the leaves of Phaseolus vulgaris at different calcium supply in relation to calcium oxalate crystal formation. *J. Plant Physiol.* **2001**, *158*, 139–144. [CrossRef]
45. Hönow, R.; Hesse, A. Comparison of extraction methods for the determination of soluble and total oxalate in foods by HPLC-enzyme-reactor. *Food Chem.* **2002**, *78*, 511–521. [CrossRef]
46. Tecimen, H.B.; Gorres, J.H.; Melnichuk, R.D. Effect of Lumbricus rubellus and Amynthas agrestis earthworms on soil biogeochemistry at the aggregate scale in northern hardwood forests. *J. Sustain. For.* **2020**, *40*, 1–16. [CrossRef]

47. Richardson, J.T. Eta squared and partial eta squared as measures of effect size in educational research. *Educ. Res. Rev.* **2011**, *6*, 135–147. [CrossRef]
48. Price-Christenson, G.J.; Johnston, M.R.; Herrick, B.M.; Yannarell, A.C. Influence of invasive earthworms (*Amynthas* spp.) on Wisconsin forest soil microbial communities and soil chemistry. *Soil Biol. Biochem.* **2020**, *149*, 107955. [CrossRef]
49. White, P.J.; Broadley, M.R. Calcium in plants. *Ann. Bot.* **2003**, *92*, 487–511. [CrossRef]
50. Poschenrieder, C.; Gunsé, B.; Corrales, I.; Barceló, J. A glance into aluminum toxicity and resistance in plants. *Sci. Total Environ.* **2008**, *400*, 356–368. [CrossRef]
51. Cromack, K., Jr.; Sollins, P.; Todd, R.; Fogel, R.; Todd, A.; Fender, W.; Crossley, M.; Crossley, D., Jr. The role of oxalic acid and bicarbonate in calcium cycling by fungi and bacteria: Some possible implications for soil animals. *Ecol. Bull.* **1977**, *25*, 246–252.
52. Parkin, T.B. Soil microsites as a source of denitrification variability. *Soil Sci. Soc. Am. J.* **1987**, *51*, 1194–1199. [CrossRef]
53. Sexstone, A.J.; Revsbech, N.P.; Parkin, T.B.; Tiedje, J.M. Direct measurement of oxygen profiles and denitrification rates in soil aggregates. *Soil Sci. Soc. Am. J.* **1985**, *49*, 645–651. [CrossRef]
54. Görres, J.H.; Savin, M.C.; Neher, D.A.; Weicht, T.R.; Amador, J.A. Grazing in a porous environment: 1. The effect of soil pore structure on C and N mineralization. *Plant Soil* **1999**, *212*, 75–83.
55. Görres, J.H.; Savin, M.C.; Amador, J.A. Soil micropore structure and carbon mineralization in burrows and casts of an anecic earthworm (*Lumbricus terrestris*). *Soil Biol. Biochem.* **2001**, *33*, 1881–1887. [CrossRef]
56. Amador, J.A.; Görres, J.H.; Savin, M.C. Carbon and Nitrogen Dynamics in (L.) Burrow Soil. *Soil Sci. Soc. Am. J.* **2003**, *67*, 1755–1762. [CrossRef]
57. Burtelow, A.E.; Bohlen, P.J.; Groffman, P.M. Influence of exotic earthworm invasion on soil organic matter, microbial biomass and denitrification potential in forest soils of the northeastern United States. *Appl. Soil Ecol.* **1998**, *9*, 197–202. [CrossRef]
58. Parkin, T.B.; Berry, E.C. Microbial nitrogen transformations in earthworm burrows. *Soil Biol. Biochem.* **1999**, *31*, 1765–1771. [CrossRef]
59. Amador, J.A.; Potts, D.A.; Savin, M.C.; Tomlinson, P.; Görres, J.H.; Nicosia, E.L. Mesocosm-scale evaluation of faunal and microbial communities of aerated and unaerated leachfield soil. *J. Environ. Qual.* **2006**, *35*, 1160–1169. [CrossRef]
60. Li, X.; Yang, J.; Morris, J.; Hester, A.; Nakata, P.A.; Hirschi, K.D. Genetically modified Medicago truncatula lacking calcium oxalate has increased calcium bioavailability and partially rescues vitamin D receptor knockout mice phenotypes. *J. Bioequiv. Bioavailab.* **2013**, *5*, 47–52. [CrossRef]
61. Korth, K.L.; Doege, S.J.; Park, S.-H.; Goggin, F.L.; Wang, Q.; Gomez, S.K.; Liu, G.; Jia, L.; Nakata, P.A. Medicago truncatula mutants demonstrate the role of plant calcium oxalate crystals as an effective defense against chewing insects. *Plant Physiol.* **2006**, *141*, 188–195. [CrossRef] [PubMed]
62. Ward, D.; Spiegel, M.; Saltz, D. Gazelle herbivory and interpopulation differences in calcium oxalate content of leaves of a desert lily. *J. Chem. Ecol.* **1997**, *23*, 333–346. [CrossRef]
63. Konno, K.; Inoue, T.A.; Nakamura, M. Synergistic defensive function of raphides and protease through the needle effect. *PLoS ONE* **2014**, *9*, e91341. [CrossRef]

soil systems

MDPI

Article

Decadal Changes of Organic Carbon, Nitrogen, and Acidity of Austrian Forest Soils

Robert Jandl *, Ernst Leitgeb and Michael Englisch

Austrian Forest Research Center, Seckendorff Gudent Weg 8, A-1131 Vienna, Austria;
ernst.leitgeb@bfw.gv.at (E.L.); michael.englisch@bfw.gv.at (M.E.)
* Correspondence: robert.jandl@bfw.gv.at; Tel.: +43-664-826-99-07

Abstract: Repeated soil surveys provide opportunities to quantify the effect of long-term environmental change. In recent decades, the topics of forest soil acidification as a consequence of acidic deposition, the enrichment of forest ecosystems with nitrogen, and the loss of carbon due to climate change have been discussed. We used two forest soil surveys that were 20 years apart, in order to establish the direction and magnitude of changes in soil carbon, nitrogen, and soil acidity. Soils have been initially sampled in the late 1980s. The plots were revisited twenty years later. Archived soil samples from the first survey were reanalyzed with the same protocol as the new samples. We found changes in the stocks of soil organic carbon, soil nitrogen, and soil pH. However, the changes were inconsistent. In general, as many sites have gained soil organic carbon, as sites have lost carbon. Most soils have been slightly enriched with nitrogen. The soil pH has not changed significantly. We conclude that changes in the evaluated soil chemical properties are mainly driven by forest management activities and ensuing forest stand dynamics, and atmospheric deposition. We have no convincing evidence that climate change effects have already changed the soil organic carbon stock, irrespective of bedrock type.

Keywords: forest soil chemistry; forest soil survey; soil organic carbon; soil nitrogen; soil acidity; Austrian forest soils

Citation: Jandl, R.; Leitgeb, E.; Englisch, M. Decadal Changes of Organic Carbon, Nitrogen, and Acidity of Austrian Forest Soils. *Soil Syst.* **2022**, *6*, 28. https://doi.org/10.3390/soilsystems6010028

Academic Editor: Klaus von Wilpert

Received: 15 February 2022
Accepted: 11 March 2022
Published: 17 March 2022

Publisher's Note: MDPI stays neutral with regard to jurisdictional claims in published maps and institutional affiliations.

1. Introduction

Soil chemical properties are remarkably inert and are known to be slow responders to changing site conditions. Despite rapid changes in seasonal climate conditions, changes in the above- and below-ground litterfall density from the herbaceous vegetation, shrubs, trees, and soil organic carbon stocks remain stable over time. Soil pH is buffered due to the interaction of soil water with the soil matrix. The stocks of soil organic carbon and nitrogen are huge, as compared to the annual fluxes of these elements.

The reliable detection of changes of soil chemical properties and the identification of the main drivers of change are challenging. Soil sampling is a destructive process, and a previously sampled spot cannot be re-sampled. Yet, soils are spatially variable. Hence, the 'signal' of temporal change of soil chemical soil properties is overlain by 'noise', due to spatial variability [1,2]. In order to corroborate whether or not a soil chemical property has changed, it is either possible to analyze a large number of replicates, or to re-sample soils after a long time. Based on data from the German Forest Soil Survey, it was expected that a significant enrichment of soils with nitrogen would be detectable after 20 years, whereas the detection of significant changes in the stocks of soil organic carbon would remain elusive [3].

There are both scientific and political reasons for interest in changes of soil properties. Soil pH has received a lot of attention in the context of forest decline in the Northern hemisphere, particularly in the 1980s [4–6]. With respect to soil acidification, the main issue is an irreversible soil detoriation, mostly due to the destruction of clay minerals at very acidic sites. Soil properties can be affected long term by adverse impacts on the nutrient and water retention capacity.

With respect to soil nitrogen, a major concern was the induction of imbalanced tree nutrition, due to elevated rates of the deposition of atmospheric compounds containing nitrogen. When plants are growing according to an increasing supply of nitrogen, the supply of other nutrients may not keep up, and nutrient deficiencies may develop [7]. The topic of nitrogen eutrophication and nitrogen saturation triggered major research efforts, with quite controversial opinions about the benefits and problems on nitrogen enrichment, depending on the metrics that have been used. Many forests have responded to elevated nitrogen deposition with higher growth rates, yet nitrate leaching from soils to aquifers and a decline is plant species richness remained a concern [8–12]. The debate on the nitrogen enrichment of forests due to atmospheric deposition is quite vivid, and continues in the discussion on biodiversity losses.

The most recent soil-related topic is the impact of climate change on soil organic carbon. The sheer size of the soil organic carbon stock raises the question as to whether the stock is stable over time [13]. Two main lines of argument are brought forward. Firstly, increases in tree productivity due to global warming increase the rate of biomass production. More above-ground and below-ground litterfall can increase the soil organic carbon stock. On the contrary, increased temperatures are stimulating soil decomposition processes and lead to a decrease in soil organic carbon stocks. An important question for managed forest ecosystems is whether forest management strategies or land management in general can possibly affect the direction of changes in soil organic carbon stocks [14,15], thereby incorporating forest soils in climate change mitigation strategies. A well known effort is the 4-per-mil concept that implies that minor changes in the soil organic carbon stock may have a relevant impact on climate change mitigation [16,17]. Yet, numerous socio-economic and political impediments may severly limit achievable soil carbon sequestration, thereby reducing the role of soils for climate change mitigation [18].

The mechanisms of soil organic carbon storage in soils are well investigated [19–22]. Yet, it is still elusive to define a reference level for soil organic carbon stock for forests under given site and management characteristics [23,24]. Climatic parameters, soil texture, and soil oxides are identified as valuable predictors for agricultural soils [25,26]. The biogeochemical cycle of nitrogen is tightly linked to the cycle of organic carbon. Resolving soil organic carbon dynamics is an integral part of understanding soil nitrogen and involves both anthropogenic and natural drivers of ecosystem dynamics.

Recently, the stocks of soil organic carbon in Austria have been assessed as part of the CarboSeq project of FAO [27]. The study included all types of land use. It was confined to the upper 30 cm of the soil and focussed on carbon in order to maximize the number of participating countries. The Austrian contribution is explained in detail in a separate publication. Accordingly, Austrian forest soils hold 128 t C/ha in the organic surface layer and the upper 30 cm of the mineral soil [28].

In this paper, we use the available data of the Austrian Forest Soil Survey and the BioSoil project in order to identify changes in soil acidity and soil organic carbon and nitrogen [29,30]. We evaluate the data from a repeated forest soil inventory and include the organic surface layer and the mineral soil to a depth of 50 cm. We hypothesize that soil nitrogen has increased significantly within the 20 years of our study, whereas changes in soil pH are small. The hypothesis is supported by an analysis of atmospheric deposition trends. Already before the first soil survey in 1989, effective measures had been adopted to reduce the emissions of sulphur dioxide, which was a main cause for soil acidification. Yet, the emissions of nitrogen oxides remained at high levels [31–33]. We further hypothesize that eventual changes in soil organic carbon will not be statistically significant. The expected decline due to higher decomposition rates of soil organic matter is partially compensated by increased carbon inputs to the soils due to increased forest productivity. In addition, numerous measured and unmeasured parameters contribute to the considerable spatial heterogeneity of soil organic carbon stocks. This hypothesis is supported by experiences from several other national forest soil inventories [29,34–36].

2. Materials and Methods

2.1. Sites

The sites of the soil investigation are located on the regular grid of the Austrian Forest Inventory [37]. On more than 500 sites, the initial forest soil survey has taken place, and soil analysis was finished in 1989 [30]. Twenty years after the initial assessment, a repetition was made within the BioSoil project (http://icp-forests.net, accessed on 15 February 2022). Within the BioSoil, only 139 sites were selected, and soil samples for chemical analysis were collected. Rather than sampling soils from pedogenetic soil horizons, in both surveys, samples were taken from fixed depth steps of the mineral soil. The sampling protocol has been changed between the surveys in order to support a European harmonization effort. Whereas the Austrian Forest Soil Survey used the separation of 0–10, 10–20, 20–30, 30–50, and 50–80 cm, the BioSoil survey used 0–5, 5–10, 10–20, 20–40, and 40–80 cm. At each sampling point, four soil pits were opened, and the collected samples were pooled, in order to obtain one representative sample per soil horizon and plot. The rock content of the soil horizons was visually estimated in the field. The organic surface layer was sampled separately. A quadratic steel frame with 30 cm side length was put on the surface, and the organic material inside the frame was collected. Samples were dried and weighed, and chemically analyzed with the same protocol as samples of the mineral soil.

The site characteristic for subsetting the dataset in our analysis is bedrock, distinguishing between soils derived from calcareous or silicatic bedrock. Pragmatically, every soil profile where carbonate was detected in the field test (fizzing when applying diluted HCl) was grouped to 'calcareous soils', otherwise to the group of 'silicatic soils'. This dichotomy was chosen because the geological characteristics, as shown in a geological map, incompletely reflect pedological site conditions at some sites. Examples are sites in Upper Austria where the geological map shows silicatic schists. Yet, in some places, the bedrock is overlain by fluvially transported calcareous quarternary material.

The climatic characteriziation of the sites is available by the mean annual temperature and precipitation for the period 1960–90. The climate data were provided by the Zentralanstalt für Meteorologie und Geodynamik (ZAMG; http://www.zamg.ac.at, accessed on 15 February 2022) in Vienna. The climate data for the sampling sites were interpolated from the network of climate stations. An elevation–correction of measured data was necessary, as mountain regions are insufficiently represented by climate stations [38].

2.2. Soil Analysis

Soil samples of both soil surveys were delivered to the lab of the Austrian Forest Research Center, and were air dried. A part of the samples was stored in a soil archive in order to allow later re-analysis. The samples were analyzed according to the ICP manual [39]. Briefly, concentrations of organic carbon and nitrogen were analyzed with a Carlo-Erba combustion analyzer. Soil pH was determined in a 0.01 m $CaCl_2$ slurry. Data integrity was ensured by benchmarking the applied measurement protocols within the international interlaboratory comparison that was organized by ICP-Forests [40]. The laboratory protocols were updated whenever its data deviated from benchmark values of the round-robin tests. A potential bias of old and new soil chemical data is caused by changes in the technical infrastructure of the laboratory, and due to changing laboratory staff. In order to avoid this bias, archived soil samples were re-analyzed. For each of the plots of the BioSoil project, the soil samples were analyzed together with retrieved soil samples from the initial survey. An unpublished comparison of the data showed that the concentrations of soil organic carbon and nitrogen and the pH_{CaCl_2} have not changed in the air-dried archived samples during 20 years of storage. Therefore, we are confident that eventually, detected differences in soil chemical properties will reflect soil changes and will not be consequences of a bias due to inconsistencies in sample stability and laboratory protocols.

Soil texture was assessed with the pipette method ([39] Part X, Method SA03). In the soil survey of the year 1989, soil texture at each site was measured for the deepest sampled horizon of the mineral soil. A preliminary unpublished project has shown that the deepest layer of the mineral soil is representative of the entire soil profile. The obtained particle size distribution was taken as a stable site property, and was used for the calculation of organic carbon and nitrogen stocks for both soil surveys.

Soil bulk density ρ was estimated with a function that has been derived from a database of Austrian forest soils. The predictors of ρ are the concentration of soil organic carbon and soil texture classes [41].

In order to account for different soil horizon depths that were used in the sampling protocols of the two soil surveys, we split each horizon in 1 cm-slices and assigned to each slice the soil chemical and physical properties of the respective soil horizon. With rock content, *eho*, and the depth of soil horizons the mass of fine soil, i.e., particles < 2 mm, per area (kg fine soil/m^2/cm) was calculated. Fine soil mass was multiplied with the concentrations of organic carbon and nitrogen, in order to obtain the masses of organic carbon and nitrogen. The stocks of organic carbon and nitrogen in the mineral soil are represented by the cumulated values of the 1 cm-slices. The stocks of organic carbon and nitrogen in the organic surface layer were obtained from the multiplication of the mass of the organic layer and the respective concentrations of carbon and nitrogen. Annual changes of the stocks of organic carbon and nitrogen for each site were calculated as differences between the respective stocks in the two surveys, divided by the time between surveys, i.e., 20 years.

Neither the initial forest soil survey of 1989 nor the BioSoil project of 2009 included soil ecological parameters.

2.3. Data Evaluation

Soil data were statistically evaluated. We confined our analysis to the organic surface layer and the upper 50 cm of the mineral soil. The number of sites where the mineral soil extend below 50 cm was small in both soil surveys and 50 cm was a reasonable cut-off.

A comparison of the concentrations and stocks of organic carbon and nitrogen, and the pH-value, respectively, in both soil surveys was made by pairwise *t*-tests. The data were than stratified further according to 'soil survey' and 'geological bedrock'. Differences between strata were analyzed by ANOVA and a subsequent multiple comparison of means (Tukey-HSD test). For data processing, statistical analysis, and graphics, we used R v.4.1.1 ('Kick Things') and the packages AQP, agricolae, MASS, lattice, dplyr, and ggplot [42–47]. Our analysis uses the subset of data, where both initial and repeated chemical soil analyses are available, and where the required ancillary data are available. For different parameters, the sample size therefore varies slightly.

We were interested as to whether our data reveal convincing predictors of soil organic carbon sequestration. Mean annual air temperature and soil texture were, among others, proposed as candidate predictors [26]. We calculated correlations between individual sites (R functions pairs and cor.test) and a best-fit model was obtained with a stepwise, multiple forward regression in order to scrutinize the predictors with data that are available from our sites (R package MASS, functions lm and stepAIC) [45].

3. Results

The differences in the concentrations of organic carbon, nitrogen, and pH in soil that have developed over the course of 20 years are shown in Table 1. The comparison is confined to the upper 30 cm of the soil, where comparable sampling depths have been used in both surveys. The organic carbon and nitrogen concentrations are significantly higher in the survey of 2009, whereas soil pH has not significantly changed. In the upper 10 cm of the mineral soil, a highly significant increase in the concentrations of organic carbon and nitrogen took place. The HSD-Tukey test indicated differences between sites on calcareous and silicatic bedrock, respectively. In general, the comparison of means distinguished only a few groups, due to the high variability within strata. The difference between bedrock types prevails in the entire soil profile, whereas differences in concentrations of organic carbon and nitrogen between the two soil surveys are not significant in the depths 10–20, and 20–30 cm, respectively. The pH-data show consistently the expected trend of higher values in deeper horizons of the mineral soil. The entire data set shows the statistically significant separation between sites on silicatic vs. calcareous bedrock, respecitvely. No temporal trend was detected.

Table 1. Comparison of concentrations of carbon and nitrogen (mg g^{-1}) and the pH value in a repeated soil survey. The values are the arithmetic mean ($\overline{x}$) and the standard deviation (*sd*) of a sample size of n = 119 sites, with 90 sites on silicatic and 29 sites on calcareous bedrock. Statistical differences between the two soil surveys are given by the *p*-values; differences between the strata 'survey' and 'bedrock' are shown by letters that indicate the grouping according to a comparison of means (Tukey test).

	Survey 1989	**Survey 2009**	**Survey 1989**	**Survey 2009**	**Survey 1989**	**Survey 2009**
	All Data		**Silicatic Bedrock**		**Calcareous Bedrock**	
Carbon						
Forest floor	330.1± 70.4	450.9 ± 80.9	320.1 ± 70.4	442.7 ± 88.2	353.4 ± 65.4	469.9 ± 58.8
	<0.001		bc	a	b	a
0–10 cm	79.1 ± 65.5	103.8 ± 72.5	56.8 ± 33.3	82.3 ± 50.5	130.4 ± 89.3	155.4 ± 89.9
	<0.001		c	b	a	a
10–20 cm	42.8 ± 41.2	47.2 ± 50.8	29.1 ± 21.3	31.4 ± 23.0	76.8 ± 56.8	86.6 ± 75.1
	n.s.		b	b	a	a
20–30 cm	28.9 ± 29.1	30.6 ± 33.1	20.8 ± 17.0	20.1 ± 17.5	41.9 ± 35.6	58.6 ± 46.7
	n.s.		b	b	a	a
Nitrogen						
Forest floor	12.1 ± 2.2	14.3 ± 2.8	11.8 ± 2.1	13.9 ± 2.7	12.9 ± 2.1	15.1 ± 2.9
	<0.001		c	b	b	a
0–10 cm	4.0 ± 3.0	5.6 ± 3.6	2.8 ± 1.5	4.4 ± 2.7	6.7 ± 1.5	8.5 ± 4.1
	<0.001		d	c	b	a
10–20 cm	2.3 ± 2.1	2.7 ± 2.7	1.5 ± 0.9	1.7 ± 1.2	4.5 ± 0.9	5.1 ± 3.6
	n.s.		b	b	a	a
20–30 cm	1.6 ± 1.4	1.8 ± 2.0	1.1 ± 0.7	1.1 ± 0.8	2.9 ± 1.9	3.6 ± 2.8
	n.s.		b	b	a	a
pH_{CaCl_2}						
Forest floor	4.2 ± 0.9	4.3 ± 1.0	3.9 ± 0.7	3.9 ± 0.7	5.1 ± 0.8	5.3 ± 0.8
	n.s.		b	b	a	a
0–10 cm	4.6 ± 1.4	4.5 ± 1.3	3.8 ± 0.5	3.7 ± 0.5	6.4 ± 1.0	6.2 ± 1.1
	n.s.		b	b	a	a
10–20 cm	4.9 ± 1.3	4.9 ± 1.3	4.2 ± 0.5	4.1 ± 0.4	7.2 ± 0.6	6.6 ± 1.0
	n.s.		b	b	a	a
20–30 cm	5.0 ± 1.3	5.1 ± 1.3	4.4 ± 0.5	4.3 ± 0.5	7.0 ± 0.8	6.9 ± 0.9
	n.s.		b	b	a	a

Small differences have led to slightly higher stocks of soil organic carbon and nitrogen in the forest floor and the upper 50 cm of the mineral soil over the course of 20 years. The differences are highly significant for the forest floor, and statistically insignificant for the mineral soil. Within the surveys, significant differences between sites with calcareous vs. silicatic bedrock were identified (Table 2). The accumulation of organic carbon and nitrogen is larger on the forest floor than in the mineral soil. Visual inspection suggests that, at most sites, the changes are smaller than the range of the standard deviation around the mean. Yet, a shift towards higher stocks of organic carbon and nitrogen took place within 20 years between the surveys. Silicatic and calcareous soils have changed over time in a similar way, as indicated by the proximity of the linear regression functions (Figure 1).

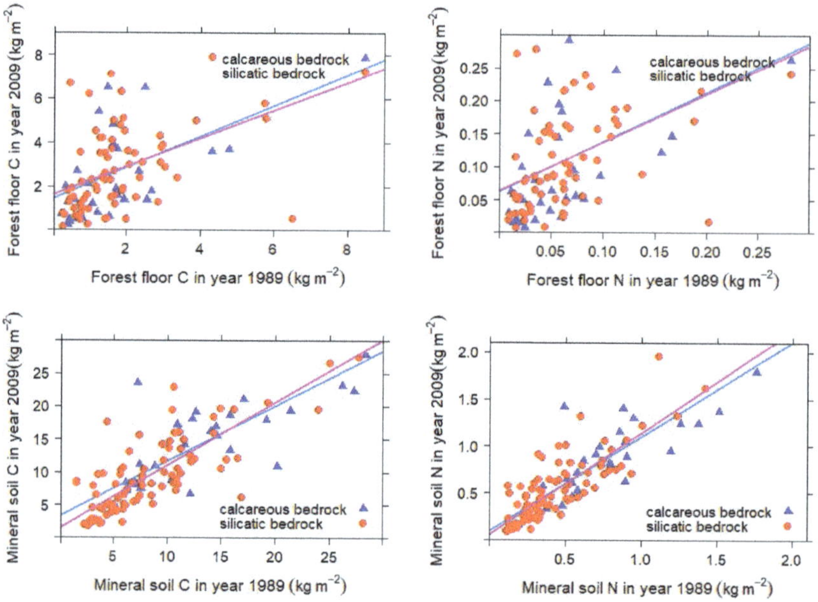

Figure 1. Change of stocks of soil organic carbon and nitrogen in Austrian forest soils within 20 years. Upper panel: Change in the litter layer. Lower panel: Change in the mineral soil to a depth of 50 cm (**left**: organic carbon, **right**: nitrogen). The blue triangles show sites on calcareous bedrock, the red circles sites on silicatic bedrock, respectively. The lines indicate the linear regression functions.

Table 2. Stocks of soil organic carbon and nitrogen (kg m^{-2}) in the organic surface layer and the upper 50 cm of the mineral soil. Statistical differences between the two soil surveys are given by the *p*-values; differences between the strata 'survey' and 'bedrock' are shown by letters that indicate the grouping according to a comparison of means (Tukey test).

	Survey 1989 All Data	Survey 2009	Survey 1989 Silicatic Bedrock	Survey 2009	Survey 1989 Calcareous Bedrock	Survey 2009
Carbon						
Forest floor	1.8 ± 1.4	2.7 ± 2.0	1.9 ± 1.5	2.7 ± 1.9	1.5 ± 1.1	2.5 ± 2.3
	<0.001		b	a	b	ab
Mineral soil	10.0 ± 7.4	11.1 ± 7.8	8.5 ± 5.4	9.7 ± 6.5	14.7 ± 10.3	15.6 ± 9.6
	n.s.		b	b	a	a
Nitrogen						
Forest floor	0.06 ± 0.05	0.11 ± 0.09	0.07 ± 0.05	0.11 ± 0.08	0.05 ± 0.04	0.10 ± 0.10
	<0.001		bc	a	bc	ab
Mineral soil	0.53 ± 0.41	0.63 ± 0.47	0.43 ± 0.26	0.53 ± 0.35	0.85 ± 0.59	0.95 ± 0.64
	n.s.		b	b	a	a

An evaluation of individual sites, stratified by bedrock material (soils derived from silicatic vs. calcareous bedrock), is shown in Figure 2. The sites were ordered according to the size of the change on stocks of soil organic carbon and nitrogen over time. The figure shows very few sites with surprisingly large changes. Most sites have almost negligible annual changes of the stocks of organic carbon and nitrogen, indicating that soils have overall been enriched with organic carbon and nitrogen. Very large gains and losses of either element cannot be explained with ecological processes, and are possibly artefacts due to peculiar local conditions that are not reflected in the data. Yet, we had no evidence for measurement errors or other sources of bias.

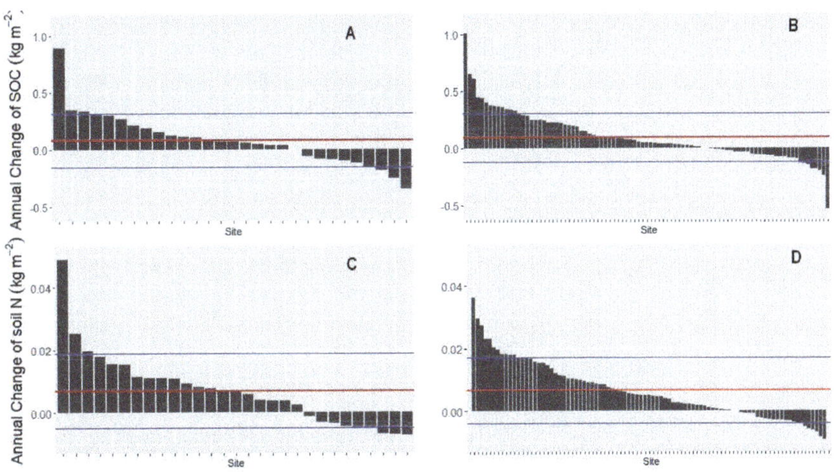

Figure 2. Annual change of organic carbon and nitrogen stocks in Austrian forest soils on 139 sampled sites ordered according to size of the change in the respective stocks. The thick horizontal red line shows the mean annual change, the fine horizontal blue lines show the mean ± the standard deviation. The annual change of organic carbon is shown in (**A**) for sites on calcareous bedrock and (**B**) for sites on silicatic bedrock. The annual change of nitrogen is shown in (**C**) for sites on calcareous bedrock and (**D**) for sites on silicatic bedrock.

The extent of soil acidfication is shown in Table 1 and in Figure 3. There is no indication that forest soils have acidified within the 20 years between the soil surveys. A de-acidification is not evident either. The range between the 25- and 75 percentiles is narrower for soils derived from silicatic bedrock. The wider range for calcareous soils can is explained by soils that are superficially acidified, but carbonate is still present in the subsoil. Such soils are encountered in Upper Austria, where quarternary deposits have been accumulated after the last glaciation and that have acidified since then. The forest soils are often acidic in the organic surface layer and the upper mineral soil.

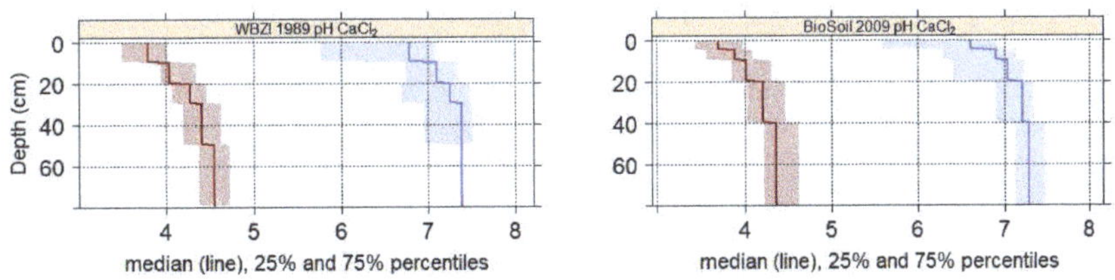

Figure 3. Extent of change in soil acidity over 20 years. Depth gradient of the median and the 25 and 75 percentiles of the pH in the mineral soil. Red lines and shades represent sites on silicatic bedrock, blue lines and shades represent sites on calcareous bedrock. **Left** graph: Forest soil survey in year 1989; **right** graph: BioSoil survey in year 2009.

Soil process understanding tells that climatic factors and mineralogical properties are indicative of the long-term carbon storage in soils [26]. Of major relevance is the content of clay minerals and silty materials that provide coupling sites for organic molecules in the mineral soil. Such considerations are relevant for the quantification of the carbon sequestration potential in soils. An obvious factor for the assessment of soil organic carbon stock changes is air temperature, because warmer sites allow for higher soil microbial activities and may trigger a depletion soil organic carbon stocks. In Figure 4, we investigate whether soil organic carbon, both expressed as concentration and stock, is related to the clay or clay-plus-silt content, or to the annual mean air temperature, respectively, at our investigated sites. Figure 4 does not indicate a strong correlation between the independent variables soil texture and air temperature, and soil organic carbon as dependent variable. The correlation coefficients and their respective statistical significances are shown in Table 3. A multiple stepwise regression with the concentration of organic C in the upper 30 cm of the mineral soil as dependent variable and the annual mean air temperature, and percentage of clay and silt, respectively, as independent variables entered temperature and silt and yielded a modest correlation ($R^2 = 0.12$). The respective correlation for the organic carbon stock in the upper 30 cm of the mineral soil had an R^2 of 0.10. The relevance of air temperature for the prediction of organic carbon was much higher than the relevance of soil texture. Overall, mean annual air temperature and soil texture are predictors of the concentration and stock soil organic carbon with only moderate predictive power (Table 4).

Table 3. Correlation between soil organic carbon concentrations (mean carbon concentration of upper 30 cm of the mineral soil) and soil organic carbon stocks (sum of organic carbon in upper 30 cm of the mineral soil) with the clay content (%), the sum of clay and silt (%), and the mean annual temperature ($\overline{T}$ (°C)), average of the years 1960 to 1990), respectively. The analysis is based on data from 429 sites of the Austrian Forest Soil Survey 1989.

	Concentration of Organic Carbon		**Stock of Organic Carbon**	
	Correlation Coefficient	***p*-Value**	**Correlation Coefficient**	***p*-Value**
clay	−0.03	0.47	−0.06	0.22
Σ(silt + clay)	0.08	0.10	0.05	0.26
$\overline{T}$	−0.30	<0.001	−0.28	<0.001

Table 4. Equations for estimating the soil organic carbon concentration (mg C/g) and the soil organic carbon stock (kg C/m^2), respectively, from mean annual temperature (°C) and the sum of silt and clay (%). The coefficients have been derived from a multiple stepwise regression. 'DF' ... degrees of freedom, R^2 ... coefficient of determination, RSE ... relative standard error. The statistical significance of the intercepts and coefficients is given by ⋆⋆⋆ ... $p < 0.001$, ⋆ ... $p < 0.05$ and 'n.s.' ... 'not significant'. The analysis is based on data of the Austrian Forest Soil Survey of 1989.

Sites	**Intercept**	**Temperature**	**Silt + Clay(%)**	**DF**	**R^2**	**RSE**
Soil organic carbon concentration						
all sites	100.7 ± 15.6	−7.1 ± 1.0	−0.09 ± 0.2	475	0.09	37.7
	⋆⋆⋆	⋆⋆⋆	n.s.			
silicatic sites	115.6 ± 16.8	−8.5 ± 0.9	−0.3 ± 0.2	340	0.22	27.5
	⋆⋆⋆	⋆⋆⋆	n.s.			
calcareous sites	48.0 ± 29.4	−4.9 ± 2.7	0.7 ± 0.3	132	0.07	49.3
	n.s.	n.s.	⋆			
Soil organic carbon stock						
all sites	12.4 ± 1.9	−0.7 ± 0.1		473	0.07	4.6
	⋆⋆⋆	⋆⋆⋆				
silicatic sites	14.9 ± 2.4	−0.9 ± 0.1		339	0.13	3.9
	⋆⋆⋆	⋆⋆⋆				
calcareous sites	7.0 ± 3.3	−0.5 ± 0.3	0.1 ± 0.0	131	0.06	5.5
	⋆	n.s.	⋆			

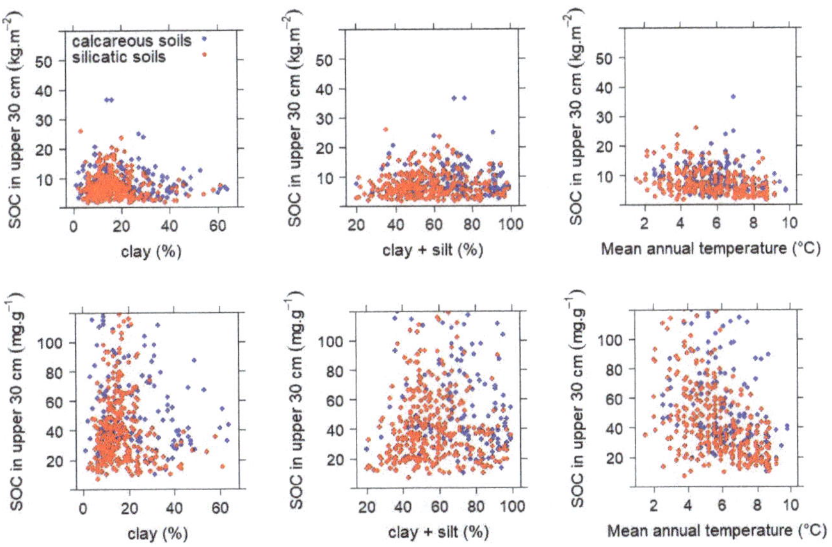

Figure 4. Relation of soil organic carbon stock, and concentration, respectively, soil texture and air temperature in Austrian forest soils. Upper panel: Relation of organic carbon stocks with the clay content (**left**), the sum of clay and silt (**center**), and air temperature (**right**). Lower panel: Relation of organic carbon concentrations with the clay content (**left**), the sum of clay and silt (**center**), and air temperature (**right**). Air temperature represents the mean of the years 1960–1990. The different color represent the bedrock material (blue: soils derived from calcareous bedrock; red: soils derived from silicatic bedrock.

4. Discussion

Soils are an integral compartment of biogeochemistry and are reflecting environmental change. Soils receive element input from the atmosphere, the vegetation, and from geochemical processes such as rock weathering. Necessarily, changes in the rates of biogeochemical processes eventually affect soil chemical properties [19,48]. Yet, changes in chemical soil processes are often slow. In particular, changes of large stocks such as soil organic carbon and nitrogen are responding slowly to external processes except for extreme events such as massive erosion or accumulation of soil material. With many anthropogenic influences on soils, the detection of soil changes receives increasing attention [49].

The availability of large-scale forest soil surveys in many countries, particularly in Europe, has fuelled the question whether anthropogenic environmental change is already evident from the analysis of soil chemical properties. An obvious approach is the repetition of soil surveys. Results of changes in soil chemical properties obtained from field data are required to corroborate the results of simulation experiments. Yet, the high spatial variability makes the distinction of signal and noise difficult. Examples of the successful identification of changes in soil organic carbon, nitrogen, and pH are available from soil monitoring projects in the United Kingom, Switzerland, Germany, and Sweden [50–53]. Evidence from ground truthing is required when anthropogenically induced changes of soil properties are addressed in conceptual studies.

We investigated the change in soil pH, soil organic carbon, and soil nitrogen. Soil pH was under scrutiny when combustion processes have enriched the atmosphere with compounds that acidified forest soils. A successful emission reduction for sulfuric compounds was implemented already in the 1990ies and sulfur dioxide emissions in Austria were greatly reduced within a short time [33]. The shared awareness for potential forest detoriation among European countries has enabled the implementation of efforts towards massive reductions of SO_2 emissions and the load of acidity was greatly reduced. Yet, forest soil acidification was by no mean a new phenomenon of the late 20th century. Biomass harvest for the benefit of agricultural production and bioenergy for a growing human population and evolving cottage industries that led to a slow, yet uni-directional degradation of forest soils [54–56]. Historical changes in agricultural practices and the use of fossil fuels instead of bioenergy alleviated the pressure on forest ecosystems. The question remains, in which time span forest soils can recover from these earlier effects. After all, the only natural de-acidification process is chemical rock weathering [57], which works at a rather slow rate.

In the time span between our two soil surveys no major anthropogenic large-scale soil acidification processes were at work. On the contrary, acidic emissions were reduced, and presumably soil-acidifying tree species such as Norway spruce are gradually replaced by forests that are dominated by deciduous tree species. Rather than expecting further soil acidification, a de-acidification was deemed possible. However, the data do not indicate such a tendency, yet (Table 1 and Figure 3). Table 1 shows that soil pH values are almost identical in both surveys. The pH values show the expected depth gradient and difference between soils derived from calcareous and silicatic bedrock, respectively. No detectable soil acidification or de-acidification has taken place in the last 20 years. The natural recovery of acidified soils is obviously a slow process that could be accelerated by liming at sites where soil acidification is deemed critical. Evidence is given by a large-scale liming campaign in SW-Germany [58,59]. Yet, some of the effects of liming may be transient and further long-term research for its full evaluation is warranted [10,60].

Research on soil organic carbon was process-based for a long time [21,22,61,62]. The consideration that soils represent a huge organic carbon stock and increasing the stock by a minute fraction may be part of successful climate-change mitigation, because capturing and sequestering carbon dioxide from the atmosphere brought the research on soil organic carbon to the forefront of attention [63,64]. Greenhouse gas emission inventories on the national, the European, and the global scale have shown that forest ecosystems act as a sink for carbon dioxide [13,65–67]. However, the role of forest soils is not entirely clear. Where the productivity of the forest is increased due to climate change or where an increase of the forest area takes place, a temporary CO_2 sink can be expected. At sites where warming accelerates the decomposition of soil organic carbon a CO_2 source is likely [68]. In cases where forest soils are CO_2 sinks, the pressure on the implementation of technological reductions of green-house gas emissions is alleviated. Carbon sequestration soils are by no means the ultimate solution, but it can still buy time for the development of technical solutions. In the Green Deal of the European Union soil organic carbon is a recognized key element of climate change mitigation [69]. Programmatic approaches such as the 4-per-mil-initiative are married with conceptual approaches on the soil organic carbon sequestration potential, and are also part of the widely used Roth-C simulation programme [17,26,70]. Yet, some expectations on the role of forest soils towards carbon sequestration are overly optimistic and may serve as a reason to further delay action on sustainable climate change mitigation.

In our analysis we found that even 20 years (approximately a fifth of the rotation period of an average Austrian forest) the changes in soil chemical properties were small and partially statistically insignificant (Table 2, Figures 1 and 2). This finding corroborates the concept that elevated input of organic carbon, not necessarily increasing the soil stock size of organic carbon, to a large extent. Instead, an ecological theory shows that biogeochemical cycles are rather accelerated. A higher availability of soil organic matter due to litterfall increases the decomposition rate of organic compounds. The net effect on the soil organic stock is small [71]. The organic soil layer is rather enriched in organic carbon (Tables 1 and 2). Organic material that is not chemically bound to mineral substances is probably more easily decomposed than organic material that is associated with minerals [62]. The organic carbon stocks in the mineral soil, that hold approximately 80% of soil organic carbon, were affected to a lesser degree (Figure 1). Figure 2 shows, on average, very small positive and negative annual changes of the carbon stocks. Some large differences cannot be explained on the basis of the data analysis, but could depend on personal biases by the field crew or forest stand dynamics that are not captured in the recorded data. The unclear pattern indicates that soil organic stock changes are influenced by several factors, e.g., numerous effects of forest management, that may have a stronger immediate effect on soil organic carbon stocks than climate change.

We had hoped to identify a metric for the carbon sequestration potential of Austrian forest soils. Such a benchmark would be instrumental to constrain the expectations on the potential and technically achievable contribution of forest soil organic carbon in climate change mitigation. In case studies, climatic factors and soil texture have been identified as good predictors of the carbon sequestration potential [26]. These parameters are also key in the widely used RothC model [72]. In Figure 4 we show that neither concentrations nor stocks of organic carbon are correlated with soil texture nor mean air temperature (Table 4). An upper limit for the expectable increase in soil organic carbon, i.e., a benchmark for potential carbon stocks, is not available and cannot be derived from the available data of the two soil surveys.

The biogeochemical fluxes of carbon and nitrogen are closely interlinked. In our survey, we found a clear signal towards increased nitrogen stocks. Again, the signal is stronger in the forest floor material than in the mineral soil (Tables 1 and 2, and Figure 1). Obviously, deposited nitrogen was withheld in the soil effectively. This is a consequence of the prevailing nitrogen limitation in Austria's forests. Despite decades of high nitrogen deposition the essential nutrient is still effectively retained. Centuries of exploitative forest use have reduced the nitrogen stocks that are still not fully replenished [73]. Therefore, nitrogen is incorporated in the biogeochemical cycles, and is readily re-absorbed by plants once it is released by the decomposition of soil organic matter. Nitrogen eutrophication as a threat to forest ecosystems is still discussed. However, the threat of ground water pollution due to nitrate leaching is only locally an issue, whereas biodiversity issues are more critical [8,74].

An emerging important aspect of soil monitoring is the assessment of soil ecological parameters. Microbial activity and the community structure of soil organisms are essential for understanding soil processes. The field of soil microbial ecology is quickly evolving and has yielded new insights in soil functioning [75–77]. However, at the time of planning, the forest soil surveys the confidence in robust and expeditive methods for soil biological parameters was low and the surveys were confined to soil physical and soil chemical parameters and site characteristics.

Author Contributions: R.J. analyzed the raw data and wrote the paper. M.E. conceptualized the initial soil survey. E.L. coordinated the BIOSOIL project. All authors have read and agreed to the published version of the manuscript.

Funding: The project was funded by the Austrian Government and Forest Focus. It is a contribution to the ACRP project CASAS.

Institutional Review Board Statement: The study did not involve humans or animals.

Informed Consent Statement: Not applicable for studies not involving humans.

Data Availability Statement: Data can be requested from the author. The requests require scrutinization against the data sharing policy of the Austrian Forest Research Center.

Acknowledgments: We are grateful for the commitment of the field and lab personnel of the Austrian Forest Research Center and to Ambros Berger and Georg Kindermann for data crunching advice. We also acknowledge the contributions of Franz Mutsch and Robert Hacker, who had passed away too early. We also thank the reviewers for highly constructive comments.

Conflicts of Interest: The authors declare no conflict of interest.

References

1. Goidts, E.; Van Wesemael, B.; Crucifix, M. Magnitude and sources of uncertainties in soil organic carbon (SOC) stock assessments at various scales. *Eur. J. Soil Sci.* **2009**, *60*, 723–739. [CrossRef]
2. Kravchenko, A.N.; Robertson, G.P. Whole-Profile Soil Carbon Stocks: The Danger of Assuming Too Much from Analyses of Too Little. *Soil Sci. Soc. Am. J.* **2011**, *75*, 235–240. [CrossRef]
3. Mellert, K.; Weis, W.; Rücker, G. *Ermittlung der (Potenziell) zu Erwartenden Signalstärke von Bodenveränderungen—Grundlagen für die Bewertung und Interpretation von Ergebnissen der BZE—Ergebnisse einer Literaturstudie im Auftrag der Bundesforschungsanstalt für Forst und Holzwirtschaft*; Technical Report. Available online: https://www.yumpu.com/de/document/read/19086710/ermittlung-der-potenziell-zu-erwartenden-signalstarke-von- (accessed on 15 March 2022).
4. Johnson, D.; Cole, D. Nutrient cycling in conifer forests. In *Ecosystems of the World*; Coniferous Forests; Andersson, F., Ed.; Elsevier: Amsterdam, The Netherlands, 2005; Volume 6, Chapter 427–450, pp. 221–292.
5. Reuss, J.O.; Johnson, D.W. *Acid Deposition and the Acidification of Soils and Waters*; Ecological Studies; Springer: New York, NY, USA, 1986; Volume 59.
6. Ulrich, B. Stability, elasticity, and resilience of terrestrial ecosystems with respect to matter balance. *Ecol. Stud.* **1987**, *61*, 11–49.
7. Schulze, E.D. *Carbon and Nitrogen Cycling in European Forest Ecosystems*; Ecological Studies; Springer: Berlin/Heidelberg, Germany, 2000; Volume 142.
8. Dirnböck, T.; Grandin, U.; Bernhardt-Römermann, M.; Beudert, B.; Canullo, R.; Forsius, M.; Grabner, M.T.; Holmberg, M.; Kleemola, S.; Lundin, L.; et al. Forest floor vegetation response to nitrogen deposition in Europe. *Glob. Chang. Biol.* **2014**, *20*, 429–440. [CrossRef]
9. Sutton, M.A.; Nemitz, E.; Skiba, U.M.; Beier, C.; Butterbach-Bahl, K.; Cellier, P.; de Vries, W.; Erisman, J.W.; Reis, S.; Bleeker, A.; et al. *The Nitrogen Cycle and Its Influence on the European Greenhouse Gas Balance*; Centre for Ecology & Hydrology: Bailrigg, UK, 2011.
10. Binkley, D.; Högberg, P. Tamm Review: Revisiting the influence of nitrogen deposition on Swedish forests. *For. Ecol. Manag.* **2016**, *368*, 222–239. [CrossRef]
11. Rothe, A.; Huber, C.; Kreutzer, K.; Weis, W. Deposition and soil leaching in stands of Norway spruce and European beech: Results from the Höglwald research in comparison with other European case studies. *Plant Soil* **2002**, *240*, 33–45. [CrossRef]
12. Scherer-Lorenzen, M.; Schulze, E.D.; Don, A.; Schumacher, J.; Weller, E. Exploring the functional significance of forest diversity: A new long-term experiment with temperate tree species (BIOTREE). *Perspect. Plant Ecol. Evol. Syst.* **2007**, *9*, 53–70. [CrossRef]
13. Pan, Y.; Birdsey, R.A.; Fang, J.; Houghton, R.; Kauppi, P.E.; Kurz, W.A.; Phillips, O.L.; Shvidenko, A.; Lewis, S.L.; Canadell, J.G.; et al. A Large and Persistent Carbon Sink in the World's Forests. *Science* **2011**, *333*, 988–993. [CrossRef] [PubMed]
14. Smith, P.; Davis, S.J.; Creutzig, F.; Fuss, S.; Minx, J.; Gabrielle, B.; Kato, E.; Jackson, R.B.; Cowie, A.; Kriegler, E.; et al. Biophysical and economic limits to negative CO_2 emissions. *Nat. Clim. Chang.* **2016**, *6*, 42–50. [CrossRef]
15. Millar, C.I.; Stephenson, N.L. Temperate forest health in an era of emerging megadisturbance. *Science* **2015**, *349*, 823–826. [CrossRef]
16. Minasny, B.; Malone, B.P.; McBratney, A.B.; Angers, D.A.; Arrouays, D.; Chambers, A.; Chaplot, V.; Chen, Z.S.; Cheng, K.; Das, B.S.; et al. Soil carbon 4 per mille. *Geoderma* **2017**, *292*, 59–86. [CrossRef]
17. Rumpel, C.; Amiraslani, F.; Koutika, L.S.; Smith, P.; Whitehead, D.; Wollenberg, E. Put more carbon in soils to meet Paris climate pledges. *Nature* **2018**, *564*, 32–34. [CrossRef] [PubMed]
18. Amundson, R.; Biardeau, L. Opinion: Soil carbon sequestration is an elusive climate mitigation tool. *Proc. Natl. Acad. Sci. USA* **2018**, *115*, 11652–11656. [CrossRef]
19. Amundson, R. *Introduction to the Biogeochemistry of Soils*; Cambridge University Press: Cambridge, UK, 2021.
20. Crow, S.E.; Swanston, C.W.; Lajtha, K.; Brooks, J.R.; Keirstead, H. Density fractionation of forest soils: Methodological questions and interpretation of incubation results and turnover time in an ecosystem context. *Biogeochemistry* **2007**, *85*, 69–90. [CrossRef]
21. Sollins, P.; Swanston, C.; Kramer, M. Stabilization and destabilization of soil organic matter—A new focus. *Biogeochemistry* **2007**, *85*, 1–7. [CrossRef]
22. Six, J.; Conant, R.T.; Paul, E.A.; Paustian, K. Stabilization mechanisms of soil organic matter: Implications for C-saturation of soils. *Plant Soil* **2002**, *241*, 155–176. [CrossRef]
23. Baritz, R.; Seufert, G.; Montanarella, L.; Van Ranst, E. Carbon concentrations and stocks in forest soils of Europe. *For. Ecol. Manag.* **2010**, *260*, 262–277. [CrossRef]
24. De Vos, B.; Cools, N.; Ilvesniemi, H.; Vesterdal, L.; Vanguelova, E.; Carnicelli, S. Benchmark values for forest soil carbon stocks in Europe: Results from a large scale forest soil survey. *Geoderma* **2015**, *251*, 33–46. [CrossRef]
25. Wenzel, W.W.; Duboc, O.; Golestanifard, A.; Holzinger, C.; Mayr, K.; Reiter, J.; Schiefer, A. Soil and land use factors control organic carbon status and accumulation in agricultural soils of Lower Austria. *Geoderma* **2022**, *409*, 115595. [CrossRef]
26. Wiesmeier, M.; Urbanski, L.; Hobley, E.; Lang, B.; von Lützow, M.; Marin-Spiotta, E.; van Wesemael, B.; Rabot, E.; Liess, M.; Garcia-Franco, N.; et al. Soil organic carbon storage as a key function of soils—A review of drivers and indicators at various scales. *Geoderma* **2019**, *333*, 149–162. [CrossRef]
27. FAO and ITPS. *Global Soil Organic Map V1.5*; FAO: Rome, Italy, 2020. [CrossRef]
28. Baumgarten, A.; Haslmayr, H.P.; Schwarz, M.; Huber, S.; Weiss, P.; Obersteiner, E.; Aust, G.; Englisch, M.; Horvath, D.; Leitgeb, E.; et al. Organic soil carbon in Austria–Status quo and foreseeable trends. *Geoderma* **2021**, *402*, 115214. [CrossRef]

29. Mutsch, F.; Leitgeb, E.; Hacker, R.; Amann, C.; Aust, G.; Herzberger, E.; Pock, H.; Reiter, R. Projekt BioSoil–Europäisches Waldboden-Monitoring (2006/07) Datenband Österreich. Band I: Methodik, Standort- und Bodenbeschreibung, Bodendaten aus Burgenland, Kärnten, Niederösterreich und Oberösterreich; Band II: Bodendaten aus Salzburg, Steiermark, Tirol und Vorarlberg, Deskriptive Statistik. *BFW Berichte* **2013**, *145*, 300.
30. Englisch, M.; Karrer, G.; Mutsch, F. Methodische Grundlagen. *Mitteilungen Der Forstl. Bundesversuchsanstalt* **1992**, *168*, 5–22.
31. Jandl, R.; Herman, F.; Smidt, S.; Butterbach-Bahl, K.; Englisch, M.; Katzensteiner, K.; Lexer, M.; Strebl, F.; Zechmeister-Boltenstern, S. Nitrogen dynamics of a mountain forest on dolomitic limestone–A scenario-based risk assessment. *Environ. Pollut.* **2008**, *155*, 512–516. [CrossRef] [PubMed]
32. Moldanová, J.; Grennfelt, P.; Jonsson, Å. Nitrogen as a threat to European air quality. In *The European Nitrogen Assessment*; Cambridge University Press: Cambridge, UK, 2011; Chapter 18.
33. Smidt, S.; Spangl, W.; Nagl, C. Trends von Luftschadstoffeinträgen in österreichischen Waldgebieten. *Cent. Gesamte Forstwes* **2009**, *127*, 1–24.
34. Grüneberg, E.; Ziche, D.; Wellbrock, N. Organic carbon stocks and sequestration rates of forest soils in Germany. *Glob. Chang. Biol.* **2014**, *20*, 2644–2662. [CrossRef]
35. Hounkpatin, K.O.L.; Stendahl, J.; Lundblad, M.; Karltun, E. Predicting the spatial distribution of soil organic carbon stock in Swedish forests using a group of covariates and site-specific data. *Soil* **2021**, *7*, 377–398. [CrossRef]
36. Ortiz, C.A.; Liski, J.; Gärdenäs, A.I.; Lehtonen, A.; Lundblad, M.; Stendahl, J.; Ågren, G.I.; Karltun, E. Soil organic carbon stock changes in Swedish forest soils—A comparison of uncertainties and their sources through a national inventory and two simulation models. *Ecol. Model.* **2013**, *251*, 221–231. [CrossRef]
37. Hauk, E.; Schadauer, K. *Instruktionen für die Feldarbeit der Österreichischen Waldinventur 2016+*, Fassung 2016 ed.; Bundesforschungs- und Ausbildungszentrum für Wald: Wien, Austria, 2016; p. 203S.
38. Auer, I.; Böhm, R.; Jurkovic, A.; Lipa, W.; Orlik, A.; Potzmann, R.; Schöner, W.; Ungersböck, M.; Matulla, C.; Briffa, K.; et al. HISTALP–historical instrumental climatological surface time series of the Greater Alpine Region. *Int. J. Climatol.* **2007**, *27*, 17–46. [CrossRef]
39. Cools, N.; de Vos, B. *Manual on Methods and Criteria for Harmonized Sampling, Assessment, Monitoring and Analysis of the Effects of Air Pollution on Forests*; Chapter Part X: Sampling and Analysis of Soil; Thünen Institute of Forest Ecosystems: Eberswalde, Germany, 2020; p. 29 + Annex.
40. Jakovjevic, T. *Soil Interlaboratory Test Programme 10th Test 2021*; Croatian Forest Research Institute: Jastrebarsko, Croatia, 2021.
41. Foldal, C.; Bohner, A.; Jandl, R.; Berger, A. Deriving regional pedotransfer functions to estimate soil bulk density in Austria. *J. Land Manag. Food Environ.* **2021**, *71*, 241–252. [CrossRef]
42. R Core Team. *R: A Language and Environment for Statistical Computing*; R Foundation for Statistical Computing: Vienna, Austria, 2020.
43. Beaudette, D.; Roudier, P.; O'Geen, A. Algorithms for quantitative pedology: A toolkit for soil scientists. *Comput. Geosci.* **2013**, *52*, 258–268. [CrossRef]
44. De Mendiburu, F.; Yaseen, M. Agricolae: Statistical Procedures for Agricultural Research; R Package Version 1.4.0. 2020. Available online: https://github.com/myaseen208/agricolae (accessed on 15 March 2022).
45. Venables, W.N.; Ripley, B.D. *Modern Applied Statistics with S*, 4th ed.; Springer: New York, NY, USA, 2002; ISBN 0-387-95457-0.
46. Sarkar, D. *Lattice: Multivariate Data Visualization with R*; Springer: New York, NY, USA, 2008; ISBN 978-0-387-75968-5.
47. Wickham, H. *ggplot2: Elegant Graphics for Data Analysis*; Springer: New York, NY, USA, 2016.
48. Schlesinger, W.E.; Bernhardt, E.S. *Biogeochemistry—An Analysis of Global Change*, 3rd ed.; Academic Press: Amsterdam, The Netherlands, 2013.
49. Binkley, D. Forest soils in the Anthropocene. In *Global Change and Forest Soils*; Developments in Soil Science; Busse, M., Giardina, C.P., Morris, D.M., Page-Dumroese, D.S., Eds.; Elsevier: Amsterdam, The Netherlands, 2019; Volume 36, Chapter 2, pp. 9–26. [CrossRef]
50. Bellamy, P.H.; Loveland, P.J.; Bradley, R.I.; Lark, R.M.; Kirk, G.J.D. Carbon losses from all soils across England and Wales 1978–2003. *Nature* **2005**, *437*, 245–248. [CrossRef] [PubMed]
51. Pluess, A.; Brang, P.; Augustin, S. (Eds.) *Wald im Klimawandel*; Haupt Verlag: Bern, Switzerland, 2016.
52. Wellbrock, N.; Bolte, A. (Eds.) *Status and Dynamics of Forests in Germany–Results of the National Forest Monitoring*; Ecological Studies; Springer Open: Berlin/Heidelberg, Germany, 2019; Volume 237.
53. Olsson, M.T.; Erlandsson, M.; Lundin, L.; Nilsson, T.; Nilsson, A.; Stendahl, J. Organic Carbon Stocks in Swedish Podzol Soils in Relation to Soil Hydrology and Other Site Characteristics. *Scand. J. For. Res.* **2009**, *43*, 209–222. [CrossRef]
54. Glatzel, G. *Leben mit dem Wald: Österreichs Wälder im Wechsel der Zeiten*; Österreichische Akademie der Wissenschaften: Wien, Austria, 1994; pp. 289–303.
55. Glatzel, G. Historic land use and its possible implications to recently accelerated tree growth in central Europe. In *Causes and Consequences of Accelerating Tree Growth in Europe*; EFI Proceedings; Karjalainen, T., Spiecker, H., Laroussinie, O., Eds.; EFI: Joensuu, Finland, 1999; Volume 29, pp. 65–74.
56. Perlin, J. *A Forest Journey—The Role of Wood in the Development of Civilization*; Harvard University Press: London, UK, 1991.
57. Morel, F.M.; Hering, J.G. *Principles and Applications of Aquatic Chemistry*; John Wiley & Sons Ltd.: Hoboken, NJ, USA, 1993.

58. Jansone, L.; von Wilpert, K.; Hartmann, P. Natural Recovery and Liming Effects in Acidified Forest Soils in SW-Germany. *Soil Syst.* **2020**, *4*, 38. [CrossRef]
59. Jacob, F.; Andreae, H.; Eickenscheidt, N.; Augustin, N. (Eds.) *Sächsischer Waldbodenbericht—Aktueller Waldbodenzustand und Dessen Veränderung*; Schriftenreihe des Staatsbetriebs Sachsenforst; Eigenverlag: Pirna, Germany, 2018; Volume 30.
60. Binkley, D.; Fisher, R.F. *Ecology and Management of Forest Soils*, 4th ed.; John Wiley & Sons: Hoboken, NJ, USA, 2020.
61. Kononova, M.; Alexandrova, I. Formation of humic acids during plant residue humification and their nature. *Geoderma* **1973**, *9*, 157–164. [CrossRef]
62. Schmidt, M.W.I.; Torn, M.S.; Abiven, S.; Dittmar, T.; Guggenberger, G.; Janssens, I.A.; Kleber, M.; Kögel-Knabner, I.; Lehmann, J.; Manning, D.A.C.; et al. Persistence of soil organic matter as an ecosystem property. *Nature* **2011**, *478*, 49–56. [CrossRef] [PubMed]
63. FAO; ITPS. (Eds.) *Recarbonizing Global Soils: A Technical Manual of Recommended Management Practices*; Forestry, Wetlands and Urban Soils—Practices Overview; FAO: Rome, Italy, 2021; Volume 5. [CrossRef]
64. Mayer, M.; Prescott, C.E.; Abaker, W.E.; Augusto, L.; Cécillon, L.; Ferreira, G.W.; James, J.; Jandl, R.; Katzensteiner, K.; Laclau, J.P.; et al. Influence of forest management activities on soil organic carbon stocks: A knowledge synthesis. *For. Ecol. Manag.* **2020**, *466*, 118127. [CrossRef]
65. Anderl, M.; Friedrich, A.; Gangl, M.; Haider, S.; Kampel, E.; Köther, T.; Kriech, M.; Lampert, C.; Matthews, B.; Pazdernik, K.; et al. *Austrias National Inventory Report 2019. Submission under the United Nations Framework Convention on Climate Change and the Kyoto Protocol*; Technical Report REP-0677; Umweltbundesamt: Vienna, Austria, 2019.
66. Janssens, I.A.; Freibauer, A.; Ciais, P.; Smith, P.; Nabuurs, G.J.; Folberth, G.; Schlamadinger, B.; Hutjes, R.W.A.; Ceulemans, R.; Schulze, E.D.; et al. Europe's terrestrial biosphere absorbs 7 to 12% of European anthropogenic CO_2 emissions. *Science* **2003**, *300*, 1538–1542. [CrossRef]
67. Houghton, R.A.; Nassikas, A.A. Global and regional fluxes of carbon from land use and land cover change 1850–2015. *Glob. Biogeochem. Cycles* **2017**, *31*, 456–472. [CrossRef]
68. Schindlbacher, A.; Zechmeister-Boltenstern, S.; Jandl, R. Carbon losses due to soil warming: Do autotrophic and heterotrophic soil respiration respond equally? *Glob. Chang. Biol.* **2009**, *15*, 901–913. [CrossRef]
69. Montanarella, L.; Panagos, P. The relevance of sustainable soil management within the European Green Deal. *Land Use Policy* **2021**, *100*, 104950. [CrossRef]
70. Smith, P.; Davies, C.A.; Ogle, S.; Zanchi, G.; Bellarby, J.; Bird, N.; Boddey, R.M.; McNamara, N.P.; Powlson, D.; Cowie, A.; et al. Towards an integrated global framework to assess the impacts of land use and management change on soil carbon: Current capability and future vision. *Glob. Chang. Biol.* **2012**, *18*, 2089–2101. [CrossRef]
71. Hungate, B.A.; Jackson, R.B.; Field, C.B.; Chapin, F.S., III. Detecting changes in soil carbon in CO_2 enrichment experiments. *Plant Soil* **1996**, *187*, 135–145. [CrossRef]
72. Coleman, K.; Jenkinson, D. RothC-26.3—A Model for the turnover of carbon in soil. In *Evaluation of Soil Organic Matter Models*; NATO ASI Series (Series I: Global Environmental Change); Powlson, D.S., Smith, P., Smith, J.U., Eds.; Springer: Berlin/Heidelberg, Germany, 1996; Volume 38. [CrossRef]
73. Jandl, R.; Smidt, S.; Mutsch, F.; Fürst, A.; Zechmeister, H.; Bauer, H.; Dirnböck, T. Acidification and Nitrogen Eutrophication of Austrian Forest Soils. *Appl. Environ. Soil Sci.* **2012**, *2012*, 632602. [CrossRef]
74. Dirnböck, T.; Kobler, J.; Kraus, D.; Grote, R.; Kiese, R. Impacts of management and climate change on nitrate leaching in a forested karst area. *J. Environ. Manag.* **2016**, *165*, 243–252. [CrossRef] [PubMed]
75. Hall, E.K.; Maixner, F.; Franklin, O.; Daims, H.; Richter, A.; Battin, T. Linking Microbial and Ecosystem Ecology Using Ecological Stoichiometry: A Synthesis of Conceptual and Empirical Approaches. *Ecosystems* **2011**, *14*, 261–273. [CrossRef]
76. Xu, X.; Li, D.; Cheng, X.; Ruan, H.; Luo, Y. Carbon: Nitrogen stoichiometry following afforestation: A global synthesis. *Sci. Rep.* **2016**, *6*, 19117. [CrossRef] [PubMed]
77. Zechmeister-Boltenstern, S.; Keiblinger, K.M.; Mooshammer, M.; Peñuelas, J.; Richter, A.; Sardans, J.; Wanek, W. The application of ecological stoichiometry to plant-microbial-soil organic matter transformations. *Ecol. Monogr.* **2015**, *85*, 133–155. [CrossRef]

soil systems

MDPI

Article

Characteristics of Soil Structure and Greenhouse Gas Fluxes on Ten-Year Old Skid Trails with and without Black Alders (*Alnus glutinosa* (L.) Gaertn.)

Hannes Warlo [1,2,*], Stephan Zimmermann [3], Friederike Lang [1] and Helmer Schack-Kirchner [1,*]

1 Chair of Soil Ecology, University of Freiburg, Bertoldstrasse 17, D-79085 Freiburg, Germany; friederike.lang@bodenkunde.uni-freiburg.de
2 Global Change Research Institute of the Czech Academy of Sciences, 4a Belidla, CZ-60300 Brno, Czech Republic
3 Soil Functions and Soil Protection, Swiss Federal Research Institute WSL, Zürcherstrasse 111, CH-8903 Birmensdorf, Switzerland; stephan.zimmermann@wsl.ch
* Correspondence: hannes.warlo@bodenkunde.uni-freiburg.de (H.W.); helmer.schack-kirchner@bodenkunde.uni-freiburg.de (H.S.-K.)

Abstract: Forest soil compaction caused by heavy machines can cause ecosystem degradation, reduced site productivity and increased greenhouse gas (GHG) emissions. Recent studies investigating the plant-mediated alleviation of soil compaction with black alder showed promising results (*Alnus glutinosa*). This study aimed to measure soil recovery and GHG fluxes on machine tracks with and without black alders in North-East Switzerland. In 2008, two machine tracks were created under controlled conditions in a European beech (*Fagus sylvatica*) stand with a sandy loam texture. Directly after compaction, soil physical parameters were measured on one track while the other track was planted with alders. Initial topsoil bulk density and porosity on the track without alders were 1.52 g cm^{-3} and 43%, respectively. Ten years later, a decrease in bulk density to 1.23 g cm^{-3} and an increase in porosity to 57% indicated partial structure recovery. Compared with the untreated machine track, alder had no beneficial impact on soil physical parameters. Elevated cumulative N_2O emission (+30%) under alder compared with the untreated track could result from symbiotic nitrogen fixation by alder. Overall, CH_4 fluxes were sensitive to the effects of soil trafficking. We conclude that black alder did not promote the recovery of a compacted sandy loam while it had the potential to deteriorate the GHG balance of the investigated forest stand.

Keywords: soil compaction; soil structure; skid trails; greenhouse gas fluxes; black alder; soil recovery

Citation: Warlo, H.; Zimmermann, S.; Lang, F.; Schack-Kirchner, H. Characteristics of Soil Structure and Greenhouse Gas Fluxes on Ten-Year Old Skid Trails with and without Black Alders (*Alnus glutinosa* (L.) Gaertn.). *Soil Syst.* **2022**, *6*, 43. https://doi.org/10.3390/soilsystems6020043

Academic Editor: Thomas Baumgartl

Received: 18 February 2022
Accepted: 22 April 2022
Published: 25 April 2022

Publisher's Note: MDPI stays neutral with regard to jurisdictional claims in published maps and institutional affiliations.

1. Introduction

The compaction of forest soils by fully mechanized logging with machine weights of up to 50 tons [1] leads to reduced soil aeration by a loss of macro porosity and the disruption of pore continuity [2]. As a consequence, a decline in site productivity [3] and ecosystem degradation are possible [4]. Furthermore, limited soil aeration reduces the sink function of forest soils for greenhouse gases. Methane, which is consumed by methanotrophic bacteria in upland forest soils, is produced by methanogenic archaea when organic matter is decomposed under anaerobic conditions [5] and N_2O is produced as an intermediate product of denitrification under anoxic conditions [6]. For these reasons, it is necessary to minimize the area affected by compaction and to recover the initial soil structure where no further disturbance is expected. As mechanical soil loosening, e.g., with agricultural techniques in forests is inconvenient [7] and natural soil recovery is slow [8,9], recent studies evaluated the potential of trees and herbs to alleviate soil compaction by root penetration and the promotion of biotic activity.

The recovery of forest soils is frequently studied on skid trails [10], which serve as machine tracks within the forest stand during logging operations. Soil structure formation by plants is commonly linked to root penetration, the promotion of soil fauna by the introduction of organic C and soil shrinkage by water extraction [11]. Carminati et al. [12] demonstrated via X-ray tomography and image analysis that gaps evolve between roots and the surrounding soil at dry conditions. Meyer et al. [13,14] observed that black alder (*Alnus glutinosa* (L.) Gaertn.) planted on skid trails in combination with compost application created air-conducting porosity to a depth of 70 cm after seven years, presumably by root penetration. Additionally, Flores Fernandez et al. [15] found alder species in combination with liming and mulching suitable to improve soil aeration on a skid trail. Apart from physical penetration, roots can promote soil structuring and the stabilization of aggregates by releasing exudates [16]. Vergani and Graf [17] found evidence that roots of grey alder (*Alnus incana* (L.) Moench) improved aggregate stability and water permeability. Furthermore, biotic activity, especially earthworm activity, positively affects soil structure recovery [7]. Ebeling et al. [18] observed the recovery of forest soils with high biological activity within 10–20 years after compaction, while a forest soil with low biological activity was not completely recovered after 40 years. Likewise, superficial soil recovery of a skid trail in a black alder stand reported by Warlo et al. [19] was presumably the result of high biotic activity under alder.

Among the examined studies dealing with the plant-mediated recovery of forest soils, the most promising results were obtained with alder species. Black alder seems to be particularly suitable as its roots are supplied with atmospheric oxygen via an aerenchyma, enabling root growth under anaerobic conditions [20]. With this physiological adaption to waterlogging, black alder naturally occurs on floodplains and riparian areas from mid-Scandinavia to the Mediterranean [21]. Planted on skid trails with often anaerobic soil conditions, black alder could contribute to the recovery of soil structure by root penetration. Additionally, symbiotic N fixation by *Frankia alni* with N inputs of up to 160 kg ha^{-1} yr^{-1} [22] promotes biotic activity under alder, resulting in further positive effects on structure formation. On the other hand, N fixation can cause increased N_2O emissions [23], deteriorating the greenhouse gas balance of forest stands regardless of possibly improved soil aeration.

The primary objective of this study is to quantify the natural and alder-enhanced recovery of soil physical parameters on machine tracks. Long-term studies of forest soil recovery after soil trafficking are scarce and studies providing detailed information on initial conditions and machine configurations are even scarcer. In 2008, a compaction experiment was conducted by the Swiss Federal Research Institute WSL in a beech stand [24]. A treatment with black alders planted in 2008 on compacted soil was included to evaluate the potential of alders to promote soil regeneration. The aim was therefore to evaluate soil recovery based on changes in soil physical parameters acquired directly after compaction compared with the results of a re-analysis in 2018. To assess the possible undesired effects of alders on the greenhouse gas budget, we measured fluxes of CO_2, N_2O and CH_4 on a monthly basis for ten months prior to sampling for soil physical analyses.

2. Materials and Methods

2.1. Site Description

The studied site (9°05′05″ N, 47°38′41″ E, 550 m asl) was located in a beech (*Fagus sylvatica* L.) stand on the Swiss Plateau close to Ermatingen. The terrain was slightly sloped (<5 degrees) and the soil was classified as Luvic Cambisol [25] with an L-mull in humus form. The pH value ($CaCl_2$) in the mineral soil was 4.6 and the texture was a sandy loam. Mean annual temperature was 8.4 °C and mean annual precipitation was 900 mm [24].

In 2008, a soil compaction experiment was conducted by the Swiss Federal Research Institute WSL [24]. Four machine passages with a fully loaded Forwarder (Valmet 840.2) weighing 26 t equipped with 71 cm-wide tires (Trelleborg Twin 428 LS2 710/45 26.5) were applied to create several skid trail-like machine tracks. The skid trails have not been trafficked since 2008. We chose two "moderately compacted" [26] situations: one in the beech stand, and the other one in direct neighborhood to the beech stand, planted with black alder in 2008 (planting depth 15 cm, spacing between alders 120 cm, no addition of fertilizer or compost). Along five replicate transects (each of ca. 30 m length) through both skid trails, three strata were defined according to Schäffer et al. [27] and Teepe et al. [28]: (i) undisturbed control in the middle between both skid trails, (ii) the wheel tracks of both skid trails and (iii) the center bulges between the wheel tracks (Figure 1). The aim of stratification was to provide comparability between sub-plots with equal initial soil disturbance. As forest soil properties and soil–atmosphere greenhouse gas fluxes vary widely between sites but even on a plot scale [29,30], the two skid trails and the five replicate strata were judged as independent from each other.

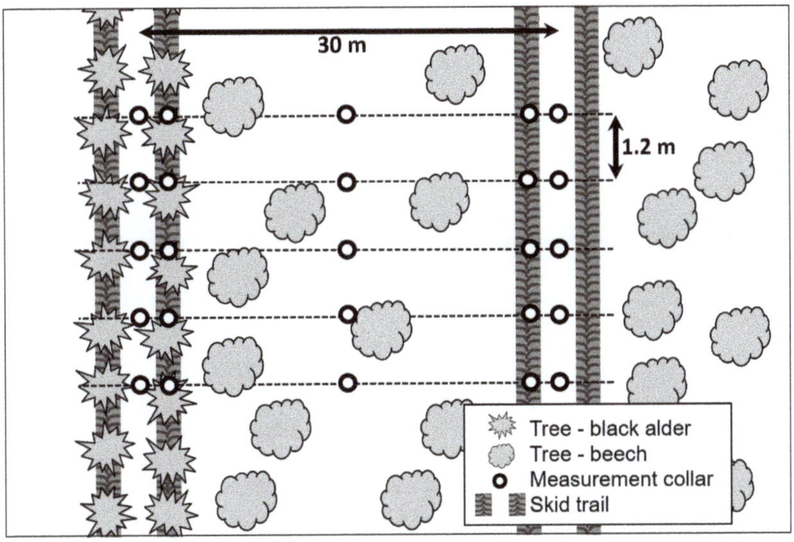

Figure 1. Sketch of the experimental design with the two investigated skid trails. Measurement collars were installed along five replicate transects in the wheel tracks of each skid trail, in the center bulges and in the undisturbed stand between the skid trails.

2.2. Filed Measurements

In 2008 and in 2012, CO_2, N_2O and CH_4 fluxes were measured with static chambers as described in Hartmann et al. [31]. In September 2017, new PVC collars (inner diameter 15.5 cm, height 9 cm) were installed to a depth of 5 cm in every stratum. The collars served as a permanent anchor for monthly measurements of soil gas fluxes (CO_2, CH_4 and N_2O) with closed chambers (closure time 30 min, 6 gas samples in vacutainers) during the period between October 2017 and August 2018. Gas analysis was conducted with a gas chromatograph (8000 series, Fisons PLC, Loughborough, UK) equipped with a flame ionization detector (FID) for CH_4 measurements and an electron capture detector (ECD) for N_2O and CO_2 measurements [32]. Gas fluxes were calculated according to Hutchinson and Livingston [33] using robust linear regressions [34] of gas concentration change over time within the chambers. A Frequency Domain probe (ML1, Delta-T Devices Ltd., Cambridge, United Kingdom) was used to measure volumetric soil moisture θ close to each collar at 5 cm depth. Chamber air temperature and soil temperature at 5 cm depth were recorded.

2.3. Soil Sampling and Laboratory Anayses

Following the compaction experiment in 2008, bulk density, total porosity and macro porosity were measured in 0–10 and 20–35 cm depth. A detailed description of the used methodology is given in Frey et al. [35]. In August 2018, soil rings (200 cm^3) were taken at the positions of the gas measurement collars in 10 cm intervals down to 50 cm depth, after the organic layer was removed (altogether 125 soil rings). Total porosity was determined with vacuum pycnometry. Water retention characteristics and pore size distribution was determined after water saturation and subsequent equilibration to a water potential of −160 hPa (pF 2.2) on a filter bed [36]. Measurement of soil gas diffusivity D_s/D_0 at field fresh moisture state and pF 2.2 was conducted according to Kühne et al. [37]. Topsoil water-filled pore space (WFPS) and air-filled porosity (ε) at the gas sampling dates were calculated based on total porosity Φ (from lab measurements) and soil moisture at field measurements. Topsoil D_s/D_0 at sampling dates was modeled with a transfer function using ε and an empirical model for forest soils proposed by Schack-Kirchner et al. [38]:

$$D_s/D_0 = 0.496 \times [\varepsilon_{calc}/100]^{1.661} \quad (1)$$

The measured D_s/D_0 was fitted well through the used transfer model. Diffusion efficiency E, which aggregates the tortuosity and discontinuity of the pore system [39], was calculated by

$$E = D_s/D_0 \times \varepsilon^{-1} \quad (2)$$

Bulk density of the mineral soil was calculated as the fraction of the soil sample dried at 105 °C and the volume of the soil sample V. Fine root (roots <2 mm diameter) mass density was calculated as the quotient of the mass of fine roots after drying at 105 °C and V. After grinding and drying sample aliquots at 105 °C, carbon (C) and nitrogen (N) contents were quantified with an elemental analyzer (Vario EL cube, Elementar, Langenselbold, Germany).

2.4. Statistical Anayses

Calculations and statistical analyses were performed with R version 3.2.3 (R Foundation for Statistical Computing, Vienna, Austria, 2015). The package "dunn.test" [40] was used to test for significant differences between treatments with Dunn's test for multiple comparisons. The package "robustbase" [34] was used for robust regression models to determine soil gas fluxes. The 95% confidence intervals of medians were approximated according to McGill et al. [41].

3. Results

3.1. Soil Physical Parameters

Ten years after soil trafficking, bulk density in the wheel tracks of both skid trails was still higher than in the undisturbed stand (Figure 2 left). The highest values were found in a depth of 20–30 cm (1.7 g cm^{-3}) and decreased in the depths below. However, except for the depth of 40–50 cm, no significant differences in bulk density between the skid trail treated with alders and untreated skid trails were evident. Center bulges generally exhibited intermediate behavior between the undisturbed stand and wheel tracks (data not shown).

The general patterns of fine root distribution (Figure 2 right) and measured relative gas diffusivity at pF 2.2 (Figure 3 left) were the reverse of the bulk density trend. This applies generally also for diffusion efficiency (Figure 3 right) but in the wheel tracks the depth gradient was very weak.

3.2. Comparison of Soil Physical Parameters in 2008 and 2018

In Table 1 are listed the soil physical parameters of the control and the skid trails for the sampling in 2008 and 2018. It has to be considered that the sampled depth in 2008 was 5 cm deeper. In 2008, the impact of soil trafficking was most pronounced at 0–15 cm depth, and was less distinct at 20–35 cm depth. Comparing soil physical parameters between 2008 and 2018 reveals decreasing bulk density and increasing macro porosity in all treatments, including the control. As already mentioned, on the skid trail planted with alders, soil physical parameters did not differ significantly from those on the untreated skid trail in 2018.

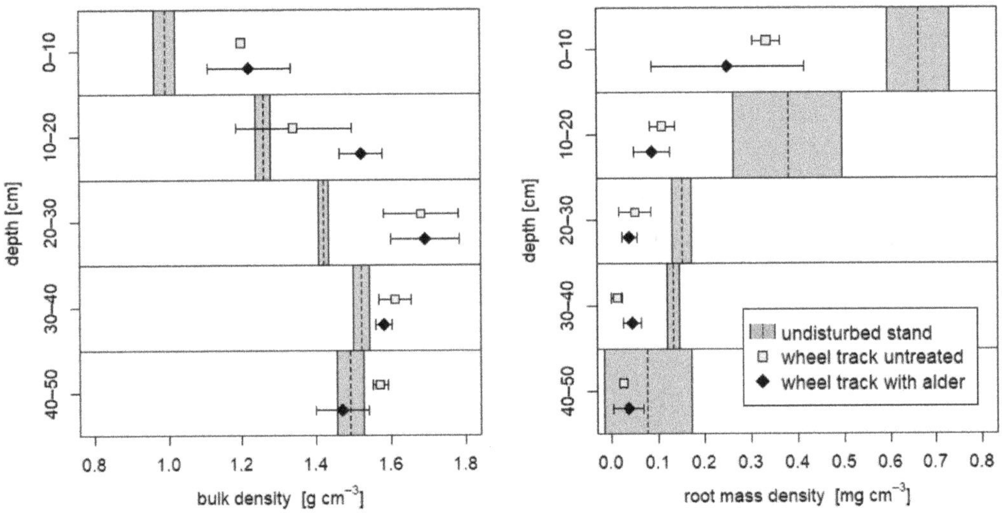

Figure 2. Median values of bulk density (**left**) and fine root mass density (**right**) in the undisturbed beech stand (dashed lines) and on the wheel tracks of the skid trails 10 years after soil trafficking. The grey shading and error bars indicate the 95% confidence intervals of median values.

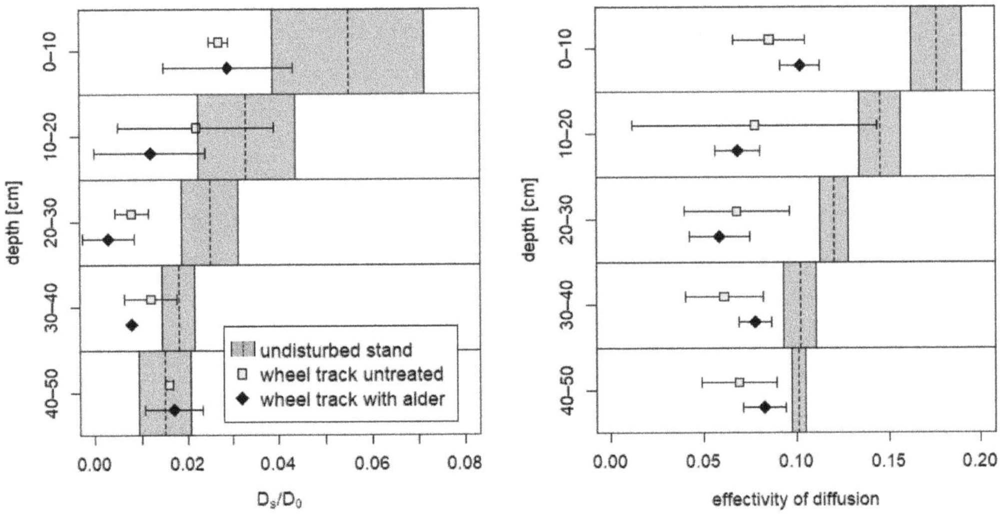

Figure 3. Median values of the relative diffusion coefficient (D_s/D_0) at pF 2.2 (**left**) and efficiency of diffusion at pF 2.2 (**right**) in the undisturbed beech stand (dashed lines) and on the wheel tracks of the skid trails 10 years after soil trafficking. The grey shading and error bars indicate the 95% confidence intervals of median values.

3.3. Soil Carbon and Nitrogen Contents

A tendency towards higher C and N contents was found in 0–10 cm depth on the wheel track of the skid trail with alders (C: 44.6 g kg^{-1}; N: 3.2 g kg^{-1}) compared with the untreated skid trail (C: 29.5 g kg^{-1}; N: 2.3 g kg^{-1}) and the undisturbed stand (C: 38.4 g kg^{-1}; N: 2.8 g kg^{-1}). In the same depth, C/N ratios ranged between 13 and 14 and declined to <10 in the depths below.

Table 1. Median values of bulk density, total porosity and macro porosity and their 95% confidence interval in the undisturbed beech stand (control) and on the wheel track of the untreated skid trail and the skid trail planted with black alders. Different letters indicate significant median differences (Dunn's test) between 2008 and 2018.

	Sampling Depth *	Beech Stand		Skid Trail (Untreated)		Skid Trail (Alder)
		2008	2018	2008	2018	2018
Bulk density [g m^{-3}]	1	1.23 ± 0.04 [a]	0.99 ± 0.03 [b]	1.52 ± 0.04 [a]	1.23 ± 0.01 [b]	1.22 ± 0.11 [b]
	2	1.49 ± 0.03 [a]	1.42 ± 0.01 [b]	1.56 ± 0.04 [a]	1.68 ± 0.10 [a]	1.69 ± 0.10 [a]
Total porosity [%]	1	50.1 ± 0.7 [a]	65.2 ± 1.8 [b]	43.3 ± 2.0 [a]	57.1 ± 3.0 [b]	57.9 ± 8.7 [b]
	2	42.4 ± 1.0 [a]	47.7 ± 0.1 [b]	39.8 ± 0.3 [a]	40.9 ± 4.2 [a]	37.7 ± 2.0 [a]
Macro porosity [%]	1	13.3 ± 2.4 [a]	15.8 ± 1.6 [a]	3.3 ± 0.6 [a]	19.8 ± 3.1 [b]	18.3 ± 8.7 [b]
	2	9.8 ± 1.6 [a]	8.5 ± 0.5 [a]	4.2 ± 0.5 [a]	11.1 ± 3.3 [b]	3.6 ± 7.7 [b]

* sampling depth 1 in 2008 = 0–15 cm; sampling depth 1 in 2018 = 0–10 cm; sampling depth 2 in 2008 = 20–35 cm and sampling depth 2 in 2018 = 20–30 cm.

3.4. Environmental Conditions during Greenhous Gas Measurements

Throughout the field measurements, WFPS was higher and D_s/D_0 was lower on both skid trails compared with the undisturbed stand (Figure 4). No treatment effect of black alder was evident.

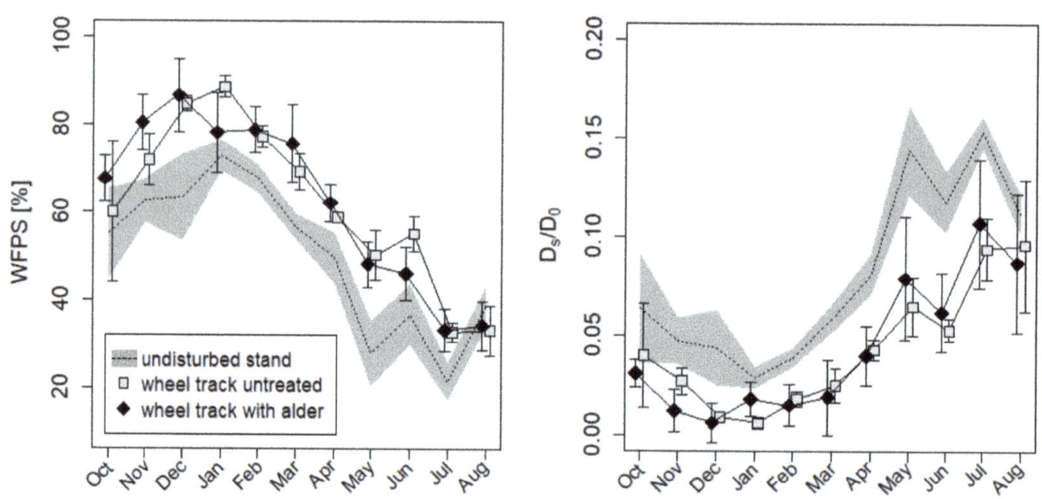

Figure 4. Monthly median values of WFPS (**left**) and modeled relative diffusion coefficient (**right**) in the undisturbed beech stand (dashed line) and on wheel tracks 10 years after soil trafficking (2017/18). The grey shading and error bars indicate the 95% confidence intervals of median values.

3.5. Greenhouse Gas Fluxes

GHG fluxes for the whole skid trail were obtained by weighting gas fluxes on the wheel tracks with 54% (2 times 71 cm tire width) and fluxes on the center bulges with 46% (120 cm width between the tires). Mostly negative CH_4 fluxes (i.e., CH_4 oxidation) were observed, with the lowest values in the undisturbed beech stand (up to −19.4 g ha^{-1} d^{-1}) and the highest values, but mostly still negative, on the skid trail with black alders (up to 0.2 g ha^{-1} d^{-1}; Figure 5). Overall, fluxes did not differ significantly between treatments; however, there was a tendency towards stronger CH_4 uptake in the undisturbed stand than on the skid trails. CH_4 fluxes were positively correlated with WFPS (r^2 = 0.51) and negatively correlated with D_s/D_0 (r^2 = 0.47). Cumulative CH_4 oxidation was strongest in the undisturbed beech stand (Table 2). On the skid trail with alders, cumulative CH_4

oxidation was significantly lower compared with the undisturbed stand but not lower than on the untreated skid trail.

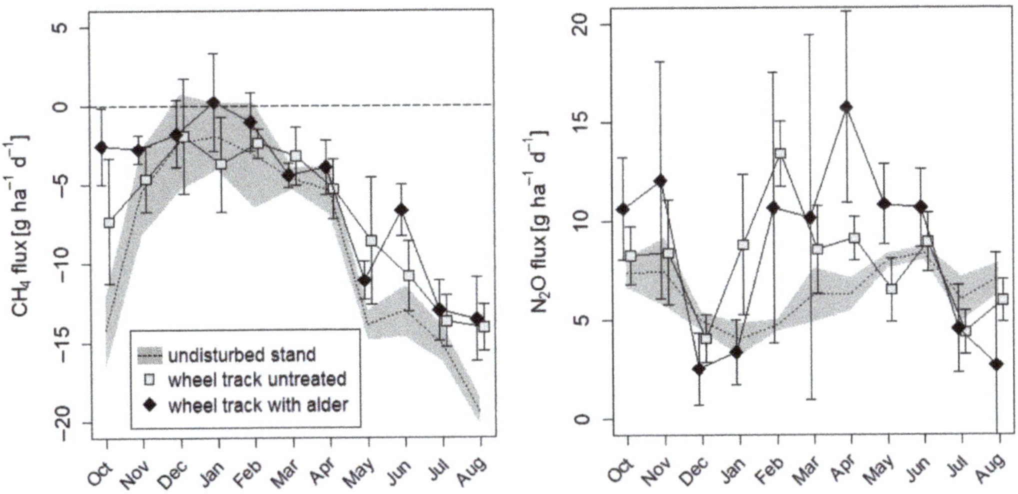

Figure 5. Medians of monthly N_2O and CH_4 flux measurements in the undisturbed beech stand (dashed line) and on wheel tracks 10 years after soil trafficking (2017/18). The grey shading and error bars indicate the 95% confidence intervals of median values.

Table 2. Median values of cumulative CO_2, N_2O and CH_4 fluxes and their 95% confidence interval. Different letters indicate significant differences between treatments.

	Beech Stand	Skid Trail Untreated	Skid Trail Alder
CH_4 (kg ha^{-1} yr^{-1})	-2.53 ± 0.79 [a]	-2.14 ± 0.62 [ab]	-1.67 ± 0.56 [b]
N_2O (kg ha^{-1} yr^{-1})	2.51 ± 0.23 [a]	2.49 ± 0.35 [a]	3.23 ± 0.70 [b]
CO_2 (Mg ha^{-1} yr^{-1})	26.1 ± 5.5 [a]	26.7 ± 5.3 [a]	28.0 ± 6.5 [a]

Median values of N_2O fluxes throughout the year were positive under all treatments, ranging from 2.5 to 15.8 g ha^{-1} d^{-1} (Figure 5). None of the tested parameters (WFPS, D_s/D_0, soil temperature, C and N contents) were significantly correlated with N_2O fluxes. Temporal variability was high on both skid trails compared with the undisturbed stand. In some cases, significantly higher N_2O emissions were found on the skid trails compared with the uncompacted control. In April and May 2018, the highest N_2O emissions were observed on the skid trail with alders. Cumulative fluxes under alder were significantly higher than in both other treatments (Table 2). CO_2 fluxes were positively correlated with soil temperature ($r^2 = 0.43$). Annual and cumulative CO_2 fluxes did not differ significantly between treatments.

4. Discussion

4.1. Soil Recovery on the Skid Trail without Treatment

This study of soil recovery after forest machine movement relies on soil physical measurements directly after soil trafficking with a forwarder and 10 years later. To our knowledge, this is unique as vehicle impact on forest soils is mostly studied by space-for-time replacement. This means that soil parameters in wheel tracks at a certain time after soil trafficking are compared with those of supposedly unaffected soil, e.g., [42,43]. What was in theory an advantage of our study turned out to be problematic due to confusion concerning the undisturbed reference which revealed an unexpected increase in macro porosity and a decrease in bulk density. Depending on whether the data from the undisturbed stand in 2008 or in 2018 are chosen as reference, soil recovery is more or less pronounced. In parts this can be an effect of the 5 cm extended depth range in the earlier sampling that brought a stronger part of the natural depth gradient of bulk density into the measurements. Furthermore, in managed forests prior compacting can be barely excluded, which makes the definition of an undisturbed control problematic. Ampoorter et al. [42] found a locally unexpected low impact of soil trafficking on a silt loam and hypothesized that a high degree of initial compaction, caused by historic uncontrolled machine traffic, could have prevented further compaction. Based on a systematic survey at 302 sites in South-West Germany, Schäffer et al. [43] reported compact or platy soil structure in 30% of the forest area outside skid trails, which they attributed to uncontrolled vehicle movement. In order to exclude such an effect, we chose control plots in 2018 explicitly along a transect between trees, inaccessible for vehicles. Because we had no other hypothesis than inherited compaction to explain the reduced bulk densities in 2018, we will hereafter use the data from 2018 as overall reference for the discussion. On the well-defined area of the skid trail itself however, we assume that potential differences in original conditions are equalized by several machine passages, such that data from 2008 and 2018 are considered sufficiently comparable. It should be kept in mind that soil sampling is always destructive, i.e., sampling at exactly the same location is not possible, which inevitably integrates spatial variability as an unknown factor.

Even though the compaction level in our study was classified as only moderate according to Lüscher et al. [26], aeration-relevant macro porosity was dramatically reduced directly after compaction (−79%), while bulk density increased by 54%. Ten years later, bulk density in the topsoil of the skid trail was less than 24% higher than in the reference, and total porosity also indicated recovery. However, a recovery of macro porosity is not enough to recover soil aeration by gas diffusion, which depends also on pore diffusion efficiency. Pore diffusion efficiency is defined as the relative diffusivity of a given volume with straight continuous pores. Deviations from these ideal pores can be caused by three geometric features [37] which can be strongly impacted by soil deformation: pores can be interrupted by shearing (connectivity), diffusion pathways can be lengthened (tortuosity), and cross sections of continuous pores can be reduced (restrictivity). In short, the observed recovery of macro porosity without a respective increase in gas diffusivity indicates a less effective pore system on the skid trails. Considering that fine roots are highly sensitive to the soil aeration status [44], limited soil aeration could explain lower fine root densities on the skid trail compared with the undisturbed stand. However, Hildebrand [45] showed for a loess loam that if bulk density was higher than 1.25 g cm^{-3}, the root growth of beech seedlings was inhibited. In our study, this threshold was exceeded in all depths below 10 cm, potentially representing a mechanical barrier for root growth.

In the literature, the relevance of structure formation by root penetration is controversial. On the one hand, roots are able to create macro pores at high soil densities by widening preexisting cracks and pores [46] and when roots dehydrate, evolving spaces between the root and surrounding soil might enable gas transport [12]. On the other hand, the radial growth pressure of roots can lead to a compaction of the surrounding soil, resulting in the opposite effect [47]. Since we found no correlation between root mass density and macro porosity or D_s/D_0, we argue that roots were not the main driver for the observed soil recovery in the present study. Alternatively, a generally high biotic activity at the site, indicated by soil respiration rates of up to 160 kg CO_2 ha^{-1} d^{-1} in the undisturbed stand and up to 170 kg CO_2 ha^{-1} d^{-1} on the skid trail, could have promoted soil recovery [7,18]. A narrow C/N ratio can be seen as beneficial for biological activity and a pH of 4.6 could offer favorable conditions for anecic and endogeic earth worms [48]. Several studies showed that earthworms are able to penetrate compacted soil with bulk densities of more than 1.7 g cm^{-3} [49,50], creating macro pores and stable soil aggregates [51]. By doing so, soil particles are ingested rather than pushed aside, causing no compaction of the surrounding soil, but instead earthworm casts at the soil surface [52]. Annual soil displacement by the burrowing activity of earthworms can exceed 100 kg m^{-2} yr^{-1} [53], involving great potential for structural restoration.

Abiotic structure formation by the shrinking and swelling of clay particles [54] or by freeze–thaw cycles [55] probably played a minor role in soil structure formation, since clay content was rather low (17%) and soil freezing at the site is rare due to a mild climate.

4.2. Soil Recovery on the Skid Trail Planted with Black Alders

The comparison of soil structure between the untreated skid trail and the skid trail planted with black alders revealed no evidence for a beneficial effect of alder on soil structure. This stands in contrast to other studies, where black alders [13,14] or grey alders [15] alleviated soil compaction, presumably due to root penetration. However, one should bear in mind that besides varying the experimental setups of studies, the interaction of numerous site-specific characteristics makes inter-study comparisons nearly impossible. This said, the results of the present study are valid for a limited area with specific site conditions.

Structure improvement in the above-cited studies was most distinct when planting alders was combined with the application of compost [13] or mulching and liming [15]. A sharp differentiation between root penetration and additional treatment effects on soil structure was therefore not always possible. However, nutrient input and reduced acidity through compost addition or liming could have ameliorated living conditions for soil-structuring biota. In addition, N fixation by alder species is frequently accompanied by an increase in soil C and N contents [56]. This fertilizing effect can additionally increase biological activity and thereby structure formation [57]. In accordance with this, Warlo et al. [19] attributed the full recovery of soil structure in the topsoil of a skid trail located in an alder stand after 17 years to generally increased biological activity due to N fixation rather than to increased root penetration only. Likewise, Ebeling et al. [18] reported the complete recovery of a compacted soil without any additional treatment but with inherent high biological activity, 20 years after compaction. In the present study, there were neither signs of higher fine root mass density under alder compared to the untreated skid trail nor of increased biological activity. The latter might be due to the fact that alders were limited to the area of the skid trail, which might be too marginal to achieve a significant boost of soil biota through N fixation.

The physiological adaption of black alders to oxygen deficiency [20] would have been expected to cause higher root mass density on the skid trail with alders compared with the untreated skid trail. As this was not the case, the above-mentioned hypothesis of root growth suppression by high mechanical impedance as opposed to limited soil aeration seems more likely.

4.3. Greenhouse Gas Fluxes

At the sampling site, Frey et al. [24] measured CH_4 fluxes of −4.2 g ha^{-1} d^{-1} at non-compacted control plots (averaged over 7, 30, 180 and 360 days after soil compaction). This is in the same range of our observations 9–10 years later. In contrast, on the skid trail Frey et al. [24] measured net positive CH_4 emission, whereas 9–10 years later, CH_4 uptake—though lower than at control plots—was observed on the skid trails. This coincides with the observed regeneration of air-conducting porosity (pores > 50 μm), which is critical for the oxygen diffusion into the soil required for CH_4 oxidation [5]. However, a trend towards less CH_4 oxidation on the skid trail planted with alders compared with the skid trail without treatment was surprising, since soil aeration and WFPS were similar throughout the year. CH_4 oxidation was possibly inhibited by elevated contents of nitrate (NO_3) or ammonium (NH_4) under alder as a result of N fixation by *Frankia alni*. Measuring NO_3 and NH_4 contents was, however, out of the scope of the present study.

Significantly elevated cumulative N_2O emissions on the skid trail with alders could likewise be the result of higher N availability under alder, as found by Mogge et al. [23]. Similar cumulative N_2O fluxes between the untreated skid trail and control suggest that soil structure, i.e., soil aeration on the skid trail, was sufficiently recovered to prevent excessive N_2O production due to anaerobic conditions. This was presumably not the case in 2008 and 2012, when N_2O fluxes on the untreated skid trail measured by Hartmann et al. [31] at the same site were 150% (light compaction) and 270% (severe compaction) higher compared with the non-compacted control. A decline in soil respiration resulting from reduced biotic activity after soil compaction [58] was not apparent and further supports the assumption of a well-advanced recovery of soil structure on both skid trails.

5. Conclusions

The main goal of this study was to evaluate the progression of soil structure regeneration on skid trails. The impact of root penetration on soil structure formation was presumably lower than in other studies and the importance of soil organisms should be taken into account for the recovery of soil compaction. This study underlines a general issue concerning the measurement of changes in soil structure over time. The destructive nature of soil physical sampling inevitably introduces uncertainties to the results due to the impossibility of sampling exactly the same point twice. As gas fluxes are sensitive to the aeration status of the soil, it might be applicable to use them as an indicator for soil structural parameters to overcome this problem. Moreover, they are easy to determine compared with soil physical parameters, such that higher area coverage and better temporal resolution of measurements could be achieved.

Author Contributions: Funding acquisition, H.S.-K.; investigation, H.W.; methodology, H.W. and H.S.-K.; project administration, H.W. and H.S.-K.; supervision, H.S.-K.; resources, F.L. and H.S.-K.; visualization, H.W.; writing—original draft, H.W.; writing—review and editing, S.Z., F.L. and H.S.-K. All authors have read and agreed to the published version of the manuscript.

Funding: This research was funded by the German Federal Ministry of Food and Agriculture, grant number FNR-22029114.

Institutional Review Board Statement: Not applicable.

Informed Consent Statement: Not applicable.

Data Availability Statement: The data presented in this study are available on request from the corresponding author.

Acknowledgments: We are grateful to Marie Gerner and Veronika Sturm for their assistance in field and laboratory work. We also thank Sebastian Bänteli for the access to the study site.

Conflicts of Interest: The authors declare no conflict of interest.

References

1. Riggert, R.; Fleige, F.; Kietz, B.; Gaertig, T.; Horn, R. Stress distribution under forestry machinery and consequences for soil stability. *Soil Sci. Soc. Am. J.* **2016**, *80*, 38–47. [CrossRef]
2. Horn, R.; Vossbrink, J.; Peth, S.; Becker, S. Impact of modern forest vehicles on soil physical properties. *For. Ecol. Manag.* **2007**, *248*, 56–63. [CrossRef]
3. Gebauer, R.; Martinková, M. Effects of pressure on the root systems of Norway spruce plants (*Picea abies* [L.] Karst.). *J. For. Sci.* **2005**, *51*, 268–275. [CrossRef]
4. Ampoorter, E.; Goris, R.; Cornelis, W.M.; Verheyen, K. Impact of mechanized logging on compaction status of sandy forest soils. *For. Ecol. Manag.* **2007**, *241*, 162–174. [CrossRef]
5. Le Mer, J.; Roger, P. Production, Oxidation, Emission and Consumption of Methane by Soils: A Review. *Eur. J. Soil Biol.* **2001**, *37*, 25–50. [CrossRef]
6. Dobbie, K.E.; Smith, K.A. Nitrous Oxide Emission Factors for Agricultural Soils in Great Britain: The Impact of Soil Water-filled Pore Space and Other Controlling Variables. *Glob. Change Biol.* **2003**, *9*, 204–218. [CrossRef]
7. Ampoorter, E.; De Schrijver, A.; De Frenne, P.; Hermy, M.; Verheyen, K. Experimental assessment of ecological restoration options for compacted forest soils. *Ecol. Eng.* **2011**, *37*, 1734–1746. [CrossRef]
8. von Wilpert, K. Forest Soil—What' s Their Peculiarity? *Soil Syst.* **2022**, *6*, 5. [CrossRef]
9. von Wilpert, K.; Schäffer, J. Ecological effects of soil compaction and initial recovery dynamics: A preliminary study. *Eur. J. For. Res.* **2006**, *125*, 129–138. [CrossRef]

10. Garland, J.J. Designated Skid Trails Minimize Soil Compaction. In *The Woodland Workbook*; Oregon State University Extension Service: Corvalis, OR, USA, 1983; pp. 1–7.
11. Angers, D.A.; Caron, J. Plant-induced changes in soil structure: Processes and feedbacks. *Biogeochemistry* **1998**, *42*, 55–72. [CrossRef]
12. Carminati, A.; Vetterlein, D.; Weller, U.; Vogel, H.; Oswald, S.E. When roots lose contact. *Vadose Zone J.* **2009**, *8*, 805–809. [CrossRef]
13. Meyer, C.; Lüscher, P.; Schulin, R. Recovery of forest soil from compaction in skid tracks planted with black alder (*Alnus glutinosa* (L.) Gaertn.). *Soil Tillage Res.* **2014**, *143*, 7–16. [CrossRef]
14. Meyer, C.; Lüscher, P.; Schulin, R. Enhancing the regeneration of compacted forest soils by planting black alder in skid lane tracks. *Eur. J. For. Res.* **2014**, *133*, 453–465. [CrossRef]
15. Flores Fernández, J.L.; Hartmann, P.; Schäffer, J.; Puhlmann, H.; von Wilpert, K. Initial recovery of compacted soil—planting and technical treatments decrease CO2concentrations in soil and promote root growth. *Ann. For. Sci.* **2017**, *74*, 1–12. [CrossRef]
16. Bardgett, R.D.; Mommer, L.; De Vries, F.T. Going underground: Root traits as drivers of ecosystem processes. *Trends Ecol. Evol.* **2014**, *29*, 692–699. [CrossRef]
17. Vergani, C.; Graf, F. Soil permeability, aggregate stability and root growth: A pot experiment from a soil bioengineering perspective. *Ecohydrology* **2016**, *9*, 830–842. [CrossRef]
18. Ebeling, C.; Lang, F.; Gaertig, T. Structural recovery in three selected forest soils after compaction by forest machines in Lower Saxony, Germany. *For. Ecol. Manag.* **2016**, *359*, 74–82. [CrossRef]
19. Warlo, H.; von Wilpert, K.; Lang, F.; Schack-Kirchner, H. Black Alder (*Alnus glutinosa* (L.) Gaertn.) on Compacted Skid Trails: A Trade-off between Greenhouse Gas Fluxes and Soil Structure Recovery? *Forests* **2019**, *10*, 726. [CrossRef]
20. Kätzel, R. Zum physiologischen Anpassungspotenzial der Schwarz-Erle (*Alnus glutinosa* [L.] GAERTN.). *Eberswalder Forstl Schr.* **2003**, *17*, 39–46.
21. Claessens, H.; Oosterbaan, A.; Savill, P.; Rondeux, J. A review of the characteristics of black alder (*Alnus glutinosa* (L.) Gaertn.) and their implications for silvicultural practices. *Forestry* **2010**, *83*, 163–175. [CrossRef]
22. Klingensmith, K.M.; Van Cleve, K. Denitrification and nitrogen fixation in floodplain successional soils along the Tanana River, interior Alaska. *Can. J. For. Res.* **1993**, *23*, 956–963. [CrossRef]
23. Mogge, B.; Kaiser, E.A.; Munch, J.C. Nitrous oxide emissions and denitrification N-losses from forest soils in the bornhoved lake region (Northern Germany). *Soil Biol. Biochem.* **1998**, *30*, 703–710. [CrossRef]
24. Frey, B.; Niklaus, P.A.; Kremer, J.; Lüscher, P.; Zimmermann, S. Heavy-machinery traffic impacts methane emissions as well as methanogen abundance and community structure in oxic forest soils. *Appl. Environ. Microbiol.* **2011**, *77*, 6060–6068. [CrossRef] [PubMed]
25. IUSS Working Group WRB. *World Reference Base for Soil Resources 2014, Update 2015, International Soil Classification System for Naming Soils and Creating Legends for Soil Maps; Rome World Soil Resour*; Rep. No 106; Food and Agriculture Organization of the United Nations (FAO): Rome, Italy, 2015; p. 203.
26. Lüscher, P.; Sciacca, S.; Thees, O. Bestrebungen zur Verbesserung des Bodenschutzes in der Schweiz. *LWF Aktuell* **2008**, *67*, 19–21.
27. Schäffer, J.; von Wilpert, K.; Kublin, E. Analysis of fine rooting below skid trails using linear and generalized additive models. *Can. J. For. Res.* **2009**, *39*, 2047–2058. [CrossRef]
28. Teepe, R.; Brumme, R.; Beese, F.; Ludwig, B. Nitrous Oxide Emission and Methane Consumption Following Compaction of Forest Soils. *Soil Sci. Soc. Am. J.* **2004**, *68*, 605. [CrossRef]
29. Maier, M.; Longdoz, B.; Laemmel, T.; Schack-Kirchner, H.; Lang, F. 2D profiles of CO2, CH4, N2O and gas diffusivity in a well aerated soil: Measurement and Finite Element Modeling. *Agric. For. Meteorol.* **2017**, *247*, 21–33. [CrossRef]
30. Muukkonen, P.; Häkkinen, M.; Mäkipää, R. Spatial variation in soil carbon in the organic layer of managed boreal forest soil-implications for sampling design. *Environ. Monit. Assess.* **2009**, *158*, 67–76. [CrossRef]
31. Hartmann, M.; Niklaus, P.A.; Zimmermann, S.; Schmutz, S.; Kremer, J.; Abarenkov, K.; Lüscher, P.; Widmer, F.; Frey, B. Resistance and resilience of the forest soil microbiome to logging-associated compaction. *ISME J.* **2014**, *8*, 226–244. [CrossRef] [PubMed]
32. Loftfield, N.; Flessa, H.; Augustin, J.; Beese, F. Automated Gas Chromatographic System for Rapid Analysis of the Atmospheric Trace Gases Methane, Carbon Dioxide, and Nitrous Oxide. *J. Environ. Qual.* **1997**, *26*, 560–564. [CrossRef]
33. Hutchinson, G.L.; Livingston, G.P. Soil-Atmosphere Gas Exchange. In *Methods of Soil Analysis: Part 4 Physical Methods*; Soil Science Society of America Book Ser. 5: Madison, WI, USA, 2002; pp. 1159–1182.
34. Maechler, M.; Rousseeuw, P.; Croux, C.; Todorov, V.; Ruckstuhl, A.; Salibian-Barrera, M.; Verbeke, T.; Koller, M.; Conceicao, E.L.; di Palma, M.A. *robustbase: Basic Robust Statistics*; R Package Version 0.93-7; R Foundation for Statistical Computing: Vienna, Austria, 2022.
35. Frey, B.; Kremer, J.; Rüdt, A.; Sciacca, S.; Matthies, D.; Lüscher, P. Compaction of forest soils with heavy logging machinery affects soil bacterial community structure. *Eur. J. Soil Biol.* **2009**, *45*, 312–320. [CrossRef]
36. Hartge, K.H.; Horn, R. *Die Physikalische Untersuchung von Böden*, 3rd ed.; Ferdinand Enke Verlag: Stuttgart, Germany, 2011.
37. Kühne, A.; Schack-Kirchner, H.; Hildebrand, E.E. Gas diffusivity in soils compared to ideal isotropic porous media. *J. Plant Nutr. Soil Sci.* **2012**, *175*, 34–45. [CrossRef]
38. Schack-Kirchner, H.; Gaertig, T.; von Wilpert, K.; Hildebrand, E.E. A modified McIntyre and Phillip approach to measure top-soil gas diffusivity in-situ. *J. Plant Nutr. Soil Sci.* **2001**, *164*, 253–258. [CrossRef]

39. Perdok, U.D.; Kroesbergen, B.; Hoogmoed, W.B. Possibilities for modelling the effect of compression on mechanical and physical properties of various Dutch soil types. *Soil Tillage Res.* **2002**, *65*, 61–75. [CrossRef]
40. Dinno, A. *dunn.test: Dunn's Test of Multiple Comparisons Using Rank Sums*; R Package Version 1.3.5; R Foundation for Statistical Computing: Vienna, Austria, 2017.
41. McGill, R.; Tukey, J.W.; Larsen, W.A. Variation of box plots. *Am. Stat.* **1978**, *32*, 12–16.
42. Ampoorter, E.; Van Nevel, L.; De Vos, B.; Hermy, M.; Verheyen, K. Assessing the effects of initial soil characteristics, machine mass and traffic intensity on forest soil compaction. *For. Ecol. Manag.* **2010**, *260*, 1664–1676. [CrossRef]
43. Schäffer, J.; Buberl, H.; von Wilpert, K. Deformation damages in forest topsoils—An assessment based on Level-I soil monitoring data from Baden-Württemberg (SW Germany). *J. Plant Nutr. Soil Sci.* **2012**, *175*, 24–33. [CrossRef]
44. Qi, B.; Marshall, J.D.; Mattson, K.G. High soil carbon dioxide concentrations inhibit root respiration of Douglas fir. *New Phytol.* **1994**, *128*, 435–442. [CrossRef]
45. Hildebrand, E.E. Der Einfluß der Bodenverdichtung auf die Bodenfunktionen im forstlichen Standort. *Forstwiss. Cent.* **1983**, *102*, 111–125. [CrossRef]
46. Gao, W.; Hodgkinson, L.; Jin, K.; Watts, C.W.; Ashton, R.W.; Shen, J.; Ren, T.; Dodd, I.C.; Binley, A.; Phillips, A.L.; et al. Deep roots and soil structure. *Plant Cell Environ.* **2016**, *39*, 1662–1668. [CrossRef]
47. Dexter, A.R. Compression of soil around roots. *Plant Soil* **1987**, *97*, 401–406. [CrossRef]
48. Sommer, M.; Ehrmann, O.; Friedel, J.K.; Martin, K.; Vollmer, T.; Turian, G. Böden als Lebensraum für Organismen-Regenwürmer, Gehäuseschnecken und Bodenmikroorganismen in Wäldern Baden-Württembergs. *Hohenh. Bodenkd. Hefte* **2002**, *63*, 98–100.
49. Capowiez, Y.; Cadoux, S.; Bouchand, P.; Roger-Estrade, J.; Richard, G.; Boizard, H. Experimental evidence for the role of earthworms in compacted soil regeneration based on field observations and results from a semi-field experiment. *Soil Biol. Biochem.* **2009**, *41*, 711–717. [CrossRef]
50. Ponder, F.; Li, F.; Jordan, D.; Berry, E.C. Assessing the impact of Diplocardia ornata on physical and chemical properties of compacted forest soil in microcosms. *Biol. Fertil. Soils.* **2000**, *32*, 166–172. [CrossRef]
51. Larink, O.; Werner, D.; Langmaack, M.; Schrader, S. Regeneration of compacted soil aggregates by earthworm activity. *Biol. Fertil. Soils.* **2001**, *33*, 395–401.
52. Greacen, E.L.; Sands, R. Compaction of Forest Soils. A Review. *Aust. J. Soil Res.* **1980**, *18*, 163–189. [CrossRef]
53. Lavelle, P.; Barot, S.; Blouin, M.; Decaens, T.; Jimenez, J.J.; Jouquet, P. Earthworms as key factor in self-organized soil systems. In *Ecosystem Engeneers: Plants to Protists*; Cuddington, K., Byers, J.E., Wilson, W.G., Hastings, A., Eds.; Elsevier: Amsterdam, The Netherlands, 2007; pp. 77–106.
54. Fisher, R.F.; Binkley, D. *Ecology and Management of Forest Soils*, 3rd ed.; Wiley: New York, NY, USA, 2000.
55. Othman, M.A.; Benson, C.H. Effect of freeze–thaw on the hydraulic conductivity and morphology of compacted clay. *Can. Geotech J.* **1993**, *30*, 236–246. [CrossRef]
56. Uri, V.; Aosaar, J.; Varik, M.; Becker, H.; Ligi, K.; Padari, A.; Kanal, A.; Lõhmus, K. The dynamics of biomass production, carbon and nitrogen accumulation in grey alder (*Alnus incana* (L.) Moench) chronosequence stands in Estonia. *For. Ecol. Manag.* **2014**, *327*, 106–117. [CrossRef]
57. Haynes, R.J.; Naidu, R. Influence of lime, fertilizer and manure applications on soil organic matter content and soil physical conditions: A review. *Nutr. Cycl. Agroecosyst.* **1998**, *51*, 123–137. [CrossRef]
58. Epron, D.; Plain, C.; Ndiaye, F.K.; Bonnaud, P.; Pasquier, C.; Ranger, J. Effects of compaction by heavy machine traffic on soil fluxes of methane and carbon dioxide in a temperate broadleaved forest. *For. Ecol. Manag.* **2016**, *382*, 1–9. [CrossRef]

soil systems

MDPI

Article

The Influence of Tree Species on the Recovery of Forest Soils from Acidification in Lower Saxony, Germany

Bernd Ahrends *, Heike Fortmann and Henning Meesenburg

Department of Environmental Control, Northwest German Forest Research Institute (NW-FVA), D-37079 Göttingen, Germany; heike.fortmann@nw-fva.de (H.F.); henning.meesenburg@nw-fva.de (H.M.)
* Correspondence: bernd.ahrends@nw-fva.de; Tel.: +49-551-69401202

Abstract: Atmospheric acid deposition has increased sharply since the beginning of industrialization but has decreased considerably since the 1980s owing to clean-air policies. Soil acidification induced by an input of acidity has been demonstrated in numerous studies using repeated forest-soil inventories. So far, relatively few data have been sampled to analyze long-term soil trends and only a few studies show the recovery of forest soils from acidification, whereas the recovery of surface waters following declining acid deposition is a widespread phenomenon. To assess a possible recovery from acid deposition, soil resampling data from 21 forested permanent soil-monitoring sites in Lower Saxony (Germany) were evaluated. For most sites, at least three repetitions of inventories from a period of 30 to 50 years were available. Trend analyses of indicators for the acid-base status of unlimed forest soils using generalized additive mixed models (GAMM) show either a trend reversal or a stagnation of the acid-base status at a strong acidification level. The recovery, if indicated by an increase of soil pH and base saturation, of soils from plots with deciduous trees appears to have occurred faster than in coniferous forest stands. This observation may be attributed to a larger amount of temporarily stored sulfur in the soil because of the higher atmospheric input into coniferous forests. As indicators for the acid-base status still show considerable soil acidification, mitigation measures such as forest liming still appear to be necessary for accelerating the regeneration process.

Keywords: base saturation; repeated soil sampling; acidification; forest soils; recovery; sulfur deposition; generalized additive mixed models

Citation: Ahrends, B.; Fortmann, H.; Meesenburg, H. The Influence of Tree Species on the Recovery of Forest Soils from Acidification in Lower Saxony, Germany. *Soil Syst.* **2022**, *6*, 40. https://doi.org/10.3390/soilsystems6020040

Academic Editor: Sören Thiele-Bruhn

Received: 9 March 2022
Accepted: 14 April 2022
Published: 20 April 2022

1. Introduction

The Industrial Revolution in Europe resulted in the increasing emission of acidifying pollutants into the atmosphere. The deposition of acidifying and eutrophying substances drastically altered the stability, nutrient cycles, and growth of forest ecosystems for several decades [1–3]. Sulfur (S) was the major component of acid deposition since the beginning of industrialization until the 1980s. The strength of the soil acidification dynamics because of acid input is primarily determined by the ability of the soil to buffer the input of acids [4,5]. The buffer capacity of soils increases with carbonate and clay mineral contents and is lowest in sandy soils. In Lower Saxony (Germany), this is particularly the case in the forested regions of the lowlands and at Harz, Solling, and Hils mountains. As a consequence of acid deposition, forest soils experienced a severe loss of base cations (Bc: Ca^{2+}, Mg^{2+}, K^{+}) with seepage water. Significantly acidified soils with a low buffer capacity show a decline in soil pH and base saturation. A loss of base cations may lead to nutrient imbalances at base-poor sites [2]. For example, at the Harz and Solling mountains in Lower Saxony, the high sulfur deposition caused severe soil acidification and Mg-deficiency symptoms of the forest stands, and large areas were affected by forest decline [6–8]. Beginning in the 1980s, clean-air policies resulted in a considerable decrease of sulfur deposition in Europe, which continues until today [9–11]. The dynamic development of sulfur deposition raises questions about the recovery of these soils from acidification [12]. This is especially true for forest ecosystems that have not yet been limed. The adsorption and release of sulfur in forest soils play decisive roles in the recovery of forest soils from acidification [13,14]. For predictions of recovery, atmospheric inputs, soil characteristics, soil S pools, and their dynamics have to be considered.

The deposition of oxidized and reduced nitrogen (N) also contributes to the acidification of forest soils [15]. However, the reductions in N deposition recorded since the early 1990s have been less pronounced than those for sulfur [9] and, accordingly, the acidification of Europe's forest soils shows a limited response to the decrease in N deposition since the 1990s [16].

From the comparison of coniferous and deciduous forest stands at comparable sites, it can be inferred that deciduous forests receive less sulfur via total deposition [17]. For the deposition process, the structure of the canopy plays an important role [18,19]. Therefore, lower deposition rates can be expected for the forest functional types that lose their leaves in autumn. In many cases, decreases in sulfur deposition have been linked directly to the degree of recovery of forest soils from acidification [20]. Accordingly, it can be assumed that because of larger stored S pools from higher sulfur deposition inputs, the recovery in conifers is significantly delayed. A long-term study at European beech and Norway spruce ecosystems in the Solling area, Germany, shows some indications of recovery of base cation to aluminum (Bc/Al) ratios in mineral soils in the beech site, whereas no recovery is observable in the spruce site [21].

The recovery of forest soils from acidification is an important topic for environmental policy with respect to future emission-reduction goals. For forest management, the planning of liming programs must be considered in view of natural regeneration. However, indications of the recovery of forest soils, including soil solution, from acidification are only sparse [20,22–24], whereas the recovery of surface waters is well documented [25–28]. In the studies of Berger et al. [29] and Reininger et al. [30], there were no indications of a recovery from acidification, or even an ongoing acidification process. Cools and De Vos [31] found a statistically significant change in the soil pH($CaCl_2$) and base saturation of European forest soils between 1994 and 2006. However, the detection of long-term trends was very different depending on the soil type, soil depth, and acid-base status of the soil.

The verification of changes from subsequent soil inventories is very difficult owing a high small-scale variability in the forest floor and the mineral soil [32–34], the low rate of change of soil properties, and the non-linearity of trends [35,36]. There are still some doubts as to whether repeated soil sampling is an efficient tool to distinguish between temporal and spatial variability [37]. Previously, soil changes were assumed to take place over timescales of centuries to millennia [22]. To detect such changes, chronosequences were frequently used to analyze a temporal development [38,39]. The high rate of soil changes after the strong increase of sulfur deposition between 1950 and 1970 offered the possibility to detect changes with repeated inventories, despite several uncertainties [21,40–42]. In contrast, the recovery of forest soils from acidification after strong reduction of the acid load appears to be delayed because of the gradual release of temporarily adsorbed sulfur from soil pools [29].

The evidence of the recovery of forest soils from acidification with repeated soil inventories at different sites also poses several statistical challenges, as most data are from observational studies rather than from factorial experiments. Due to the experimental design and the use of soil chemical response variables, the assumptions of the parametric statistical methods are not always fulfilled. Generalized additive mixed models (GAMMs) provide an approach that allows (1) the analysis of response variables with non-Gaussian distributions, such as base saturation; (2) the inclusion of random effects to account the "pseudo replicated" structure of the data (correlated errors among inventories on the same sampling site); (3) the consideration of non-linear processes; (4) to account for sampling heterogeneity across space and time; and (5) inconsistencies in the timing of soil inventories that could be addressed by modelling the within-year-variation as a covariate [43–45].

The objective of our study was to improve the detection and understanding of long-term changes of acidification of highly polluted forest soils in Lower Saxony, Germany. Specifically, we considered two main research questions. First, are there significant indications of forest soil recovery after three decades of reduced sulfur deposition? Second, are the trends of the indicators for soil acidification different for tree species owing to the lower atmospheric sulfur input in deciduous forest stands? The pH values, either measured in water (pH(H_2O)) or in calcium chloride solution (pH($CaCl_2$)) and base saturation, were used as indicators for the acid-base status of the forest soils in this study.

2. Materials and Methods

2.1. Study Sites

Included in this study were all 21 forest study plots (Figure 1) with terrestrial soils of the network of permanent soil monitoring plots in Lower Saxony (in German: BDF = Bodendauerbeobachtungsflächen [46]) existing since 1992. Some of the plots belong also to the ICP Forests Intensive Monitoring Programme Level II established under the UNECE Convention on Long-Range Transboundary Air Pollution [47].

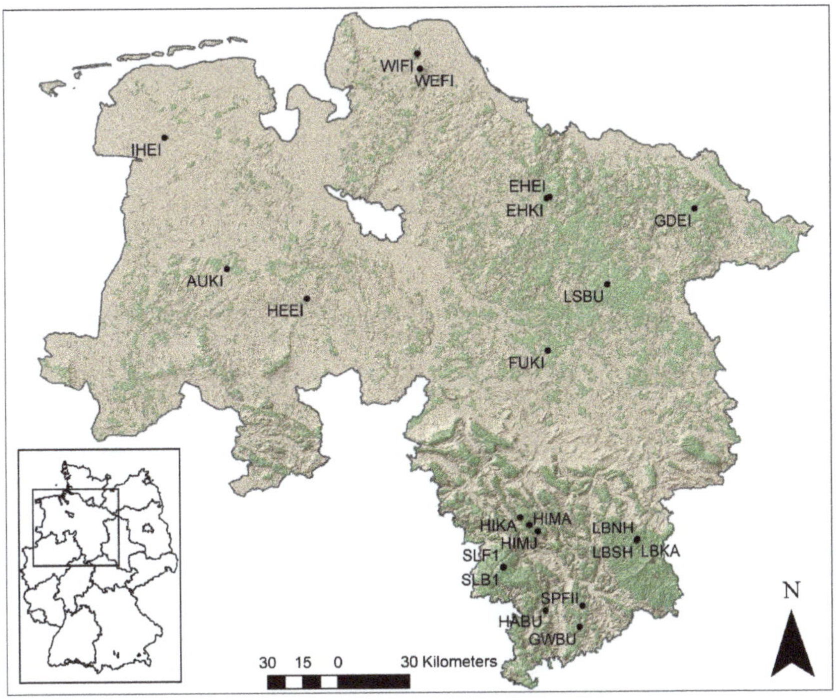

Figure 1. Location of 21 permanent soil-monitoring plots in Lower Saxony, Germany. The forested areas are represented with the color green. Abbreviations show the last four letters of the plot code (see Table 1).

The most frequent tree species is the Norway spruce (n = 10) (*Picea abies* (L.) H. Karst), followed by European beech (n = 4) (*Fagus sylvatica* L.), oak (n = 4) (*Quercus robur* L. and *Qu. petraea* (Matt.) Liebl.), and Scots pine (n = 3) (*Pinus sylvestris* L.). The sites are located at altitudes between 31 and 657 m a.s.l. Mean air temperature and mean annual precipitation (1981–2010) ranged from 6.4 to 9.7 °C and from 684 to 1443 mm, respectively. Additional information on the sites is provided in Table 1. The most frequent soil type is Podzol (n = 13), followed by Cambisol (n = 4), Luvisol (n = 2), Fluvisol (n = 1), and Planosol (n = 1).

Table 1. Location and plot characteristics for repeated soil-sampling plots in Lower Saxony.

Location	Plot Code	Lat [°]	Long [°]	MTS [-]	Alt [m]	Slope [°]	Aspect [°]	Stype [WRB]	MAT [°C]	MAP [mm]
Westerberg	F001WEFI	53.67	9.09	spruce	37	1.1	311	Podzol	8.7	903
Ehrhorn	F002EHEI	53.18	9.90	oak	110	2.9	337	Cambisol	8.5	843
	F012EHKI	53.17	9.88	pine	82	1.1	241	Podzol	8.8	826
Lüss	F003LSBU	52.84	10.17	beech	115	2.3	310	Podzol	8.5	859
Fuhrberg	F004FUKI	52.59	9.87	pine	37	0.1	107	Podzol	9.7	684
Lange Bramke	F005LBNH	51.86	10.42	spruce	613	15.8	19	Podzol	6.5	1453
	F022LBSH	51.86	10.41	spruce	603	23.0	151	Podzol	6.4	1446
	F023LBKA	51.86	10.42	spruce	656	8.1	109	Podzol	6.5	1453
Solling	F006SLB1	51.76	9.58	beech	506	1.9	207	Cambisol	7.0	1246
	F007SLF1	51.76	9.58	spruce	507	1.4	92	Podzol	7.0	1245
Harste	F008HABU	51.59	9.83	beech	250	5.5	82	Luvisol	8.5	774
Göttingen	F009GWBU	51.53	10.05	beech	421	0.3	246	Luvisol	7.5	897
Wingst	F010WIFI	53.73	9.07	spruce	34	3.7	102	Podzol	8.7	933
Ihlow	F011IHEI	53.41	7.45	oak	53	0.3	274	Podzol	9.4	846
Göhrde	F013GDEI	53.12	10.84	oak	97	1.9	127	Podzol	8.5	742
Heerenholz	F014HEEI	52.80	8.37	oak	48	0.3	204	Fluvisol	9.4	795
Hils	F016HIKA	51.95	9.69	spruce	424	24.3	253	Podzol	7.6	1196
	F017HIMA	51.92	9.74	spruce	253	6.3	244	Cambisol	8.2	1020
	F019HIMJ	51.90	9.79	spruce	317	14.0	26	Cambisol	7.8	1075
Spanbeck	F020SPFI	51.61	10.07	spruce	251	1.7	72	Planosol	8.6	830
Augustendorf	F021AUKI	52.91	7.86	pine	31	1.0	102	Podzol	9.4	843

Lat: latitude, Long: longitude; MTS: main tree species, Alt: altitude above sea level, Stype: soil type, MAT: mean air temperature, MAP: annual mean precipitation (reference period 1981–2010).

2.2. Estimation of Total Sulfur Deposition

Deposition assessments at some of the permanent monitoring plots were conducted according to the ICP Forests Manual on the sampling and analysis of deposition [48]. Fifteen continuously open bulk samplers were placed under the forest canopy. At plots with European beech, stemflow was assessed at a subset of the trees [48]. Samples were collected at least fortnightly, filtered, and stored in the dark at about 4 °C before being chemically analyzed.

As already mentioned in the introduction, we focused on sulfur deposition because the acidification of Europe's forest soils shows a limited response to the decrease in N deposition since the 1990s [16]. While sulfur deposition fluxes are only observed at a subset (9 of 21) of the permanent inventory plots, we estimated the modeled deposition fluxes of sulfur from the mapped atmospheric deposition data for Germany that were provided by the Federal Environment Agency (UBA) [49]. These data only cover the period from 2000–2015. To obtain the sulfur deposition time series from 1950 to 2020 for each monitoring site, we adapted temporal reconstruction methods. This was done with a simplified version of the model MAKEDEP by Alveteg et al. [50]. For the scaling of the sulfur deposition, we distinguished between marine and non-marine deposition. For sea spray, we assumed a constant deposition over time. For the non-marine proportion, we created a standard function for scaling. This function was based on the trend of sulfur deposition for Europe described by Engardt et al. [9]. This curve was adjusted using observed deposition data from sites in Lower Saxony. Lower Saxony received very high sulfur deposition inputs in some regions in the past, especially in the Solling and Harz mountains. More details on the methods for computing yearly deposition fluxes and the applied scaling function are given in Figure A1.

The quality of regionalized data and the reconstruction process with the same standard scaling function for all sites in Lower Saxony was evaluated at the sites with observed total S depositions (Figure 2). When comparing the observed and estimated deposition, both the site-specific level and the temporal development of the sulfur deposition is well represented with the estimated unified scaling function. An exception is the site F007SLF1 at Solling, where a systematic underestimation of the observed deposition is visible (Figure 2). Therefore, the function was adjusted by increasing the estimated deposition by a factor of 1.45 for this site (see dotted line in Figure 2).

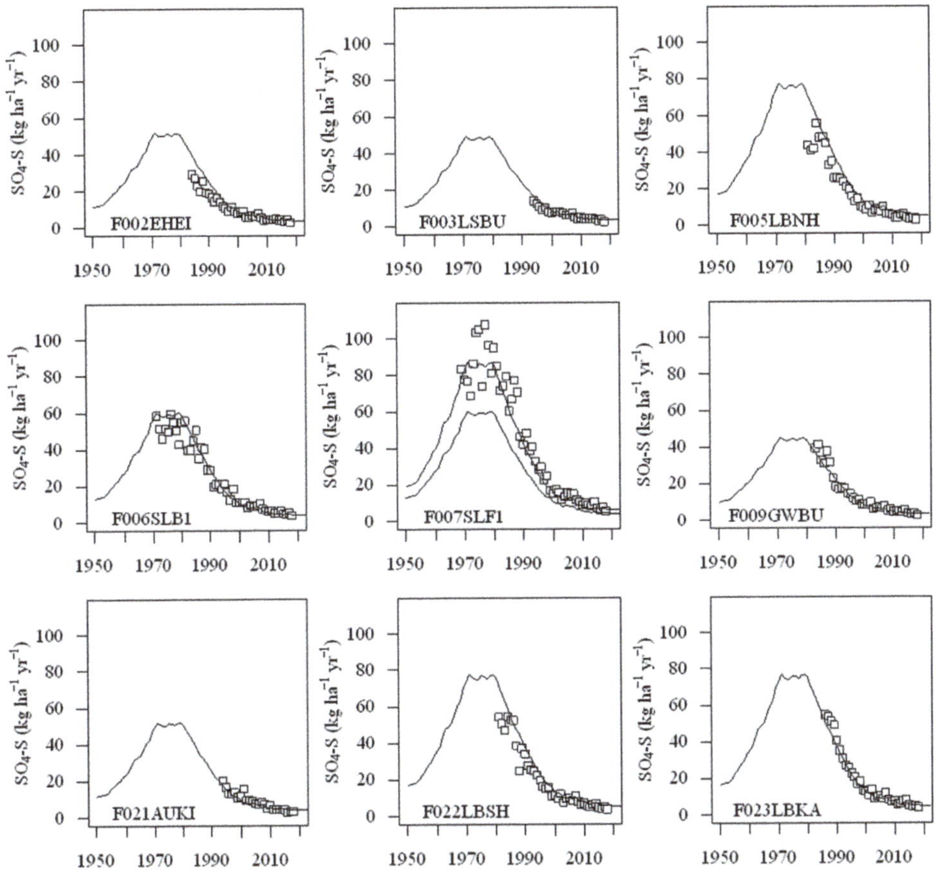

Figure 2. The time series of observed total sulfur deposition from throughfall measurements (□) at permanent soil-monitoring plots and the reconstructed time series of sulfur deposition for the period 1950 to 2020. The dotted line shows a site-specific recalibration of the observed data for the F007SLF1 plot.

2.3. Sampling Procedures and Chemical Analysis

The sampling of soils at the individual permanent soil-monitoring plots is conducted at intervals of no more than 10 years (Figure 3). At least three inventories are available for most plots (exceptions: F016HIKA, F017HIMA, and F010WIFI). The sampling is conducted in an alternating system with about two inventories each year. The sampling-site design, as well as the sampling procedures, chemical analyses, and quality checks, are documented by Barth et al. [51] and in the ICP Forests Manual on Sampling and Analysis of Soil [52]. Accordingly, soil samples were taken at 24 locations within the plot, which were aggregated to six composite samples per plot and depth interval. The inventories following this design are described by black circles in Figure 3. The organic layer was divided into a litter (L), a moderately decomposed (Of), and a highly decomposed (Oh) layer, if present. The mineral soil was sampled at fixed depths of 0–5 cm, 5–10 cm, 10–20 cm, and 20–30 cm, and at a maximum 20 cm interval for deeper soil layers. The soil samples were analyzed for their content of the elements carbon (C), N, and phosphorus, for pH, and for their effective cation exchange capacity (CEC). Exchangeable cations (Na^+, K^+, Mg^{2+},Ca^{2+}, Al^{3+}, Fe^{3+}, Mn^{2+}, and H^+) were determined after percolating the sieved (<2 mm) soil samples with NH_4Cl and the cations in the percolate were subsequently determined using ICP methods and pH measurements for H^+ [53–55]. CEC was calculated as the sum of the cation equivalents [56]. Base saturation (BS) was calculated as a percentage of the exchangeable base cations (Na^+, K^+, Ca^{2+} and Mg^{2+}) from the CEC. For the determination of pH, samples were mixed with a volume ratio of sample to solution of 1:5 with H_2O (pH(H_2O)) or 0.01 M $CaCl_2$ solution (pH($CaCl_2$)) and pH was determined with a glass electrode. The pH values and base saturation were used as indicators for the acid-base status of the soil. Sample preparation and analysis followed standard procedures [57].

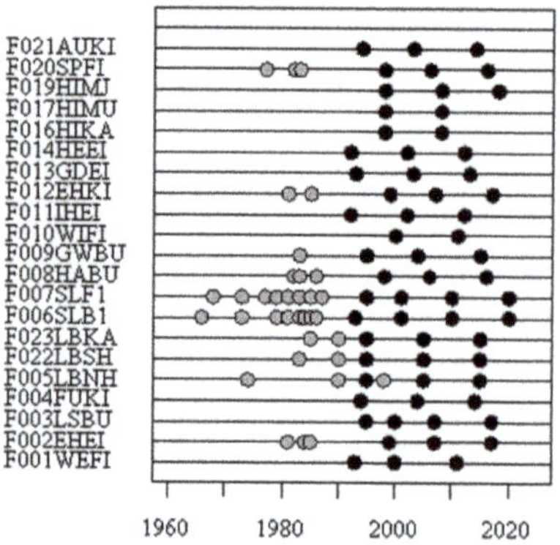

Figure 3. Sampling years of soil inventories at the permanent soil monitoring plots in Lower Saxony, Germany. (Black cycles: sampling design complies with the ICP Forests specifications; grey cycles: inventories with partly deviating sampling design, but comparable analytical methods).

For most plots, data from soil inventories from prior to the start of the permanent soil monitoring programme (BDF programme) are available, which allows for the extension of the number of replicates and the study period [14,21,58]. The analytical methods are comparable to the methods used in more recent inventories. The sampling design, however, differs with respect to the number of replicates and sampling depth. These inventories are marked with grey circles in Figure 3. Despite the methodological differences, these data bear important information with respect to the long-term dynamics of the acid-base status.

2.4. Data Handling

The average *CEC*, BS, and pH values for the different depth ranges (*d*) of 0–30 cm, 0–50 cm, and 0–100 cm were calculated as follows for *CEC*:

$$CEC_d = \frac{1}{z_d \cdot BD_d} \sum_{l=1}^{n} z_l \cdot BD_l \cdot CEC_l \quad (1)$$

for BS:

$$BS_d = \frac{1}{z_d \cdot BD_d \cdot CEC_d} \sum_{l=1}^{n} z_l \cdot BD_l \cdot CEC_l \cdot BS_l \quad (2)$$

and for pH, the aggregation was done on the basis of the H^+ concentration.

The total thickness (depth ranges) d is given by:

$$z_d = \sum_{l=1}^{n} z_l \tag{3}$$

and bulk density (*BD*) by:

$$BD_d = \frac{1}{d}\sum_{l=1}^{n} z_l \cdot BD_l \tag{4}$$

where z = thickness and l = soil layer.

The plots were stratified into sub-groups to enable a more detailed analysis, as well as to consider the different soil chemical processes (such as chemical weathering rates or cation leaching) and other general conditions (such as forest type), as shown in Figure 4.

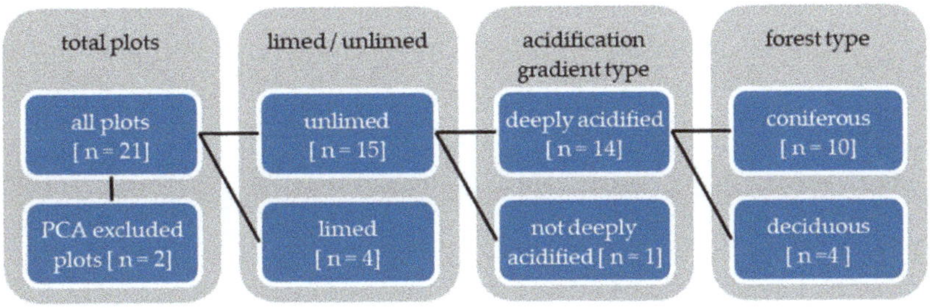

Figure 4. Stratification of the inventory plots into functional groups for detailed statistical analysis. Further details on the definition of the "vertical acidification gradient type" can be found in Hartmann and von Wilpert [59] and in Section 3.2. PCA = principal component analysis.

2.5. Derivation of Meteorological Data

For a climatic characterization of the study plots (see Table 1), observational data of the German Meteorological Service (Deutscher Wetterdienst, DWD) were used (Table 1). The regionalization of the data from the climate and precipitation stations of the DWD to the soil-monitoring sites was performed using the methods described in Dietrich et al. [60] for the standard period from 1981 to 2010.

2.6. Statistical Analysis

R 4.0.3 software (R Development Core Team 2020) was used for the statistical analyses. Depth profile plots and statistics are generated with the package 'aqp': algorithms for quantitative pedology [61].

2.6.1. Detection of "Atypical" Plots Using Principal Component Analysis (PCA)

If some plots are "atypical" (belonging to different clusters) compared to most of the other monitoring plots studied, they tend to bias the interpretation and conclusions of the analysis [62]. PCA can be regarded as a classification procedure and was thus used to detect "atypical" study plots. A given plot belongs to one of the following classes: (a) non-atypical, or (b) atypical [63]. The PCA was performed with the package 'FactoMineR' [64] and the results are displayed using the 'factoextra' package. In addition to the soil chemical acidification indicators (pH, base saturation), the cation exchange capacity (CEC) was also considered as a variable for the PCA. All depth ranges (0–30; 0–50; and 0–100) were used in one analysis to include the effects of the vertical differentiation of the parameters in the soil profiles. With one exception, for all plots the most recent soil inventory was used. At the site F009GWBU, we used the results from the inventory in the year 2004. During this inventory, much greater profile depths were achieved, which makes it easier to compare this site with the others.

2.6.2. Mixed-Effects Models

We used generalized additive mixed models (GAMM) to examine the relationship between the response and the inventory year. Our data consist of y_{it}, the base saturation or pH values measured at site i at year t for i = 1, . . . ,21 and t = 1966, . . . ,2020. Following Knape [45], we separated the smooth trend components from the random among-year variations using the following structure of a mixed effects model:

$$y_{it} \sim \text{Gamma}(\exp(a + S(t) + \varepsilon_t + s_i)),\ \varepsilon_t \sim N\left(0, \sigma^2\right), \tag{5}$$

where a is an intercept, S(t) is a smooth function of the year representing nonlinear changes in base saturation or pH values, ε_t is a random effect for the year, and s_i is a site effect to account for the among-site variations. With this formulation the temporal change is described as a combination of S(t) and ε_t.

The above formulated model structure from Knape [45] was developed for the count data of populations that often have Poisson-distributed sampling errors. Therefore, a gamma distribution with a log link was used to account for the properties of the soil's chemical variables. We performed checks for the approximate normality of random effects and heteroscedasticity (using the functions 'checkfit' and 'gam.check' [45,65]; for examples, see Figures A2–A4). For assessing unusual observations (outliers), we performed statistical outlier detection with the residual plots and the Grubbs test calculation [66] using the package 'outliers'.

An important choice is the selection of the parameter that determines the flexibility of the smooth functions, since a non-limited smooth component would be able to capture all variations of soil sampling between the different years [45]. If the flexibility is too high there is a risk of overfitting and if it is too low a part of the variation in the data may not be captured well. With an appropriate selection of the degree of smoothing, the S(t) component can be interpreted as the long-term trend while the random effect is capturing the short-term variation, e.g., due to sampling variation/uncertainty. Accordingly, the exact interpretation of the short-term and long-term components depends on the smoothing degree of the long-term component and on how the temporal random effect is modeled. In most cases, we used the automatic selection of smoothing parameters. For subgroups with small sample sizes (e.g., limed sites), it was necessary to reduce the degree of smoothing to avoid temporal overfitting. Standard software to parameterize this type of model is available from the R library 'poptrend' [45]. A major advantage of the function 'ptrend', included in the library 'poptrend', is that significant increases and decreases in the trend could be visualized for different periods of the fitted curve. In conventionally parameterized GAMMs, only the significance of the overall smoothing function can be assessed. More details on methods for computing confidence intervals for the trend estimates and changes are given elsewhere (Figure A1 in [45]).

Uncertainty estimates are an important part of the communication of the trends in soil data. In the function 'ptrend', uncertainty estimates were performed with simulations using parametric bootstrapping based on assuming the asymptotic normality of the estimates [67]. This approach considerably reduced the computational effort relative to fitting GAMs to every bootstrap resample [67], but the cost is that the confidence intervals produced in this way do not account for any uncertainty in the selection of the degree of smoothing [45].

For visualization, the long-term trend (S(t)) is standardized. As default in 'ptrend', the long-term trend is standardized with respect to the predicted base saturation or pH value at the first monitoring year. We used for standardization the average predicted base saturation or pH value (*ciBase* = *mean*) at site i, ignoring the temporal random effects. This has an advantage over the default standardization in that it is less affected by uncertainty in the observed values from the first monitoring year [45].

3. Results

3.1. Site-Specific Load and Reduction of Sulfur Deposition

For the characterization of the site-specific sulfur deposition load, the estimated annual rates were cumulated from 1950 to 2020 (Figure 5). For the 21 plots considered in this study, the accumulated sulfur deposition in the period between 1950 and 2020 ranged from 1319 to 2819 Mg ha^{-1} with a mean of 1813 Mg ha^{-1} and a median of 1670 Mg ha^{-1}. The plot with the maximum deposition load (2819 Mg ha^{-1}) is a spruce stand in the Solling area (F007SLF1). The adjacent beech stand (F006SLB1) received an accumulated deposition load of 1936 Mg ha^{-1} of sulfur.

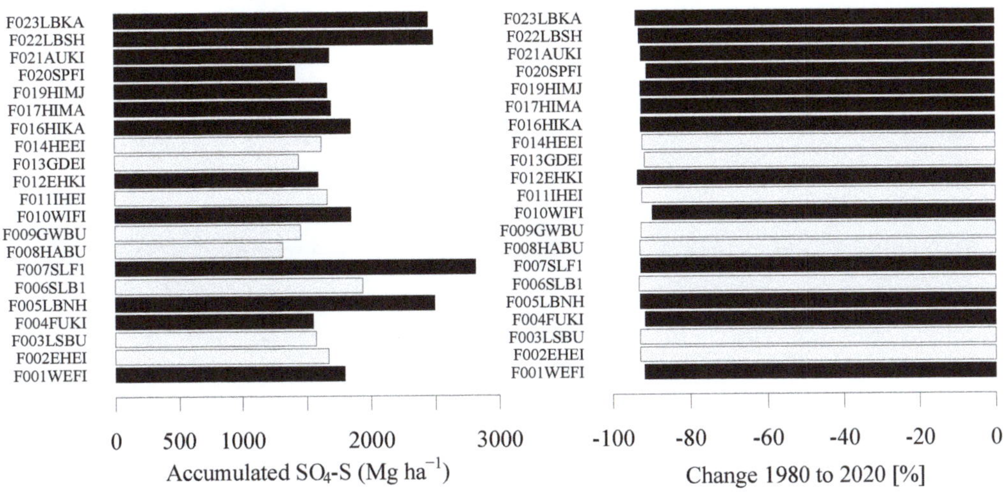

Figure 5. Estimated cumulative sulfur load (Σ1950 to 2020) and relative change of sulfur deposition between 1980 and 2020 for the permanent soil-monitoring plots in Lower Saxony, Germany. ■ coniferous forest; □ deciduous forest.

Due to the use of a uniform scaling function in the deposition reconstruction procedure, the relative reduction of the deposition at the individual plots turns out to be similar. The relative changes of the sulfur deposition between 1980 and 2020 ranged from −89 to −94% with a mean of −92%. The lower relative decreases relate to the plots near the coast with a higher sulfur deposition of marine origin.

3.2. Soil Chemical Status at the Time of the Last Soil Inventory

At most plots, observed pH($CaCl_2$) is in the acidic range between pH($CaCl_2$) 3 and 5 (Figure 6). In the topsoil, pH($CaCl_2$) is even below 3 at some plots. These soils are predominantly deeply acidified; only at a few plots can higher pH values be found at greater depths (> 70 cm). In contrast, one plot on carbonate bedrock (F009GWBU) shows pH values well above 7 below about 20 cm depth. Forest soils with low base saturation (<20%) in the entire soil profile were observed at most plots, at mountain sites with base-poor silicate bedrock, and at unconsolidated sandy substrates in the lowlands (Figure 7). To characterize the acid-base status of forest soils, the base saturation can be classified in ecologically relevant groups of vertical gradients [59,68]. Here, the statistically defined six types by Hartmann and Wilpert [59] were used as a classification scheme.

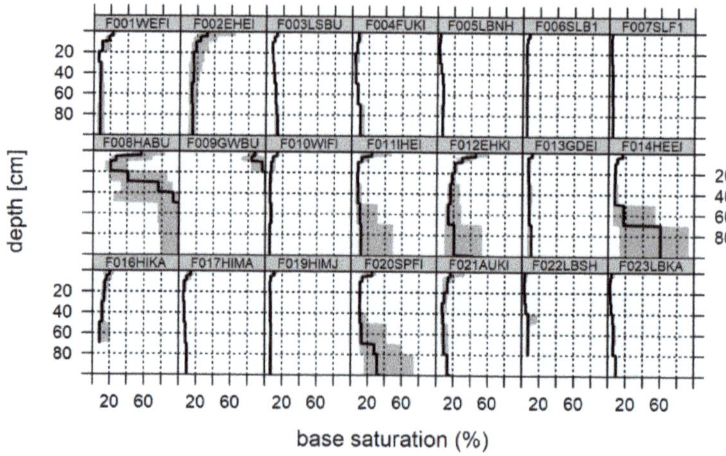

Figure 6. Depth profiles of pH($CaCl_2$) in the mineral soil for the most recent soil inventory (cf. Figure 3) at the permanent soil monitoring plots in Lower Saxony. The solid line describes the median of six composite samples taken at 24 locations and the grey areas the error range (5th and 95th percentile).

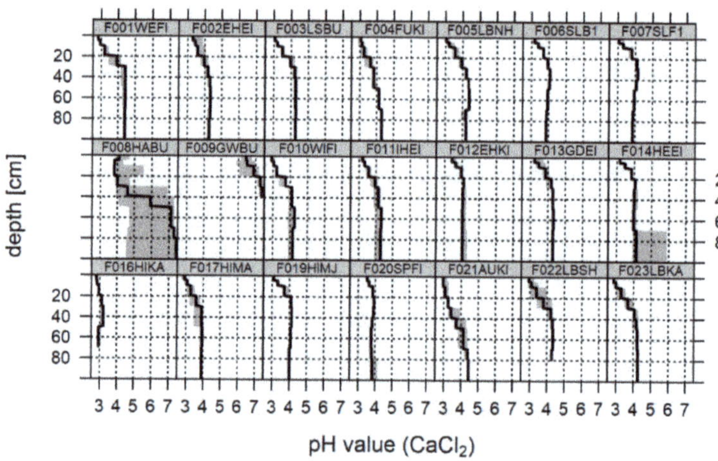

Figure 7. Depth profiles of base saturation in the mineral soil for the most recent soil inventory (cf. Figure 3) at the permanent study plots in Lower Saxony. The solid line describes the median of six composite samples taken at 24 locations and the grey areas the error range (5th and 95th percentile).

The groups 1, 2, and 3 are defined as soils with 100% base saturation in the (near) subsoil and very high (group 1—F009GWBU), high (group 2—F008HABU) or low (group 3) base saturation values in the topsoil. These sites are mainly found on limestone and some unconsolidated carbonate sediments. Group 4 is distinctly acidified in the main rooting zone with a high base saturation in the subsoil (F012EHKI, F014HEEI, and F020SPFI). All other plots can be assigned to group 5. The soils from this group are deeply acidified with very low base saturation values over the entire soil profile. Group 6 shows an increase of base saturation in the topsoil as a result of liming. The elevated base saturation in the upper 10 cm of the plot F012EHKI is typical for group 6. However, this plot was assigned to group 4 because of a high base saturation in the subsoil. Plots F011IHEI, F014HEEI, and F021AUKI were limed in the 1980s. However, the effects of liming were no longer evident at the time of the last soil inventory. Accordingly, these limed plots were assigned to group 5, but not included in the comparison of forest types.

3.3. Statistical Detection of "Atypical" Monitoring Sites (Main Cluster Groups)

According to the principal component analysis of the soil chemical variables (pH, base saturation, and CEC) at the most recent soil inventory, three clusters of the permanent soil monitoring plots can be identified. Two clusters are represented by only one plot each, namely F008HABU and F009GWBU, respectively, whereas the third cluster comprise all other plots (Figure 8). According to this result, F008HABU and F009GWBU were excluded from the following analysis of the changes in soil pH and base saturation.

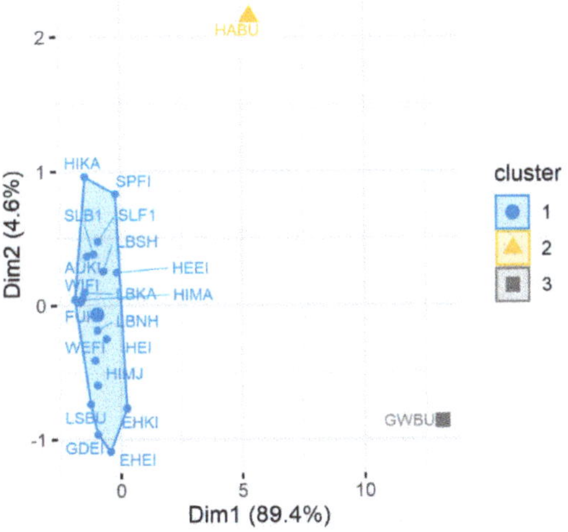

Figure 8. First two components of a principal component analysis of soil chemical variables at the most recent (Figure 3) soil inventory and designation of permanent soil monitoring plots to clusters 1 to 3. Clusters 2 and 3 consist of only one plot each. Abbreviations show the last four letters of the plot code (see Table 1).

3.4. Changes of the Acid-Base Status and Indications of Recovery

If all plots are evaluated, a significant decrease in base saturation up to about 1990 for all evaluation depths considered is visible (Figure 9). Neither significant increases nor decreases can be detected after 1990. The plots F021AUKI, F014HEEI, F011IHEI, and F012EHKI were limed between 1985 and 1992. At these plots, a significant increase in base saturation for all depth intervals occurred after liming (Figure 9; 'limed plots'). The unlimed plots showed no tendency toward recovery. Only if the deeper soil layers are included (0–100), a slight, but mostly insignificant increase of base saturation can be recognized. To delineate site and tree species effects, only the plots of base saturation group 5 were considered in the following evaluation (Figure 9; 'group 5'; 'group 5 coniferous'; 'group 5 deciduous'). All plots included in the tree species comparison have similar soil acid-base status on average. The depth-weighted average base saturations for the depth range 0–100 cm in the subgroups 'group 5', 'group 5 coniferous', and 'group 5 deciduous' are 7%, 7%, and 8%, respectively.

For the same groups, the depth-weighted average pH(H_2O) values are 4.4, 4.4, and 4.5, respectively, and the pH($CaCl_2$) values are 4.0, 4.0, and 4.0, respectively. The effective cation exchange capacities are 402 $kmol_c\ ha^{-1}$, 422 $kmol_c\ ha^{-1}$, and 371 $kmol_c\ ha^{-1}$, respectively.

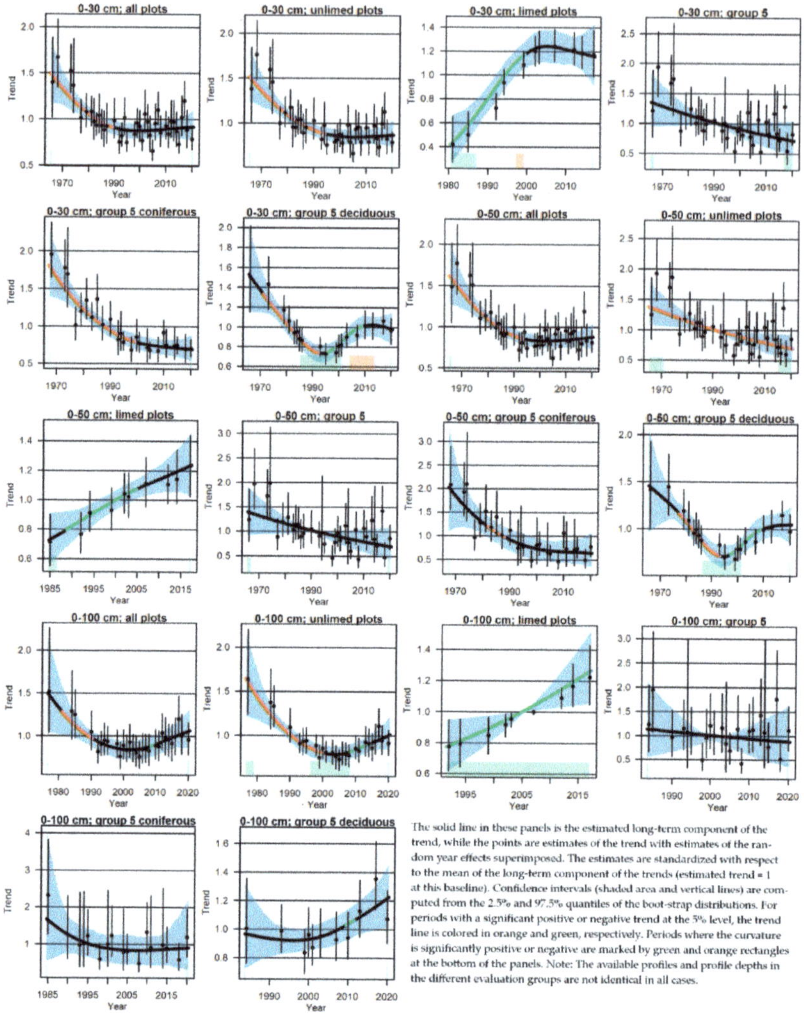

Figure 9. Estimated trends for the base saturation in mineral soil for different aggregation depths (0–30 cm, 0–50 cm, and 0–100 cm).

The difference in the dynamics between coniferous and deciduous forest sites is evident. At the deciduous forest sites, a period from 1990 to 2000 shows a significant positive curvature for the depth ranges 0–30 cm and 0–50 cm (Figure 9). A reversal of the decrease after 1990, with a significant increase of base saturation since 2010, is only visible for these two depth ranges. The lack of significance in the 0–100 cm depth range might be owing to the lower number of soil samples in this depth range. In contrast, a decrease of base saturation at coniferous plots continued until about 2000 in all depth ranges. In the subsequent period, the base saturation remained at a low level (<10%; Figure 7).

Observed pH(H_2O) is available from the early 1990s. As with base saturation, there has been a significant increase in pH(H_2O) at the plots with liming measures between 1985 and 1992 for all three depth ranges (Figure 10). There are indications of recovery at the unlimed plots between 2000 and 2010 (consistently significant for 0–50 cm and 0–100 cm). After this period, there appears to be a stabilization or even a slight decrease of the pH values.

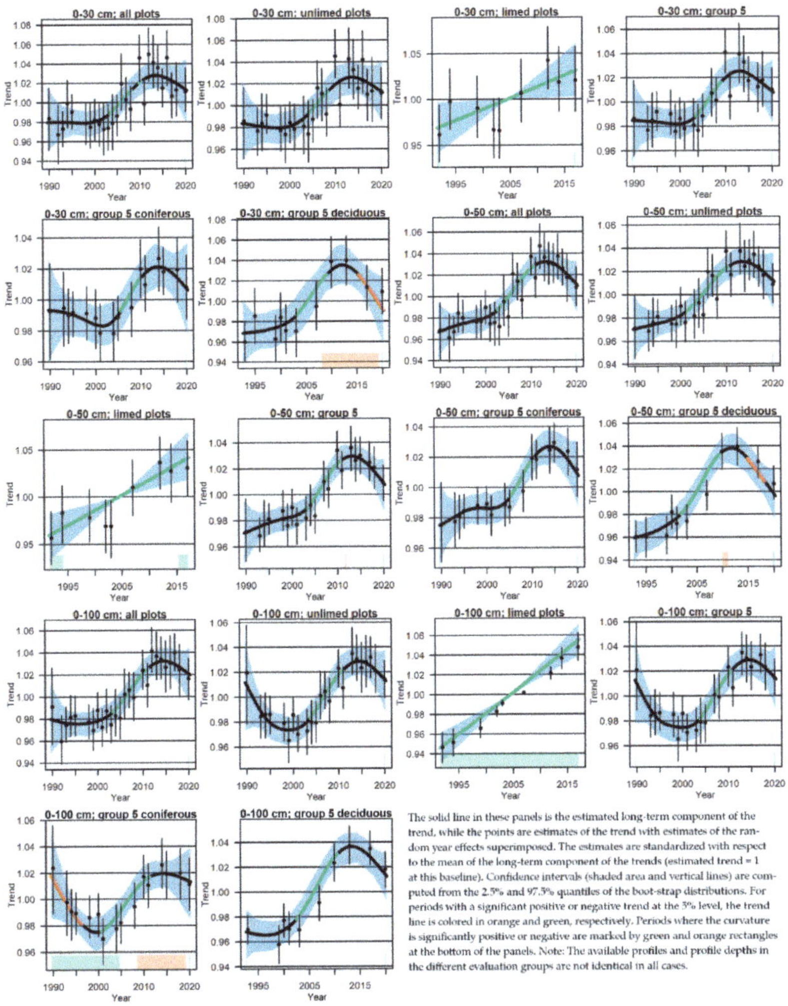

Figure 10. Estimated trends for pH(H_2O) values in mineral soil for different aggregation depths (0–30 cm, 0–50 cm, and 0–100 cm).

As compared to the deciduous forest plots, the recovery of pH(H_2O) was delayed at the coniferous plots, while the magnitude of the recovery is considerably more pronounced at the deciduous forests. This is most evident for the 0–50 cm depth interval. A decrease after 2010 should be interpreted with caution, as there are only inventories from two plots.

In contrast, the temporal patterns of pH($CaCl_2$) are much more difficult to describe (Figure 11). The effects of liming do not appear to be as pronounced as for base saturation and pH(H_2O), especially in the top 30 cm. However, for the other two depth ranges (0–50 cm and 0–100 cm), there is a significant increase in pH($CaCl_2$) after liming in the years 1985 and 1992. The temporal dynamics of pH($CaCl_2$) show a very striking sinusoidal structure for the groups 'all plots' and 'unlimed plots'. These results show that interpretations are difficult when very different groups (soil types, parent material, and forest types) are analyzed together. When deciduous and coniferous plots from the deeply acidified group 5 are evaluated separately, this structure is no longer apparent. For pH($CaCl_2$), there is no significant increase in the topsoil (depth ranges 0–30 cm and 0–50 cm), neither at the deciduous nor coniferous plots. If the deeper soil horizons are included, there is a significant increase in the pH($CaCl_2$) values, especially in the period from 2000 to 2010 for all aggregation levels of the subgroups ('all plots', 'unlimed plot', 'group 5', 'group 5 coniferous', and 'group 5 deciduous'). This is particularly interesting as sulfate deposition declined steeply until the year 2000 (Figure 2), but at a much slower rate afterward.

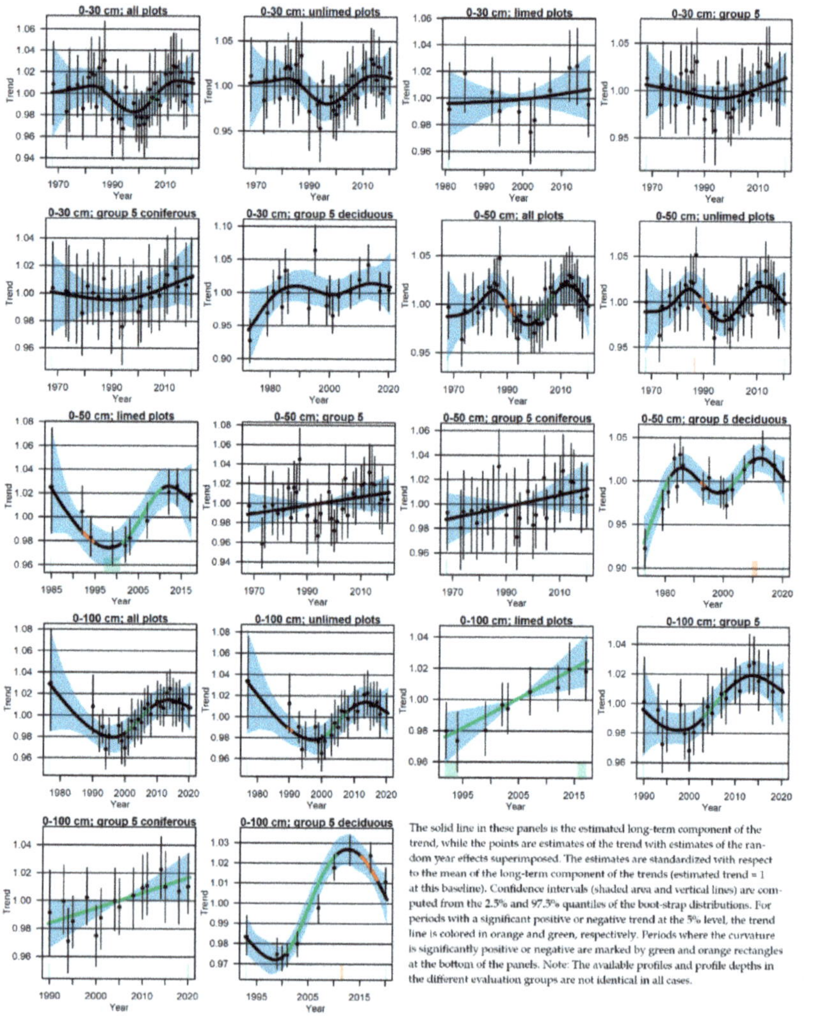

Figure 11. Estimated trends for pH($CaCl_2$) values in mineral soil for different aggregation depths (0–30 cm, 0–50 cm, and 0–100 cm).

4. Discussion

4.1. Site-Specific Sulfur Deposition Time-Series

The approach for quantifying total sulfur deposition for the years 2000 to 2015 with an emissions-based method is described in detail in Schaap et al. [49]. In short, three major calculation steps are conducted in this model: (1) the chemical transport model LOTOS-EUROS [69,70] is used to calculate dry deposition as a product of modeled ambient air concentration fields of S species and modeled deposition velocities. (2) In the next step, modeled rainwater concentrations from the LOTOS-EUROS model are used in combination with about 100 monitoring stations of precipitation chemistry in Germany. These data serve to adjust the modeled rainwater concentration distribution from the LOTOS-EUROS model using residual kriging. The generated concentration field is multiplied with high-resolution precipitation data (1 × 1 km). (3) Occult deposition is calculated from fog water concentrations. In such a model chain, there are numerous sources of uncertainties [9,71,72]. An example of the uncertainties caused by the spatial resolution of 1 × 1 km are the two sites at the Solling area (F006SLB1 and F007SLF1). While the deposition in the beech stand is well described by the modeling, considerable deviations from the observed deposition are shown for the neighboring spruce stand.

A further uncertainty arises from the reconstruction of the deposition with uniform deposition scaling functions for all sites. Despite these two main uncertainties, with the exception of site F007SLF1, there is a good agreement with observed deposition fluxes, both in terms of absolute magnitude and temporal dynamics (Figure 2). The highest S deposition rates were observed and estimated for coniferous forests (Figures 2 and 5).

The estimated S deposition for the soil inventory sites peaked in the early 1980s with maximum deposition rates of over 80 kg S ha^{-1} a^{-1}. This is comparable with other European studies [73].

4.2. Soil Chemical Status, Sampling Design and Statistical Approach

A high proportion of deeply acidified soils in our study documents the strong acidification of forest soils in Lower Saxony. Very low pH values indicate a historically high acid load and the exhausted buffer capacities of the mineral soil. The second National Forest Soil Inventory (NFSI II) in Lower Saxony revealed a base saturation in the main rooting zone (10–30 cm) of below 20% at 65% of the forest area [74]. A base saturation of at least 15–20% is regarded as a critical limit for vital growth and the sufficient regeneration ability of most tree species [56,75]. If base saturation falls below this critical limit, a significant increase of toxic Al^{3+} ions and an increasing mobilization of heavy metals into the soil solution can be suspected. While "natural" soil acidification from carbonic or organic acids is usually limited to the topsoil, most plots in this study experienced deep acidification, resulting from the transport of deposition-borne mobile anions such as sulfate and nitrate into deeper soil layers.

The detection of changes in the acid-base status of forest soils using repeated soil inventories necessitates a high accuracy of the inventories and sufficient changes of soil chemical variables between the inventories. Therefore, a high reproducibility of the sampling design and the comparability of the methods used for the repeated inventories is required [35,36,76]. Our data are arranged in a matrix of observations, in which some are missing because of the experimental design (Figure 3). Braun et al. [77] used linear mixed-effect models (LMM) with the plot as a random effect and the year as a fixed effect to analyze time trends in the soil solution. The inclusion of the plot as a random effect is necessary to account for the "pseudo-replicated" data structure (correlation among measurements on the same plot). Major challenges arise because the soil's chemical changes are often nonlinear and non-synchronous in the different soil horizons [31,36]. Additionally, soil changes are usually very slow and may be superimposed on the short-term and seasonal fluctuations [36]. Especially in forest soils, large spatial heterogeneities can contribute to a reduced detectability of soil changes [36,78]. In particular, the spatial variability of exchangeable base cations contents, which are needed for the calculation of base saturation may be very high [36,79], especially if the different depth intervals are considered [80].

The error ranges of the GAMMs reveal that the plot-scale variability is sufficiently captured given the sampling design (Figures 6 and 7). Since 1992, all study plots were sampled with an equal number of sampling points. For a few plots, the error range (in deeper soil layers) suggests that a higher number of replicates might be advisable, or even necessary. Furthermore, the ability to detect soil changes increases with an increased number of plots and sampling dates at the different plots. In our study, many plots have a very high number of sampling dates over a period of 30 to 50 years (the median for all sites are four inventories; min = 2 and max = 12). An increasing number of sampling dates for a site can provide a more accurate assessment of soil changes over time. Mobley et al. [36] concluded that the sampling scheme must incorporate more sampling dates to better capture nonlinear structures in the dynamics of soil variables. We assume that in our study this advantage outweighs the disadvantages of a slightly different sampling design (lower number of repetitions) before 1992 (Figure 3). Soil inventories at the study plots prior to 1992 mainly document an advancing acidification process under high loads of acidic deposition. The course of a potential recovery from acidification because of a reduction of atmospheric acidic deposition during the period 1992–2020 is consistently substantiated through soil inventories with the same number of replicates in approximately 10-year intervals. During the whole study period, sampling procedures and analytical methods are checked by rigorous QA/QC routines, including the analyses of several hundred samples with new and old methods, in case a new analytical method was introduced, to secure the conformity of the methods [81].

The analysis of response variables with non-Gaussian distributions, the "pseudo replicated" structure of the inventories at specific sites, the consideration of non-linear change processes, the sampling heterogeneity over space and time, and the non-equidistance in the timing of the soil inventories, has been addressed in this study using GAMMs [43–45]. Knape [45] showed that the inclusion of temporal random effects in the estimation of smooth trends makes it possible to separate long-term changes from short-term fluctuations. In our study, we assume that short-term fluctuations are mainly caused by uncertainties in the observation of a site-specific representative mean of the different acidification indicators. The GAMM framework published by Knape [45] was developed for modeling trends in the count data of populations. Changing the response distribution to Gamma, the underlying pitfalls in modeling count data and long-term soil chemical variables are very similar. Soil data, as well as count data, could have a high variation in sampling effort and detectability. Therefore, a very flexible but robust statistical approach is needed. Such an approach is also a "classic GAMM". The main advance of the package 'poptrend' is the visual interpretability of the long-term recovery trend. A distinction can be made between periods with a significant increase, a significant decrease, and/or a stagnation of acidification indicators. A direct implementation of sulfur deposition in the modeling approach is a much greater challenge because the highest sulfur output occurs many years later than the occurrence of the highest sulfur deposition loads [21].

4.3. Change in Soil Chemical Variables and Indications of Recovery

After cluster analysis, two of our sampling sites were excluded from the study, as the soil chemical status was very different from the other sites and even from each other. This was also done against the following background: Kirk et al. [82] stated that in soils in the carbonate or silicate buffer range the pH values do not necessarily increase after a decline in deposition acidity [4]. A strong increase in pH values is more likely in the aluminum buffer range, where the dissolution of Al silicates, the destruction of clay minerals, and the protolysis of Al hydroxides are the primary soil chemical processes. Accordingly, Kirk et al. [82] found the largest increase in soils with low initial pH(H_2O). In addition, other studies found a recovery in the most acidic forest soils (pH($CaCl_2$) $\leq$ 4.0 or BS $\leq$ 20%) [31,83].

Between 1970 and 1990, the pH($CaCal_2$) values and base saturation decreased in the soils of the studied permanent soil-monitoring plots. Hazlett et al. [20] summarized in their introduction results from numerous soil sampling studies across Europe and eastern North America. These resampling studies confirmed a decrease in the soil pH and base saturation in many forest soils for this period. Since the early 1980s, there has been a drastic reduction of the deposited sulfur and acidity in Europe and North America (cf. Figures 3 and 5). The comparison of 21 plots in Lower Saxony in this study showed a recovery from acidification indicated by an increase of soil pH and base saturation using repetitions of forest soil inventories. Particularly, the soil data from the Solling plots

(F006SLB1; F007SLF1) belong to the longest time series of repeated soil inventories in forest ecosystems worldwide. The observed and estimated sulfur deposition for the study plots shows a very different absolute reduction and cumulative load of sulfur deposition in the past (Figures 3 and 5). This is of great importance because some studies indicate that the intensity of the decrease in sulfur deposition could be linked directly to the degree of recovery [20].

Considering the group 'all plots', the base saturation for all aggregation depths (0–30 cm, 0–50 cm, and 0–100 cm) predominantly show a very slight increase, but the increase is not significant (Figure 9). Despite the considerable reductions in sulfur deposition in Lower Saxony (Figure 3), the possible recovery from soil acidification appears to be very slow. Meesenburg et al. [14] found for German NFSI plots that unlimed acid-sensitive sites experienced an ongoing acidification of the subsoil with corresponding losses of base cations. At liming trials in southwestern Germany, changes in untreated (unlimed) plots were only marginal and the soils remained highly acidic [84]. Major causes for a delayed recovery of forest soils from acidification are the still-substantial deposition of nitrogen species [1,16] with subsequent generation of acidity through uptake and nitrification processes, as well as the remobilization of previously retained sulfur over a longer period [29]. After a period of N accumulation with increasing nitrogen stocks, forest ecosystems may become N saturated, resulting in an increasing risk of leaching nitrate and base cations into surface waters [2]. In addition, changes in sulfur deposition were identified as key drivers in carbon stabilization and nitrogen leaching risks [3,85]. These processes, although seemingly linked to recovery, result in a significantly delayed recovery or even the further acidification of soils. Another reason for a very weak recovery is the simultaneous decrease in base cation deposition [86,87]. In our study, a trend reversal (insignificant) or stabilization at a low level can be observed at the depth intervals 0–30 cm and 0–100 cm. However, at the depth interval 0–50 cm an ongoing acidification seems to be evident. In the German NFSI, those sites, which were limed and are stocked with deciduous tree species, showed indications of recovery [14]. If the limed plots are examined individually in our study, there is a significant increase in base saturation after liming. It is well-known from numerous studies [14,84,88–90] that limed forest sites experience a significant increase in base saturation and pH values (especially in the topsoil). For example, Guckland et al. [88] found an 11% increase of base saturation up to a 40 cm depth. The German NFSI also found an increase of base saturation in the upper 30 cm of the limed plots [14].

All plots with vertical base saturation gradient type 5 were analyzed separately to reveal the potential effects of the tree functional groups (deciduous versus coniferous stands). While a recovery cannot be detected for the plots with conifers, the deciduous plots show a significant increase in base saturation after about 1995. The tree functional group-specific differences diminish with increasing soil depth. Figure 5 shows that the sulfur deposition load in the past was lower for deciduous than for coniferous forests. Accordingly, coniferous forest plots have a higher potential for the remobilization of temporarily stored sulfur. However, it should be noted that some of the coniferous permanent monitoring plots are located in highly polluted regions (Section 4.1). Therefore, the described tree functional group effect may to a certain extent be confounded by a plot-specific deposition load. Nevertheless, for the plots F006SLB1 and F007SLF1, it was demonstrated that the tree functional group effect is indeed decisive for the deposition input, retention of elements, and output fluxes [21]. However, it should be noted that this effect could not be solely because of lower deposition at deciduous stands compared to conifers. There are also many other factors and processes in the nutrient cycling and the related organic carbon dynamics of forest ecosystems that might explain differences in acidification dynamics between the forest function groups (conifers and deciduous) [22,91]. For example, litterfall, fine root turnover and decomposition, nitrogen retention, and depth and distribution of the rooting system may possibly differ significantly between the groups. Deep-rooting deciduous tree species [92] can enhance the zone, where mineral weathering contributes to the cycling of base cations, and a higher base cation content of deciduous litter [93,94] may also have an effect on the replenishment of the cation exchange sites with base cations.

A consistent significant increase of pH($CaCl_2$) can only be observed in acidification depth gradient group 5 (Figure 11). Cools and De Vos [31] found that the pH($CaCl_2$) significantly increased at plots with very acidic forest soils (pH($CaCl_2$). For forest soils with pH($CaCl_2$) above 4.0, they found a further decrease. It should be noted here that there are very few plots with pH values above 4.0 in our study. When analyzing the different dynamics of soil recovery of deciduous and coniferous forests, the mean pH($CaCl_2$) values of the selected (see Figure 4 and Section 3.2) deciduous and coniferous plots are both 4.0.

The pH(H_2O) shows a general increase after the drastic reduction of sulfur inputs since the early 1980s. However, the recovery takes place with a strong delay. In our study, the increase of pH(H_2O) clearly appears later than the decrease in sulfur deposition. These results confirm other studies, according to which a significant delay can be assumed [20,22,82]. In our study, this time lag appears to be longer in the coniferous forests than in the deciduous ones. Watmough et al. [95] point out that the release of formerly stored sulfur delays the recovery from acidification. Other studies also show that a considerable proportion of formerly deposited sulfur is temporally stored in organic sulfur pools [21,96,97].

A consistent increase in soil pH was also found in Austria [98]. In Germany, the comparison between the first and second NFSI resulted in a significant increase of pH(H_2O) in all depth intervals of the mineral soil [14]. Other studies from long-term soil monitoring programs in Europe, the United States, and Canada have shown that soil pH increases because of decreasing sulfur deposition [20,82]. In contrast, Berger et al. [29] found at the 'Vienna Wood' forest in the soil areas between trees and in the deeper soil horizons show no indications of recovery from acidification. Only within the stemflow soil area could a significant increase in pH(H_2O) be found. They conclude that the recovery in the sampled soils in 1984 and again in 2012 at 97 beech stands may be highly delayed, especially in the deeper soil horizons. For the period between 1994 and 2007, an ongoing soil acidification was found in some parts of the Hrubý Jeseník region, Czech Republic [30]. In other parts, a slight decline in acidity was noted.

Our study shows different trends for pH(H_2O) and pH($CaCl_2$). Meesenburg et al. [14] attribute this pattern to a reduction of the ionic strength in the soil solution, in particular due to the decrease in sulfur concentrations. At the time of the first soil inventories, many of our sites still had pH values at which the sorption of sulfur is particularly high [99]. Depending on the soil type, sulfur fixation often has its maximum at a pH(H_2O) of 4.0 and remains approximately constant as the pH continues to decrease. However, if a reduced atmospheric sulfur load leads to a slight increase in pH and at the same time to a reduction in the sulfur content in the soil solution, the adsorbed sulfur is dissolved again and further acidification is promoted [100]. Thus, the reduction of acid inputs over the last three decades has only led to a significant increase of pH(H_2O) in the soil solution. For pH($CaCl_2$), this development has not yet been clearly established.

Increases of base saturation and of pH($CaCl_2$) are only apparent for deciduous forests with lower historical sulfur inputs. The recovery of forest soils from acidification requires compensation for the very slow natural soil acidification [101] and other acidifying processes (e.g., timber logging and the natural accumulation of biomass) through weathering and other deacidifying processes. Different soil types with varying soil textures can have very different chemical weathering rates. Accordingly, the intensity of the acidification and recovery processes could be very different. For this reason, we divided the plots in groups with different vertical base saturation gradients and excluded the plots F008HABU and F009GWBU from the analysis.

Under certain conditions (low weathering rate, high utilization intensities), recovery is not to be expected at all in the decades to come [29]. The potential for a resupply of exchangeable base cations through weathering remains an important question because the uncertainty of the silicate mineral weathering rates estimates are very high [102,103]. Accordingly, some studies (summarized in [104]) concluded that it is unlikely that weathering rates can replenish base cations to the extent necessary to bring about recovery. Sverdrup et al. [105] postulated a delayed recovery with an increasing soil depth. In contrast, in our study, there is no dampening of the change signals by adding deeper soil layers. In some cases, the signal becomes even clearer. Cools and De Vos [31] explained this development as follows: "In the acidification process, there could be a significant delay from the topsoil, which is first affected by acid deposition, to the bottom of the soil profile. During acidification, hydrogen and Al^+ ions mobilized in the soil solution may exchange with the base cations on a cation exchangeable complex and delay the decrease in pH. During recovery, the reverse process could occur, and while the upper layers recover, simultaneously the bottom layers may still acidify."

The significance of recovery for some subgroups (e.g., deciduous trees) shows that soil resampling appears to be a valuable method to detect soil changes over varying time periods at sites with different forest types and acid deposition histories. However, the lag time between the decrease in sulfur deposition and the recovery of soil chemical indicators underlines the importance of continuing long-term studies.

5. Conclusions

We conclude that despite a reduction of sulfur deposition by about 90% in Lower Saxony, the recovery from soil acidification is slow. The most recent soil inventories show a trend reversal or a stabilization at a low level. This recovery of the soils apparently occurred faster at deciduous compared to coniferous plots. A possible explanation for this finding could be a larger amount of temporarily stored sulfur in the soil because of higher atmospheric input in the coniferous forests. While the acidification indicators are still at a critical level and recovery is very slow and delayed in the coniferous forest soils, the acceleration of the regeneration process through liming still seems to be necessary. Furthermore, high nitrogen deposition loads in the Lower Saxony forests still appear to continue, resulting in an increasing risk of base cation and nitrate leaching into surface waters. Therefore, continued monitoring of the acid-base status of forest soils at permanent soil-monitoring plots seems to be necessary to track further ecosystem responses to changing environmental conditions, such as deposition, climate change, and weathering.

Author Contributions: Conceptualization, B.A. and H.M.; methodology, B.A., H.F. and H.M.; formal analysis, B.A. and H.F.; investigation, H.F. and H.M.; statistical analysis, B.A.; writing—original draft preparation, B.A.; writing—review and editing, B.A., H.F. and H.M.; visualization, B.A.; supervision, H.M.; project administration, H.M.; funding acquisition, H.M. All authors have read and agreed to the published version of the manuscript.

Funding: The study has been funded since 1992 by the State of Lower Saxony through the Permanent Soil Monitoring Programme. Partial funding of data collection and evaluation was provided by the European Union under Council Regulation (EEC) 3528/86 on the Protection of Forests against Atmospheric Pollution, the Regulation (EC) 2152/2003 concerning monitoring of forests and environmental interactions in the community (Forest Focus), and by the project LIFE 07 ENV/D/000218 Further Development and Implementation of an EU-level Forest Monitoring System (FutMon).

Institutional Review Board Statement: Not applicable.

Informed Consent Statement: Not applicable.

Data Availability Statement: The datasets related to this article are available from the corresponding authors on reasonable request. The original datasets for regionalized deposition are available on request from UBA, Dessau, Germany.

Acknowledgments: We thank our colleagues from the Northwest German Forest Research Institute involved in operating the network of permanent soil-monitoring plots in Lower Saxony in the last decades. The constructive comments by the four anonymous reviewers, which helped to improve the manuscript significantly, are gratefully acknowledged.

Conflicts of Interest: The authors declare no conflict of interest.

Appendix A

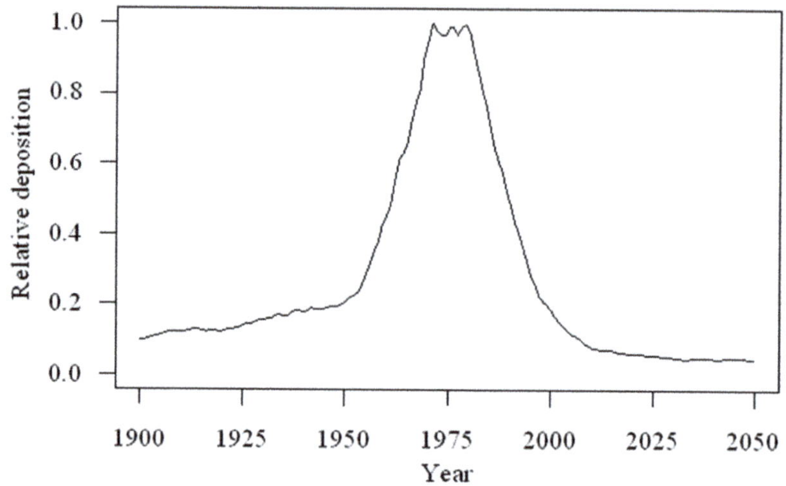

Figure A1. Standard curve for non-marine deposition in Lower Saxony, Germany. Adapted from Engardt et al. [9].

$$TD_{s(yr)} = \overline{TD}_{s(2000...2015)} \cdot SF_{(yr)} \cdot \overline{SF}_{(2000...2015)} \quad \text{(A1)}$$

where $TD_{s(yr)}$ is the annual sulfur deposition in the year yr (kg ha^{-1} yr^{-1}), $\overline{TD}_{s(2000\ldots2015)}$ is the mean sulfur deposition for period from 2000 to 2015, $SF_{(yr)}$ is the annual specific scale factor in the year yr (taken from Figure A1), and $\overline{SF}_{(2000\ldots2015)}$ the mean scale factor for the period from 2000 to 2015.

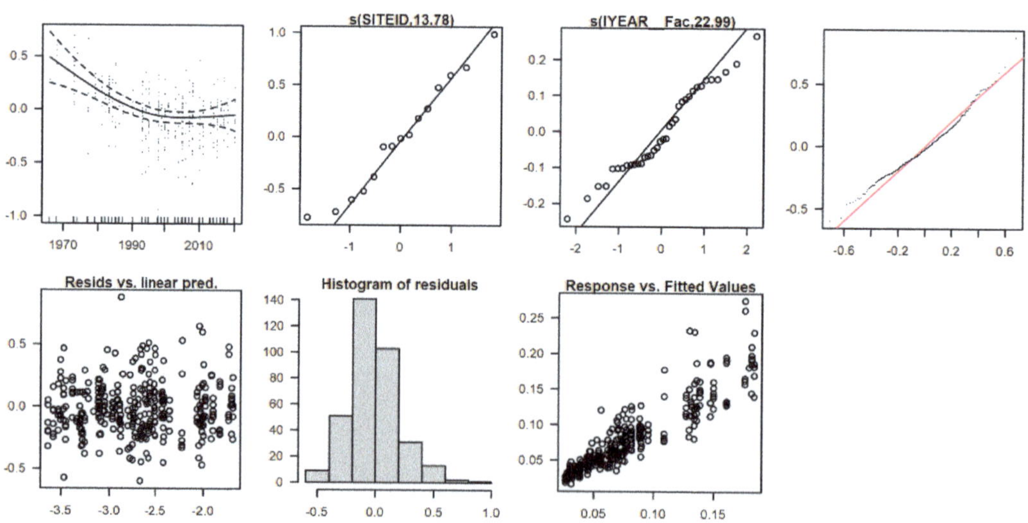

Figure A2. Example of graphical residual analysis (normality; homogeneity) of the selected generalized additive mixed model (GAMM) for the base saturation at unlimed plots in a 0–30 cm soil depth.

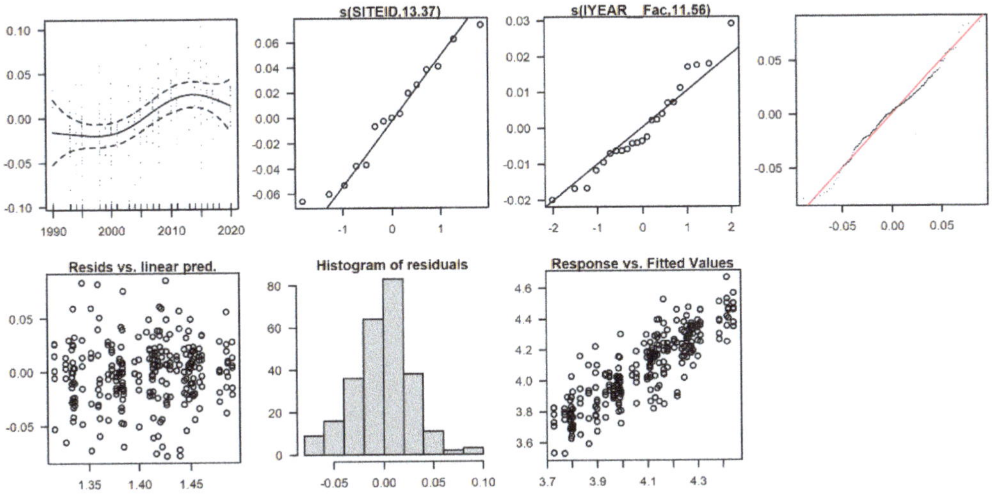

Figure A3. Example of graphical residual analysis (normality; homogeneity) of the selected generalized additive mixed model (GAMM) for pH(H_2O) values at unlimed plots in a 0–30 cm soil depth.

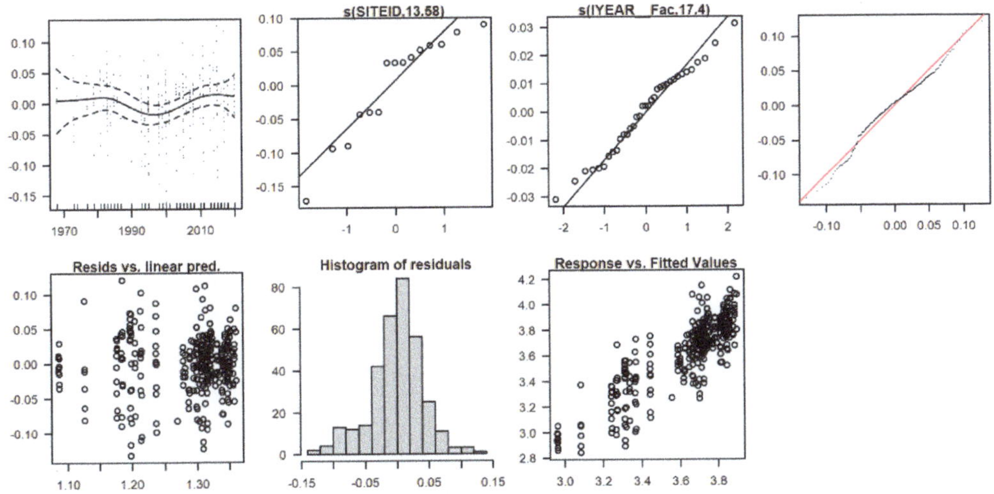

Figure A4. Example of graphical residual analysis (normality; homogeneity) of the selected generalized additive mixed model (GAMM) for pH($CaCl_2$) values at unlimed plots in a 0–30 cm soil depth.

References

1. Brumme, R.; Ahrends, B.; Block, J.; Schulz, C.; Meesenburg, H.; Klinck, U.; Wagner, M.; Khanna, P.K. Cycling and retention of nitrogen in European beech (*Fagus sylvatica* L.) ecosystems under elevated fructification frequency. *Biogeosciences* **2021**, *18*, 3763–3779. [CrossRef]
2. Wilpert, K.V. Forest Soils—What's Their Peculiarity? *Soil Syst.* **2022**, *6*, 5. [CrossRef]
3. Oulehle, F.; Evans, C.D.; Hofmeister, J.; Krejci, R.; Tahovska, K.; Persson, T.; Cudlin, P.; Hruska, J. Major changes in forest carbon and nitrogen cycling caused by declining sulphur deposition. *Glob. Chang. Biol.* **2011**, *17*, 3115–3129. [CrossRef]
4. Ulrich, B. Ökologische Gruppierung von Böden nach ihrem chemischen Bodenzustand. *Z. Pflanzenernähr. Bodenkd.* **1981**, *149*, 289–305. [CrossRef]
5. Ulrich, B. Soil Acidity and its Relations to Acid Deposition. In *Effects of Accumulation of Air Pollutants in Forest Ecosystems*; Ulrich, B., Pankrath, J., Eds.; Springer: Dordrecht, The Netherlands, 1983; pp. 127–146.

6. Hauhs, M. Lange Bramke: An Ecosystem Study of a Forested Catchment. In *Acidic Precipitation. Volume I: Case Studies*; Adriano, D.C., Havas, M., Eds.; Springer: New York, NY, USA, 1989; pp. 275–305.
7. Hauck, M.; Zimmermann, J.; Mascha, J.; Dulamsuren, C.; Bade, C.; Ahrends, B.; Leuschner, C. Rapid recovery of stem growth at reduced SO_2 levels suggests a major contribution of foliar damage in the pollutant-caused dieback of Norway spruce during the late 20th century. *Environ. Pollut.* **2012**, *164*, 132–141. [CrossRef]
8. Landmann, G.; Hunter, I.R.; Hendershot, W. Temporal and spatial development of magnesium deficiency in forest stands in Europe, North America and New Zealand. In *Magnesium Deficiency in Forest Ecosystems*; Hüttl, R.F., Schaaf, W., Eds.; Nutrients in Ecosystems; Springer: Dordrecht, The Netherlands, 1997; Volume 1, pp. 23–64.
9. Engardt, M.; Simpson, D.; Schwikowski, M.; Granat, L. Deposition of sulphur and nitrogen in Europe 1900–2050. Model calculations and comparison to historical observations. *Tellus Ser. B* **2017**, *69*, 1328945. [CrossRef]
10. Meesenburg, H.; Eichhorn, J.; Meiwes, K.J. Atmospheric Deposition and Canopy Interactions. In *Functioning and Management of European Beech Ecosystems*; Brumme, R., Khanna, P.K., Eds.; Ecological Studies; Springer: Berlin/Heidelberg, Germany, 2009; Volume 208, pp. 265–302.
11. Müller, F.; Bergmann, M.; Dannowski, R.; Dippner, J.W.; Gnauck, A.; Haase, P.; Jochimsen, M.C.; Kasprzak, P.; Kröncke, I.; Kümmerlin, R.; et al. Assessing resilience in long-term ecological data sets. *Ecol. Indic.* **2016**, *65*, 10–43. [CrossRef]
12. Lawrence, G.B.; Bailey, S.W. Recovery Processes of Acidic Soils Experiencing Decreased Acidic Deposition. *Soil Syst.* **2021**, *5*, 36. [CrossRef]
13. Lange, H.; Hauhs, M.; Schmidt, S. Long-term sulfate dynamics at Lange Bramke (Harz) used for testing two acidification models. *Water Air Soil Pollut.* **1995**, *79*, 339–351. [CrossRef]
14. Meesenburg, H.; Riek, W.; Ahrends, B.; Eickenscheidt, N.; Grüneberg, E.; Evers, J.; Fortmann, H.; Köng, N.; Lauer, A.; Meiwes, K.J.; et al. Soil Acidification in German Forest Soils. In *Status and Dynamics of Forests in Germany*; Wellbrock, N., Bolte, A., Eds.; Springer: Cham, Switzerland, 2019; Volume 237, pp. 93–120.
15. De Vries, W.; Dobbertin, M.H.; Solberg, S.; van Dobben, H.F.; Schaub, M. Impacts of acid deposition, ozone exposure and weather conditions on forest ecosystems in Europe: An overview. *Plant Soil* **2014**, *380*, 1–45. [CrossRef]
16. Schmitz, A.; Sanders, T.G.M.; Bolte, A.; Bussotti, F.; Dirnböck, T.; Johnson, J.; Peñuelas, J.; Pollastrini, M.; Prescher, A.-K.; Sardans, J.; et al. Responses of forest ecosystems in Europe to decreasing nitrogen deposition. *Environ. Pollut.* **2019**, *244*, 980–994. [CrossRef]
17. Schrijver, D.A.; Geudens, G.; Augusto, L.; Staelens, J.; Mertens, J.; Wuyts, K.; Gielis, L.; Verheyen, K. The effect of forest type on throughfall deposition and seepage flux: A review. *Oecologia* **2007**, *153*, 663–674. [CrossRef]
18. Zhang, L.; Brook, J.R.; Vet, R. A revised parameterization for gaseous dry deposition in air quality models. *Atmos. Chem. Phys.* **2003**, *3*, 2067–2082. [CrossRef]
19. Finkelstein, P.L. Deposition Velocities of SO_2 and O_3 over Agricultural and Forest Ecosystems. *Water Air Soil Pollut. Focus* **2001**, *1*, 49–57. [CrossRef]
20. Hazlett, P.W.; Emilson, C.E.; Lawrence, G.B.; Fernandez, I.J.; Ouimet, R.; Bailey, S.W. Reversal of forest soil acidification in the northeastern United States and eastern Canada: Site and soil factors contributing to recovery. *Soil Syst.* **2020**, *4*, 54. [CrossRef]
21. Meesenburg, H.; Ahrends, B.; Fleck, S.; Wagner, M.; Fortmann, H.; Scheler, B.; Klinck, U.; Dammann, I.; Eichhorn, J.; Mindrup, M.; et al. Long-term changes of ecosystem services at Solling, Germany: Recovery from acidification, but increasing nitrogen saturation? *Ecol. Indic.* **2016**, *65*, 103–112. [CrossRef]
22. Lawrence, G.B.; Hazlett, P.W.; Fernandez, I.J.; Ouimet, R.; Bailey, S.W.; Shortle, W.C.; Smith, K.T.; Antidormi, M.R. Declining Acidic Deposition Begins Reversal of Forest-Soil Acidification in the Northeastern U.S. and Eastern Canada. *Environ. Sci. Technol.* **2015**, *49*, 13103–13111. [CrossRef]
23. Pihl Karlsson, G.; Akselsson, C.; Hellsten, S.; Karlsson, P.E. Reduced European emissions of S and N—Effects on air concentrations, deposition and soil water chemistry in Swedish forests. *Environ. Pollut.* **2011**, *159*, 3571–3582. [CrossRef]
24. Johnson, J.; Graf Pannatier, E.; Carnicelli, S.; Cecchini, G.; Clarke, N.; Cools, N.; Hansen, K.; Meesenburg, H.; Nieminen, T.M.; Pihl-Karlsson, G.; et al. The response of soil solution chemistry in European forests to decreasing acid deposition. *Glob. Chang. Biol.* **2018**, *24*, 3603–3619. [CrossRef]
25. Sucker, C.; Wilpert, K.; Puhlmann, H. Acidification reversal in low mountain range streams of Germany. *Environ. Monit. Assess.* **2011**, *174*, 65–89. [CrossRef]
26. Wright, R.F.; Larssen, T.; Camarero, L.; Cosby, B.J.; Ferrier, R.C.; Helliwell, R.; Forsius, M.; Jenkins, A.; Kopáček, J.; Majer, V.; et al. Recovery of Acidified European Surface Waters. *Environ. Sci. Technol.* **2005**, *39*, 64A–72A. [CrossRef]
27. Stoddard, J.L.; Jeffries, D.S.; Lükewille, A.; Clair, T.A.; Dillon, P.J.; Driscoll, C.T.; Forsius, M.; Johannessen, M.; Kahl, J.S.; Kellogg, J.H.; et al. Regional trends in aquatic recovery from acidification in North America and Europe. *Nature* **1999**, *401*, 575–578. [CrossRef]
28. Watmough, S.A.; Eimers, M.C. Rapid Recent Recovery from Acidic Deposition in Central Ontario Lakes. *Soil Syst.* **2020**, *4*, 10. [CrossRef]
29. Berger, T.W.; Türtscher, S.; Berger, P.; Lindebner, L. A slight recovery of soils from Acid Rain over the last three decades is not reflected in the macro nutrition of beech (*Fagus sylvatica*) at 97 forest stands of the Vienna Woods. *Environ. Pollut.* **2016**, *216*, 624–635. [CrossRef]

30. Reininger, D.; Fiala, P.; Samek, T. Acidification of Forest Soils in the Hrubý Jeseník Region. *Soil Water Res.* **2011**, *6*, 83–90. [CrossRef]
31. Cools, N.; De Vos, B. Availability and evaluation of European forest soil monitoring data in the study on the effects of air pollution on forests. *Iforest—Biogeosciences For.* **2011**, *4*, 205–211. [CrossRef]
32. Grigal, D.F.; McRoberts, R.E.; Ohmann, L.F. Spatial Variation in Chemical Properties of Forest Floor and Surface Mineral Soil in the North Central United States. *Soil Sci.* **1991**, *151*, 282–290. [CrossRef]
33. Bruelheide, H.; Udelhoven, P. Correspondence of the fine-scale spatial variation in soil chemistry and the herb layer vegetation in beech forests. *For. Ecol. Manag.* **2005**, *210*, 205–223. [CrossRef]
34. Penne, C.; Ahrends, B.; Deurer, M.; Böttcher, J. The impact of the canopy structure on the spatial variability in forest floor carbon stocks. *Geoderma* **2010**, *158*, 282–297. [CrossRef]
35. Arrouays, D.; Bellamy, P.H.; Paustian, K. Soil inventory and monitoring. Current issues and gaps. *Eur. J. For. Res.* **2009**, *60*, 721–722. [CrossRef]
36. Mobley, M.L.; Yang, Y.; Yanai, R.D.; Nelson, K.A.; Bacon, A.R.; Heine, P.R.; Richter, D.D. How to Estimate Statistically Detectable Trends in a Time Series: A Study of Soil Carbon and Nutrient Concentrations at the Calhoun LTSE. *Soil Sci. Soc. Am. J.* **2019**, *83*, S133–S140. [CrossRef]
37. Lawrence, G.B.; Fernandez, I.J.; Richter, D.D.; Ross, D.S.; Hazlett, P.W.; Bailey, S.W.; Oiumet, R.; Warby, R.A.F.; Johnson, A.H.; Lin, H.; et al. Measuring environmental change in forest ecosystems by repeated soil sampling: A North American perspective. *J. Environ. Qual.* **2013**, *42*, 623–639. [CrossRef] [PubMed]
38. Griffiths, R.P.; Swanson, A.K. Forest soil characteristics in a chronosequence of harvested Douglas-fir forests. *Can. J. For. Res.* **2001**, *31*, 1871–1879. [CrossRef]
39. Bens, O.; Buczko, U.; Sieber, S.; Hüttl, R.F. Spatial variability of O layer thickness and humus forms under different pine beech-forest transformation stages in NE Germany. *J. Plant Nutr. Soil Sci.* **2006**, *169*, 5–15. [CrossRef]
40. Falkengren-Grerup, U. Long-term changes in pH of forest soils in southern Sweden. *Environ. Pollut.* **1987**, *43*, 79–90. [CrossRef]
41. Blake, L.; Goulding, K.W.T.; Mott, C.J.B.; Johnston, A.E. Changes in soil chemistry accompanying acidification over more than 100 years under woodland and grass at Rothamsted Experimental Station, UK. *Eur. J. Soil Sci.* **1999**, *50*, 401–412. [CrossRef]
42. Johnson, A.H.; Andersen, S.B.; Siccama, T.G. Acid rain and soils of the Adirondacks. I. Changes in pH and available calcium, 1930–1984. *Can. J. For. Res.* **1994**, *24*, 39–45. [CrossRef]
43. Zuur, A.F.; Ieno, E.N.; Walker, N.J.; Saveliev, A.A.; Smith, G.M. *Mixed Effects Models and Extensions in Ecology with R*; Springer: New York, NY, USA, 2009; p. 574. [CrossRef]
44. Ingersoll, T.E.; Sewall, B.J.; Amelon, S.K. Improved Analysis of Long-Term Monitoring Data Demonstrates Marked Regional Declines of Bat Populations in the Eastern United States. *PLoS ONE* **2013**, *8*, e65907. [CrossRef]
45. Knape, J. Decomposing trends in Swedish bird populations using generalized additive mixed models. *J. Appl. Ecol.* **2016**, *53*, 1852–1861. [CrossRef]
46. Höper, H.; Meesenburg, H. Das Bodendauerbeobachtungsprogramm. *GeoBerichte* **2012**, *23*, 6–18.
47. De Vries, W.; Vel, E.; Reinds, G.J.; Dellstra, H.; Klap, J.M.; Leeters, E.E.J.M.; Hendricks, C.M.A.; Kerkvoorden, M.; Landmann, G.; Herkendell, J.; et al. Intensive monitoring of forest ecosystems in Europe: 1. Objectives, set-up and evaluation strategy. *For. Ecol. Manag.* **2003**, *174*, 77–95. [CrossRef]
48. Clarke, N.; Zlindra, D.; Ulrich, E.; Mosello, R.; Derome, J.; Derome, K.; König, N.; Lövblad, G.; Draaijers, G.; Hansen, K.; et al. Part XIV: Sampling and Analysis of Deposition. Available online: https://storage.ning.com/topology/rest/1.0/file/get/9995560266?profile=original (accessed on 17 January 2022).
49. Schaap, M.; Hendriks, C.; Kranenburg, R.; Kuenen, J.; Segers, A.; Schlutow, A.; Nagel, H.-D.; Ritter, A.; Banzhaf, S. PINETI-3: Modellierung atmosphärischer Stoffeinträge von 2000 bis 2015 zur Bewertung der ökosystem-spezifischen Gefährdung von Biodiversität durch Luftschadstoffe in Deutschland. *Texte Umweltbundesamt* **2018**, *79*, 149.
50. Alveteg, M.; Walse, C.; Warfvinge, P. Reconstructing historic atmospheric deposition and nutrient uptake from present day values using MAKEDEP. *Water Air Soil Pollut.* **1998**, *104*, 269–283. [CrossRef]
51. Barth, N.; Brandtner, W.; Cordsen, E.; Dann, T.; Emmerisch, K.-H.; Feldhaus, D.; Kleefisch, B.; Schilling, B.; Utermann, J. Boden-Dauerbeobachtung—Einrichtung und Betrieb von Boden-Dauerbeobachtungsflächen. In *Bodenschutz. Ergänzbares Handbuch der Maßnahmen und Empfehlungen für Schutz, Pflege und Sanierung von Böden, Landschaft und Grundwasser*; Rosenkranz, D., Bachmann, G., König, W., Einsele, G., Eds.; Erich Schmidt Verlag: Berlin, Germany, 2000; Volume Bd3: 9152, pp. 1–127.
52. Cools, N.; De Vos, B. Part X: Sampling and Analysis of Soil. In *Manual on Methods and Criteria for Harmonized Sampling, Assessment, Monitoring and Analysis of the Effects of Air Pollution on Forests*; UNECE ICP Forests Programme Coordinating Centre; Thünen Institute of Forest Ecosystems: Eberswalde, Germany; Available online: https://storage.ning.com/topology/rest/1.0/file/get/9995584862?profile=original (accessed on 17 February 2022).
53. König, N.; Forstmann, H. *Probenvorbereitungs-Untersuchungs-und Elementbestimmungs-Methoden des Umweltanalytik-Labors der Niedersächsischen Forstlichen Versuchsanstalt und des Zentrallabor II des Forschungszentrums Waldökosysteme*; Forest Ecosystem Research Center: Göttingen, Germany, 1996; pp. 46–49.
54. König, N.; Fortmann, H. *Probenvorbereitungs-, Untersuchungs- und Elementbestimmungsmethoden des Umweltanalytiklabors der Niedersächsischen Forstlichen Versuchsanstalt und des Zentrallabor II des Forschungszentrums Waldökosysteme*; Forest Ecosystem Research Center: Göttingen, Germany, 1999; pp. 58–60.

55. König, N.; Forstmann, H.; Lüter, K.L. *Probenvorbereitungs-, Untersuchungs- und Elementbestimmungs-Methoden des Umweltanalytik-Labors der Niedersächsischen Forstlichen Versuchsanstalt*; Forest Ecosystem Research Center: Göttingen, Germany, 2009; pp. 75–78.
56. Meiwes, K.J.; Khanna, P.K.; Ulrich, B. Parameter for describing soil acidification and their relevance to the stability of forest ecosystems. *For. Ecol. Manage.* **1986**, *15*, 161–179. [CrossRef]
57. Wellbrock, N.; Ahrends, B.; Bögelein, R.; Bolte, A.; Eickenscheidt, N.; Grüneberg, E.; König, N.; Schmitz, A.; Fleck, S.; Ziche, D. Concept and Methodology of the National Forest Soil Inventory. In *Status and Dynamics of Forests in Germany*; Wellbrock, N., Bolte, A., Eds.; Ecological Studies; Springer: Cham, Schwitzerland, 2019; Volume 237.
58. Meiwes, K.J.; Meesenburg, H.; Eichhorn, J.; Jacobsen, C.; Khanna, P.K. Changes in C and N content of soils under beech forests over a period of 35 years. In *Functioning and Management of European Beech Ecosystems*; Brumme, R., Khanna, P., Eds.; Ecological Studies; Springer: Berlin/Heidelberg, Germany, 2009; Volume 208, pp. 49–63.
59. Hartmann, P.; Von Wilpert, k. Statistisch definierte Vertikalgradienten der Basensättigung sind geeignete Indikatoren für den Status und die Veränderungen der Bodenversauerung in Waldböden. *Allg. Forst-U. J.-Ztg.* **2016**, *187*, 61–69.
60. Dietrich, H.; Wolf, T.; Kawohl, T.; Wehberg, J.; Kändler, G.; Mette, T.; Röder, A.; Böhner, J. Temporal and spatial high-resolution climate data from 1961 to 2100 for the German National Forest Inventory (NFI). *Ann. For. Sci.* **2019**, *76*, 6. [CrossRef]
61. Beaudette, D.E.; Roudier, P.; O'Geen, A.T. Algorithms for quantitative pedology: A toolkit for soil scientists. *Comput. Geosci.* **2013**, *52*, 258–268. [CrossRef]
62. Jackson, D.A.; Chen, Y. Robust principal component analysis and outlier detection with ecological data. *Environmetrics* **2004**, *15*, 129–139. [CrossRef]
63. Pascoal, C.; Oliveira, M.R.; Pacheco, A.; Valadas, R. Detection of Outliers Using Robust Principal Component Analysis: A Simulation Study. In *Combining Soft Computing and Statistical Methods in Data Analysis*; Borgelt, C.E.A., Ed.; Advances in Intelligent and Soft Computing; Springer: Berlin/Heidelberg, Germany, 2010; Volume 77, pp. 499–507.
64. Lê, S.; Josse, J.; Husson, F. FactoMineR: An R Package for Multivariate Analysis. *J. Stat. Softw.* **2008**, *25*, 18. [CrossRef]
65. Wood, S.N. *Generalized Additive Models: An Introduction with R*; Chapman & Hall: Boca Raton, FL, USA, 2006; p. 410.
66. Dixon, W.J. Analysis of Extreme Values. *Comput. Geosci.* **1950**, *21*, 488–506. [CrossRef]
67. Harrison, P.J.; Buckland, S.T.; Yuan, Y.; Elston, D.A.; Brewer, M.J.; Johnston, A.; Pearce-Higgins, J.W. Assessing trends in biodiversity over space and time using the example of British breeding birds. *J. Appl. Ecol.* **2014**, *51*, 1650–1660. [CrossRef]
68. Kölling, C.; Hoffmann, M.; Gulder, H.-J. Bodenchemische Vertikalgradienten als charakteristische Zustandsgrößen von Waldökosystemen. *Z. Pflanzenernähr. Bodenk.* **1996**, *159*, 69–77. [CrossRef]
69. Schaap, M.; Timmermans, R.M.A.; Roemer, M.; Boersen, G.A.C.; Builtjes, P.J.H. The LOTOS–EUROS model: Description, validation and latest developments. *Int. J. Environ. Pollut.* **2008**, *32*, 270–290. [CrossRef]
70. Manders, A.M.M.; Builtjes, P.J.H.; Curier, L.; Denier van der Gon, H.A.C.; Hendriks, C.; Jonkers, S.; Kranenburg, R.; Kuenen, J.J.P.; Segers, A.J.; Timmermans, R.M.A.; et al. Curriculum vitae of the LOTOS–EUROS (v2.0) chemistry transport model. *Geosci. Model Dev.* **2017**, *10*, 4145–4173. [CrossRef]
71. Smith, R.I.; Fowler, D. Uncertainty in Estimation of Wet Deposition of Sulphur. *Water Air Soil Pollut. Focus* **2001**, *1*, 341–353. [CrossRef]
72. Ahrends, B.; Schmitz, A.; Prescher, A.-K.; Wehberg, J.; Geupel, M.; Andreae, H.; Meesenburg, H. Comparison of Methods for the Estimation of Total Inorganic Nitrogen Deposition to Forests in Germany. *Front. For. Glob. Change* **2020**, *3*, 103. [CrossRef]
73. Prechtel, A.; Alewell, C.; Armbruster, M.; Bittersohl, J.; Cullen, J.; Evans, C.D.; Helliwell, R.C.; Kopacek, J.; Marchetto, A.; Matzner, E.; et al. Response of sulphur dynamics in European freshwaters to decreasing sulphate deposition. *Hydrol. Earth Syst. Sci.* **2001**, *5*, 311–325. [CrossRef]
74. Evers, J.; Dammann, I.; König, N.; Paar, U.; Stüber, V.; Schulze, A.; Schmidt, M.; Schönfelder, E.; Eichhorn, J. Waldbodenzustandsbericht für Niedersachsen und Bremen. Ergebnisse der zweiten Bodenzustandserhebung im Wald (BZE II). *Beitr. aus der NW-FVA* **2019**, *19*, 498.
75. Reuss, J.O. Implications of the calcium-aluminum exchange system for the effect of acid precipitation on soils. *J. Environ. Qual.* **1983**, *12*, 591–595. [CrossRef]
76. Kirwan, N.; Oliver, M.A.; Moffat, A.J.; Morgan, G.W. Sampling The Soil In Long-Term Forest Plots: The Implications of Spatial Variation. *Environ. Monit. Assess.* **2005**, *111*, 149–172. [CrossRef]
77. Braun, S.; Tresch, S.; Augustin, S. Soil solution in Swiss forest stands: A 20 year's time series. *PLoS ONE* **2020**, *15*, e0227530. [CrossRef] [PubMed]
78. Dai, W.; Li, Y.; Fu, W.; Jiang, P.; Zhao, K.; Li, Y.; Penttinen, P. Spatial variability of soil nutrients in forest areas: A case study from subtropical China. *J. Plant Nutr. Soil Sci.* **2018**, *181*, 827–835. [CrossRef]
79. Li, J.; Richter, D.D.; Mendoza, A.; Heine, P. Effects of land-use history on soil spatial heterogeneity of macro- and trace elements in the Southern Piedmont USA. *Geoderma* **2010**, *156*, 60–73. [CrossRef]
80. Heinze, S.; Ludwig, B.; Piepho, H.-P.; Mikutta, R.; Don, A.; Wordell-Dietrich, P.; Helfrich, M.; Hertel, D.; Leuschner, C.; Kirfel, K.; et al. Factors controlling the variability of organic matter in the top- and subsoil of a sandy Dystric Cambisol under beech forest. *Geoderma* **2018**, *311*, 37–44. [CrossRef]
81. König, N.; Cools, N.; Derome, K.; Kowalska, A.; De Vos, B.; Fürst, A.; Marchetto, A.; O'Dea, P.; Tartari, G.A. Chapter 22—Data Quality in Laboratories: Methods and Results for Soil, Foliar, and Water Chemical Analyses. In *Developments in Environmental Science*; Ferretti, M., Fischer, R., Eds.; Elsevier: Amsterdam, The Netherlands, 2013; Volume 12, pp. 415–453.

82. Kirk, G.J.D.; Bellamy, P.H.; Lark, R.M. Changes in soil pH across England and Wales in response to decreased acid deposition. *Glob. Chang. Biol.* **2010**, *16*, 3111–3119. [CrossRef]
83. Vanguelova, E.I.; Benham, S.; Pitman, R.; Moffat, A.J.; Broadmeadow, M.; Nisbet, T.; Durrant, D.; Barsoum, N.; Wilkinson, M.; Bochereau, F.; et al. Chemical fluxes in time through forest ecosystems in the UK—Soil response to pollution recovery. *Environ. Pollut.* **2010**, *158*, 1857–1869. [CrossRef]
84. Jansone, L.; von Wilpert, K.; Hartmann, P. Natural Recovery and Liming Effects in Acidified Forest Soils in SW-Germany. *Soil Syst.* **2020**, *4*, 38. [CrossRef]
85. Lawrence, G.B.; Scanga, S.E.; Sabo, R.D. Recovery of Soils from Acidic Deposition May Exacerbate Nitrogen Export from Forested Watersheds. *J. Geophys. Res.* **2020**, *125*, e2019JG005036. [CrossRef]
86. Hedin, L.O.; Granat, L.; Likens, G.E.; Buishand, T.A.; Galloway, J.N.; Butler, T.J.; Rodhe, H. Steep declines in atmospheric base cations in regions of Europe and North America. *Nature* **1994**, *367*, 351–354. [CrossRef]
87. Meesenburg, H.; Meiwes, K.-J.; Rademacher, P. Long term trends in atmospheric deposition and seepage output in northwest german forest ecosystems. *Water Air Soil Pollut.* **1995**, *85*, 611–616. [CrossRef]
88. Guckland, A.; Ahrends, B.; Paar, U.; Dammann, I.; Evers, J.; Meiwes, K.J.; Schönfelder, E.; Ullrich, T.; Mindrup, M.; König, N.; et al. Predicting depth translocation of base cations after forest liming—Results from long-term experiments. *Eur. J. For. Res.* **2012**, *131*, 1869–1887. [CrossRef]
89. Lawrence, G.; Siemion, J.; Antidormi, M.; Bonville, D.; McHale, M. Have Sustained Acidic Deposition Decreases Led to Increased Calcium Availability in Recovering Watersheds of the Adirondack Region of New York, USA? *Soil Syst.* **2021**, *5*, 6. [CrossRef]
90. Greve, M.; Block, J.; Schüler, G.; Werner, W. Long Term Effects of Forest Liming on the Acid-Base Budget. *Appl. Sci.* **2021**, *11*, 955. [CrossRef]
91. Lawrence, G.B.; Shortle, W.C.; David, M.B.; Smith, K.T.; Warby, R.A.F.; Lapenis, A.G. Early indications of soil recovery from acidic deposition in U.S. red spruce forests. *Soil Sci. Soc. Am. J.* **2012**, *76*, 1407–1417. [CrossRef]
92. Czajkowski, T.; Ahrends, B.; Bolte, A. Critical limits of soil water availability (CL-SWA) in forest trees—An approach based on plant water status. *Vti Agric. For. Res.* **2009**, *59*, 87–93.
93. Augusto, L.; Ranger, J.; Binkley, D.; Rothe, A. Impact of several common tree species of European temperate forests on soil fertility. *Ann. For. Sci* **2002**, *59*, 233–253. [CrossRef]
94. Rumpf, S.; Schönfelder, E.; Ahrends, B. Biometrische Schätzmodelle für Nährelementgehalte in Baumkompartimenten. *Freibg. Forstl. Forsch.* **2018**, *101*, 33–73.
95. Watmough, S.A.; Aherne, J.; Alewell, C.; Arp, P.; Bailey, S.; Clair, T.; Dillon, P.; Duchesne, L.; Eimers, C.; Fernandez, I.; et al. Sulphate, nitrogen and base cation budgets at 21 forested catchments in Canada, the united states and Europe. *Environ. Monit. Assess.* **2005**, *109*, 1–36. [CrossRef]
96. Zhang, Y.; Mitchell, M.J.; Christ, M.; Likens, G.E.; Krouse, H.R. Stable sulfur isotopic biogeochemistry of the Hubbard Brook Experimental Forest, New Hampshire. *Biogeochemistry* **1998**, *41*, 259–275. [CrossRef]
97. Alewell, C.; Mitchell, M.J.; Likens, G.E.; Krouse, H.R. Sources of Stream Sulfate at the Hubbard Brook Experimental Forest: Long-Term Analyses Using Stable Isotopes. *Biogeochemistry* **1999**, *44*, 281–299. [CrossRef]
98. Jandl, R.; Schmidt, S.; Mutsch, F.; Fürst, A.; Zechmeister, H.; Bauer, H.; Dirnböck, T. Acidifiction and nitrogen eutrophication of austrian forest soils. *Appl. Envirionmental Soil Sci.* **2012**, *2012*, 9. [CrossRef]
99. Sokolova, T.A.; Alekseeva, S.A. Adsorption of Sulfate Ions by Soils (A Review). *Eurasian Soil Sci.* **2008**, *41*, 140–148. [CrossRef]
100. Meiwes, K.J.; Khanna, P.K.; Ulrich, B. Retention of sulphate by an acid brown earth and its relationship with the atmospheric impact of sulphur to forest vegetation. *J. Plant Nutr. Soil Sci.* **1980**, *143*, 402–411. [CrossRef]
101. Ulrich, B. Natural and anthropogenic components of soil acidification. *J. Plant Nutr. Soil Sci.* **1986**, *149*, 702–717. [CrossRef]
102. Klaminder, J.; Lucas, R.W.; Futter, M.N.; Bishop, K.H.; Köhler, S.J.; Egnell, G.; Laudon, H. Silicate mineral weathering rate estimates: Are they precise enough to be useful when predicting the recovery of nutrient pools after harvesting? *For. Ecol. Manag.* **2011**, *261*, 1–9. [CrossRef]
103. Futter, M.N.; Klaminder, J.; Lucas, R.W.; Laudon, H.; Köhler, S.J. Uncertainty in silicate mineral weathering rate estimates: Source partitioning and policy implications. *Environ. Res. Lett.* **2012**, *7*, 8. [CrossRef]
104. Fenn, M.E.; Huntington, T.G.; McLaughlin, S.B.; Eager, S.B.; Gomez, A.; Cook, R.B. Status of soil acidification in North America. *J. For. Sci.* **2006**, *52*, 3–13. [CrossRef]
105. Sverdrup, H.; Martinson, L.; Alveteg, M.; Moldan, F.; Kronnäs, V.; Munthe, J. Modeling recovery of Swedish ecosystems from acidification. *Ambio* **2005**, *34*, 25–31. [CrossRef]

soil systems

Article

Merits and Limitations of Element Balances as a Forest Planning Tool for Harvest Intensities and Sustainable Nutrient Management—A Case Study from Germany

Bernd Ahrends [1,*], Klaus von Wilpert [2], Wendelin Weis [3], Christian Vonderach [4], Gerald Kändler [4], Dietmar Zirlewagen [5], Carina Sucker [2] and Heike Puhlmann [2]

1 Department of Environmental Control, Northwest German Forest Research Institute (NW-FVA), D-37079 Goettingen, Germany
2 Department of Soil and Environment, Forest Research Institute of Baden-Wuerttemberg (FVA-BW), D-79100 Freiburg, Germany; klaus.von-wilpert@online.de (K.v.W.); carina.sucker@web.de (C.S.); heike.puhlmann@forst.bwl.de (H.P.)
3 Department of Soil and Climate, Bavarian State Institute of Forestry, D-85354 Freising, Germany; wendelin.weis@lwf.bayern.de
4 Department of Biometry and Information Sciences, Forest Research Institute of Baden-Wuerttemberg (FVA-BW), D-79100 Freiburg, Germany; christian.vonderach@forst.bwl.de (C.V.); gerald.kaendler@forst.bwl.de (G.K.)
5 INTERRA, Bureau for Environmental Monitoring, St. Peter Str. 30, D-79341 Kenzingen, Germany; info@interra.biz
* Correspondence: bernd.ahrends@nw-fva.de

Citation: Ahrends, B.; von Wilpert, K.; Weis, W.; Vonderach, C.; Kändler, G.; Zirlewagen, D.; Sucker, C.; Puhlmann, H. Merits and Limitations of Element Balances as a Forest Planning Tool for Harvest Intensities and Sustainable Nutrient Management—A Case Study from Germany. *Soil Syst.* **2022**, *6*, 41. https://doi.org/10.3390/soilsystems6020041

Academic Editor: Heike Knicker

Received: 28 February 2022
Accepted: 13 April 2022
Published: 20 April 2022

Publisher's Note: MDPI stays neutral with regard to jurisdictional claims in published maps and institutional affiliations.

Abstract: Climate change and rising energy costs have led to increasing interest in the use of tree harvest residues as feedstock for bioenergy in recent years. With an increasing use of wood biomass and harvest residues, essential nutrient elements are removed from the forest ecosystems. Hence, nutrient sustainable management is mandatory for planning of intensive forest use. We used soil nutrient balances to identify regions in Germany where the output of base cations by leaching and biomass utilization was not balanced by the input via weathering and atmospheric deposition. The effects of conventional stem harvesting, stem harvesting without bark, and whole-tree harvesting on Ca, Mg and K balances were studied. The nutrient balances were calculated using regular forest monitoring data supplemented by additional data from scientific projects. Effective mitigation management strategies and options are discussed and calculations for the compensation of the potential depletion of nutrients in the soil are presented.

Keywords: soil nutrient balance; deposition; weathering; leaching; uncertainties; harvest intensities; forest monitoring data; Germany; National Forest Inventory (NFI); National Forest Soil Inventory (NFSI)

1. Introduction

The supply of base cations like sodium (Na), potassium (K), magnesium (Mg) and calcium (Ca) into forest soils occurs through the dissolution of minerals and inputs through atmospheric deposition [1]. The decline in base cation deposition throughout large parts of Europe [2,3] will partly offset the positive effect of reduced base cation leaching due to decreased sulphur deposition [4,5], even if the sulphur emissions in Germany have decreased by more than 90% in the past decades [6]. Additionally, forests are currently and will be in the coming decades under pressure to fulfil the rising demands for timber and biomass as a sustainable energy source [7], which can contribute to reducing greenhouse gas emissions [8,9].

With the intensifying use of wood biomass and harvest residues, like tree tops, branches and bark, the associated nutrient element exports increase disproportionately [10], as the element concentrations in these tree parts are much higher than in stem wood [11,12].

Subsequently, the additional exports of base cations may have significant impacts on soil element stocks and soil quality [13], if weathering or other input fluxes cannot compensate for the resulting losses. Under these conditions, a recovery of forest soils from past acidification is not to be expected, despite a considerable ongoing decrease of sulphur deposition in Europe [4].

Accordingly, concerns have been raised about the sustainability of harvesting practices and their net impact on forest productivity, particularly during the second and subsequent rotation periods [14]. Therefore, a world-wide debate on the merits and trade-offs of additional forest biomass use is under way in the scientific community, as well as in politics, forestry practice and among certification authorities. Practical decisions in harvest intensity planning have long-term consequences on soil quality, forest growth and potentially necessary compensation measures. However, decisions are often based on intuition [15] or very different, hardly comparable methodological approaches [16]. As indicators should have a scientific basis and be applied operationally in a mapped way [17], a common method for site-specific assessment is to calculate input-output nutrient budgets [7,18,19].

In the Netherlands, soil nutrient balances were calculated in order to develop nationwide forest harvesting guidelines specific to regions of comparable deposition, tree species and soil types [7]. For each region-tree-soil combination, the maximum possible harvest intensities were calculated under equalized nutrient balance including uncertainties. The results show that on poorer sandy soils, even the current rates of timber and biomass exports lead to negative balances, particularly for phosphorus (P) and K. A long term mass balance study in Sweden showed net losses of Ca and Mg for stem harvesting and whole-tree harvesting (WTH) scenarios throughout most parts of the country [20]. In another study from Sweden, WTH reduced the base saturation in the soil [21]. Long-term base cation balances for forest soils in Finland demonstrated that WTH will lead to the depletion of base cations [22]. In 1066 Finish lake catchments, stem-only and stem-and-branches harvesting scenarios resulted in a balanced base cation budget, whilst WTH scenarios depleted the soil base cation pools [13]. In contrast, the calculations of Forsius et al. [23] for Finnish forests suggested that the input by weathering and deposition was sufficient to sustain the nutrient demand of WTH. For the British Isles, the predicted input from weathering and deposition were sufficient to compensate the losses of Ca, Mg, and K for stem-only and stem plus branch harvest scenarios [24]. Whole-tree harvesting resulted in a negative Ca balance at about half of the studied sites. Consistent with this, one of the oldest European WTH experiments in the United Kingdom also observed a significant decrease of soil base saturation after harvesting all above ground biomass [14]. When comparing these studies from different countries, it is important to note that conventional harvest schemes vary widely across countries [25] and that the considered WTH scenarios differ as well (for example in terms of the harvested tree compartments).

A review by Agate et al. [26] confirmed that high nutrient losses from soils have measureable consequences for forest ecosystems. Most studies therein revealed a negative effect on stand growth (tree diameter, tree height and tree volume) with a 3–7% reduction in the short and medium-term (up to 33 years after use), especially when canopy biomass including foliage was exported. The review by Thiffault et al. [27] also indicated medium-term ($\leq$24 years) growth reductions in intensively used stands. However, the long-term effects are largely unknown [28] and it should be noted that the negative effects on stand growth after WTH observed in some studies resulted from a temporary nitrogen deficiency [29]. The high risk of site degradation due to intensified biomass harvesting [30] makes it necessary to develop forest management strategies which will not impair forest productivity in the long term [8] and which take into account that the risks are highly dependent on soil fertility, stand and site conditions [31,32].

In Germany and many other countries, there is a great interest in sustainable nutrient management in the forestry sector. Nutrient balances are widely adopted tools to assess sustainable forestry. However, the calculation procedures in the individual federal states of Germany differ significantly from each other and the approaches have only been implemented in small-scale studies. For example, very different models and methods are used to calculate weathering [33–36] and leaching rates [19,37,38].

The objective of our study was to improve the nationwide assessment of the effects of harvest intensities on sustainability of the element budgets of forest soils. Specifically, we aimed at: (1) calculating methodologically uniform nutrient balances and their uncertainties for Germany using established, regular environmental monitoring systems as spatial data basis; (2) analysing the spatial patterns of harvesting effects on element balances to assess potential and risk of actual and intensified biomass harvesting for German forests and (3) deriving and quantifying strategic approaches for adapting harvest intensities as well as other nutrient management options to the actual nutrient availability at a regional scale.

2. Materials and Methods

2.1. Calculation of Nutrient Balances

Soil nutrient balances are commonly used indicators in both agriculture and forestry where annually or periodically aggregated balances express changes in soil nutrient stocks and soil fertility. Compared to agricultural systems, where fertilizer input dominates the soil nutrient balance, calculating soil nutrient balances for forest ecosystems poses a much greater challenge and the involved uncertainties are very high [39]. In forest soils, the relevant processes are inputs by atmospheric deposition as well as soil mineral weathering and the output fluxes by leaching and forest harvesting (Figure 1). For the assessment of different management options, the calculations should be performed in two steps: (1) calculation of nutrient balances without harvest removal (environmental part) and (2) calculation of total nutrient balances considering specific management options, e.g., different harvest scenarios or soil protective liming.

Figure 1. Schematic view of considered element fluxes for soil nutrient balances of Ca, Mg, and K in forest ecosystems and their relationships to each other. One 1000th of the plant-available nutrient stock in the soil is taken into account as an additional input (buffer).

In our study, the calculation of nutrient balances was restricted to the main nutrients Ca, Mg and K. These elements largely determine the resilience of forest soils to acidification.

Nitrogen (N) fluxes were not calculated because high nitrogen emissions in the recent past and at present have caused an oversupply of nitrogen throughout Germany [19,40]. Phosphorus, also an essential nutrient for tree growth and health, is characterized by very small fluxes in both leaching and deposition, and P concentrations are often near or below the limit of detection of conventional analytical methods [41]. In addition, part of the P transport in soils takes place in colloidal inorganic and organic P forms and can only be measured with considerable analytical effort [42]. Phosphorus was therefore not addressed in this study.

In some forest soils, the depletion rate of base cations in relation to element pools is very small [24]. To take this into account, one thousandth of the plant-available soil nutrient stock of Ca, Mg and K up to a depth of 90 cm was added to the nutrient balance and thus would be depleted in 1000 years at the earliest. Such a buffer is appropriate because small balance deficits are more tolerable in soils with high plant-available nutrient stocks and natural soil acidification is an extremely slow process [43]. All presented balances were averaged over the years 2000 to 2010 (data and conditions of the second National Forest Soil Inventory (NFSI II)) to compensate for annual fluctuations.

2.1.1. Study Sites

Around one third (32%) of Germany is covered by forest. Due to the high diversity of soils in Germany, reflecting different landscapes formed during the ice age, climate, and bedrock conditions, we stratified Germany into eight different model regions or soil landscapes (Figure 2A).

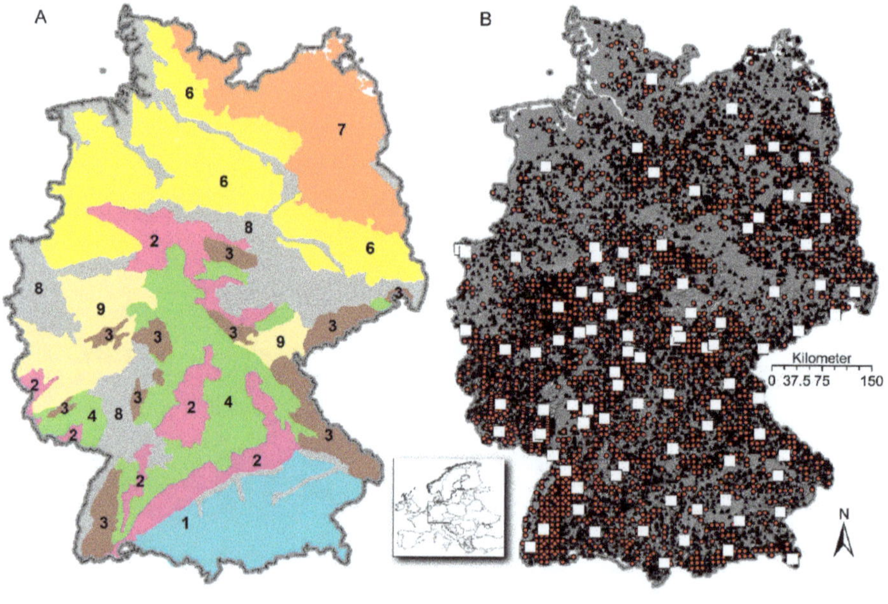

Figure 2. Stratification of Germany in soil landscapes (model regions—(**A**)) and location of intensive monitoring plots (□), National Forest Soil Inventories (NFSI) plots (•) and National Forest Inventory (NFI) tracts (▲) in Germany (**B**). Legend of left map: 1—pre-alpine moraines and limestone Alps; 2—hills on limestone bedrocks; 3—hills on crystalline bedrocks; 4—hills on sand, silt, clay bedrocks; 6—old moraines, north German lowland; 7—young moraines, north German lowland; 8—loess regions, fluvial valleys; 9—hills on clay- and silt schist bedrocks. Note: number 5 was deliberately not assigned.

For model development, application, regionalization, and evaluation, we used data from three different monitoring networks in Germany (Figure 2B). Data from Intensive

Forest Monitoring sites were used to parametrize the statistical deposition model (cf. Section 2.1.3) and for the estimation of weathering rates (cf. Section 2.1.4) and element concentrations in seepage water (cf. Section 2.1.5). Data from an 8 × 8 km grid of the NFSI II (1690 sites) were used to calculate weathering and leaching rates and then transferred from these sites to the 23,880 tracts of the third German National Forest Inventory (NFI 2012), using statistical regionalization methods (cf. Section 2.2). The German NFI is based on a systematic rectangular grid with clusters (tracts) as primary sampling units. The sample grid has a width of 4 × 4 km covering the entire forest area of Germany. The NFI data can be used to quantify different forest use scenarios and compare these to the nutrient balance without harvesting (WOH). In addition, the 16-fold higher data density of the NFI compared to the NFSI allows a more spatially differentiated identification of regions with critical nutrient balances.

2.1.2. Model Formulation

The total nutrient balance (EB) with harvest (WH) was calculated according to:

$$EB_x = DEP_x + WEA_x + LEA_x + HAR_x + \frac{S_x}{1000} \quad (1)$$

for each element x (Ca, Mg, K) separately, where DEP is the deposition, WEA the weathering rate, LEA the leaching, HAR the harvest removal of base cations and S denotes one 1000th of the plant-available nutrient stock in the soil. In scenarios without harvest (WOH), HAR was set to zero. All fluxes were calculated as kg ha^{-1} yr^{-1} and for the comparison between different charged ions, $kmol_c$ ha^{-1} yr^{-1}. Following Sverdrup et al. [44], cation exchange and the release of base cations in the decomposition process of organic matter were not taken into account, because they are internal cycles and accordingly not long-term sources.

We defined three different potentially feasible harvesting scenarios along a gradient of harvesting intensity (MIN, REAL, MAX—see Table A1). The scenarios considered different intensities of stem and crown utilization as well as the redistribution of biomass in the forest stand, its accumulation on skid trails or its export depending on the utilization technique. Scenario MIN was the most resource saving harvest intensity where only saw logs and industrial wood with diameters > 12–17 cm are harvested. Scenario REAL represented a common harvest intensity where saw logs, industrial wood, and fuel wood with diameters > 7 cm are used. In the most intense scenario MAX, all above-ground woody biomass except inevitable harvesting losses is used. More details can be found in Table A1.

2.1.3. Deposition

The estimation of total deposition (TD—sum of wet, dry and occult) of the base cations x (Ca, Mg, K) was based on combined regionalized measurements (wet deposition [45,46]) and canopy budget model calculations (dry and occult) after the 'filtering approach' [47] for the period 2000–2010. The calculations are performed as follows:

$$TD_x = BD_x + BD_x \cdot DDF \quad (2)$$

and the dry deposition factor (DDF) is calculated from Na deposition [48,49]:

$$DDF = \frac{(TF - BD)_{Na}}{BD_{Na}} \quad (3)$$

where TF = throughfall deposition and BD = bulk deposition. This approach assumes that Ca, Mg and K aerosols are deposited with equal deposition velocity as Na particles.

Generalized additive mixed models (GAMM) [50] were used to explore and quantify the impact of forest stand characteristics, topographic and atmospheric variables on DDF. For this, measured deposition data [51] from more than a hundred Intensive Forest Monitoring sites with a variety of forest stand types were analysed. The final DDF model included the predictor variables wind speed, windward and leeward effects, distance to the North Sea (as proxy for sea salt concentration), bulk deposition, tree species and stand height. Bulk open field deposition was estimated from regionalized wet deposition maps which were adjusted using correction factors for Germany from Gauger et al. [52]. The complete model with all parameters and validation results is summarized in Appendix B. The calculation of the GAMM models was performed with R 3.01 software [53], package "mgvc", landscape morphology was analysed with SAGA [54].

2.1.4. Weathering Rates

Weathering rates in the mineral soil were calculated with the geochemical model PROFILE [55], which has been frequently applied in Europe [56–58] and North America [59–61]. The particularly sensitive input variables [62,63] were parameterized as follows: The specific surface area (SSA) was calculated using the equation from Phelan et al. [60], which is a modification of the equation from the original PROFILE model [55] and is also valid for soils with clay contents of more than 20%. Dynamic soil water contents for all NFSI profiles were derived from water budget modelling using LWF-Brook90 [64]. A detailed description of the water budget model parameterization is given in a variety of recent studies [65–67]. The mean annual soil temperature was taken from regionalized climate data (cf. Section 2.1.5). As mineral analyses were not available for most of the NFSI sites, the mineralogical input to PROFILE was estimated from total geochemical soil analyses with the A2M ("Analysis to Mineralogy") model [68].

2.1.5. Leaching of Base Cations

Leaching rates were estimated by multiplying the amount of annual seepage water with an estimated element concentration in the seepage. Plot-specific soil water fluxes were estimated with the physically based hydrological model LWF-Brook90. The LWF-Brook90 model requires meteorological input data in daily resolution (precipitation, temperature, radiation, water vapour pressure, wind speed). The model was run from 2000 to 2010 using regionalized daily climate data derived from measurements at the weather stations of the German Weather Service (German: Deutscher Wetterdienst, DWD). Temperature, vapour pressure and wind speed were interpolated using GAMs, precipitation, and global radiation by kriging. Methodical details and information on model performance are given by Ahrends et al. [65].

To address the second methodological challenge—estimating element concentrations in seepage water—data on soil water extracts measured on the NFSI plots were used. An example for the estimation of seepage nitrate concentrations from water extracts measurements is described in Fleck et al. [19]. Sample preparation and analysis for the soil water extracts followed standardized procedures according to the guidelines for harmonized methodologies for laboratory analyses [69–72].

The exact procedure is described in Weis et al. [73] and includes the following main steps: (1) estimation of seepage concentrations of strong anions (sulphate, nitrate, chloride) from their concentrations in the aqueous soil extracts (water to soil ratio 2 ÷ 1); (2) estimation of inorganic dissolved carbon concentration from soil pH in water; (3) estimation of the molar fractions of the cations in the leachate from their fractions of the effective cation exchange capacity (extraction with 1 M NH_4Cl solution; for carbonate-containing soils extraction with 0.1 M $BaCl_2$ solution); (4) multiplication of the estimated cation fractions with the total anion concentration.

The approach was based on the following simplifying assumptions: the base cation leaching is mainly driven by the leaching of the anions sulphate (SO_4^{2+}), nitrate (NO_3^-) [7] and chloride (Cl^-) and the anion discharge is equal to the cation discharge; organic anions can be neglected in the discharge horizon; the cation fractions in the seepage water can be predicted with sufficient accuracy from the cation fractions at the soil exchanger; the total concentration (activity) of the ions in the seepage water plays a subordinate role. At 90% of the NFSI sites the discharge horizon was assumed to be in the mineral soil at depth of

60–90 cm; at 10% of the sites the soil development was shallower, and thus the surface of the bedrock was taken as discharge horizon there.

2.1.6. Estimation of Nutrient Export under Different Harvest Intensities

The National Forest Inventory in Germany is the primary source of national forest information and has been conducted three times so far (1987, 2002 and 2012). To project forest development and timber supply based on NFI data into the future, the empirical single-tree growth model WEHAM [74,75] was used. For this study, the resulting data were used to define initial conditions for the simulation of harvest scenarios. Normally the NFI is not to be evaluated for individual tracts, but tracts are aggregated into larger areas comprising all age classes of relevant forest types. When calculating nutrient balances for each NFI tract, we had to assume that nutrient removal by harvest corresponded to the long-term average. Hence, we used real removals during 2002 to 2012 (based on NFI data) and WEHAM projections for 2012 to 2052 encompassing 50 years of forest development. Evaluation results showed that calculated nutrient balances from the nutrient removal at the NFI tracts were independent of stand age (Figure A3).

For our nutrient balances, the exported biomass for each NFI sample tract and each scenario (Table A1) was calculated based on the available biometric tree information using additive biomass functions for different species and different components: stump, stump bark, solid wood, bark of solid wood, brushwood with a diameter of less than 7 cm and needles (if applicable; leaves for broadleaved trees were not considered) [76]. Taper curves and assortment algorithms were used including the quantification of bark share [77]. Mean contents of Ca, Mg and K in the biomass compartments "coarse wood", "bark of coarse wood" and crown biomass (brushwood plus bark and needles when coniferous) were calculated from twelve scientific studies [33,78–88] and summarized in Rumpf et al. [12]—including 451 experimental sites and 1498 trees of 5 coniferous and 6 broad-leaved tree species. The element exports with harvest were derived from the multiplication of biomass values and nutrient contents of the respective tree components based on GAM models. The pre-defined utilization scenarios (Table A1) determined the average nutrient export at each NFI sample tract.

2.2. Regionalization of Nutrient Fluxes and Soil Stocks

The data needed for calculating the soil-based components of the nutrient balances as well as the balances themselves were only available (measured or assessed by transfer functions) at the grid points of the NFSI. None of these data were available at the systematic grid with tracts of the National Forest Inventory (NFI). Therefore, results were transferred from the NFSI grid to the NFI tracts by means of regression models (stepwise multiple regression models combined with kriging of model residuals when needed) on log transformed response variables (alternatively tested boosted regression trees were outperformed according to the model validation). As predictors, we used quasi-continuously available key variables like geology, topography, soil types, climate and deposition from nationwide data bases and maps (GÜK2000 for geological overview, BÜK50 or BÜK200 for soil types, a 25 m DEM grid for deriving topographical indices and deposition maps [46]). Moreover, information on forest conditions (proportion of coniferous trees, crown condition, soil protective liming) were included in the regression models as potential co-variables.

The transfer was performed in three steps: First, the data was split in half resulting in a training and a validation data set, both randomly distributed in space. Second, the regression models were parameterized (ordinary least squares, OLS) separately for each soil landscape (cf. Figure 2A) at the NFSI grid points. An assessment of the extent to which the regionalized balances reflected the distributional characteristics of the balances based on the measured NFSI data is provided in Figure A6. Model performance was tested after back-transformation and bias correction [89] in terms of R^2 and RMSE (Root Mean Square Error) using the validation data set. Finally, the regression models were applied with their respective individual set of predictors to each of the NFI tracts with available metadata except for deposition, which was estimated directly for all NFI tracts (cf. Section 2.1.3 and Figure A7).

The completeness of measured parameters at the sampling points of the NFSI differs between the federal states, even if a big effort has been invested for harmonizing the environmental monitoring systems for all of Germany [72,90]. Data availability differed between the balance components: data for calculating weathering rates were available at 86% of the sites, for nutrient leaching with seepage water at 66% and for nutrient balances at 55%. Hence, only at around half of the NFSI points, could the complete nutrient balance including all balance components be calculated. Therefore, transfer of the nutrient balance from the NFSI points to the NFI tracts was achieved in two steps: (1) transfer of the individual balance terms from the NFSI points to the NFI tracts, (2) calculating the nutrient balance at the NFI tracts from the transferred balance components.

For the stratified model regions (Figure 2A), different regionalization models were parameterized. The regions 6 and 7 were combined for the final model selection as both are characterized by glacial till and comparable maritime climate.

Additionally, global models for whole Germany were parametrized for comparison of the model performance with the stratified models. The R^2 for the Ca, Mg and K balances were by 0.28, 0.29 and 0.18 lower in the global models than in the stratified ones. An overview on the indicators of model performance of the stratified models is given in Table 1. The model errors were highest for Ca and Mg in model regions with limestone bedrocks (region 1 and 2, Figure 2A) and amounted to a multiple of the mean balance level, whereas for K, model errors were comparatively small and quite uniform for all model regions with a CV_{RSME} (standardized RMSE) of around 0.5. All regression analyses were performed with R environment for statistical computing and the following software packages: "OLS", "randomFOREST", "stats" and "dismo". Landscape morphological indices were calculated with SAGA [54].

Table 1. Indicators of model performance for the soil-related balances (weathering + deposition—leaching with seepage) in the validation data set. OBS: number of observations; R^2: coefficient of determination; RMSE: root mean squared error [kg ha^{-1} yr^{-1}]; CV_{RMSE}: standardized RMSE.

Region	OBS	Ca Balance			Mg Balance			K Balance		
		R^2	RMSE	CV_{RMSE}	R^2	RMSE	CV_{RMSE}	R^2	RMSE	CV_{RMSE}
1	63	0.733	118.580	−8.024	0.640	35.193	−5.940	0.665	3.990	0.454
2	44	0.679	111.700	2.877	0.015	38.815	2.870	0.509	4.726	0.434
3	79	0.690	10.000	−2.421	0.527	6.809	−2.625	0.523	4.125	0.563
4	126	0.540	25.631	4.626	0.226	8.932	3.369	0.344	5.213	0.465
6|7	46	0.650	7.775	4.527	0.587	1.992	0.704	0.800	1.546	0.305
8	31	0.584	36.587	−14.405	0.339	4.168	1.447	0.382	4.604	0.611
9	61	0.623	12.438	−1.553	0.387	6.041	−1.841	0.475	3.736	0.485
Global	450	0.605	62.514	−2.054	0.447	23.164	−0.288	0.521	4.313	0.474

2.3. *Treatment of Special Sites*

On sites dominated by limestone and dolomite, the supply of Ca and Mg to forest soils and trees is usually unlimited. The high solubility of these carbonates causes large Ca and Mg fluxes from weathering and in seepage flux. In combination with equally high uncertainties, this discourages a sufficiently reliable interpretation of Ca and Mg balances for these sites. Therefore, the element balances for Ca and Mg were assumed to be even on limestone and dolomite. The occurrence of carbonate was predicted with a logistic classification model (recall accuracy >90%), which was calculated using the R packages "logistf", "stats" and "dismo".

The Ca and Mg balances are also influenced by liming, as both elements are contained in the applied dolomitic limestone. The dissolution of dolomites initially increases the input of both elements into the soil. As a result, higher leaching exports may occur temporarily, but also in the longer term [91,92]. Accordingly, liming effects should be considered in the balances of Ca and Mg. Information on past liming events is available for NFSI plots, although it is very heterogeneous with respect to timing, repetition, and amount of dolomite application in the different federal states of Germany. For the NFI tracts, reliable data on liming are missing for parts of the federal territory or are distributed very heterogeneously in some states. However, evaluations within the framework of the regionalization of input and output fluxes showed that, where liming was documented, the liming effect in the regionalization models was only rarely significant with respect to the target variables leaching, weathering and soil stocks for Ca and Mg and had very low sensitivity on model results with changing signs [93]. This shows that liming obviously has ambiguous effects in our data and, at best, only very weak influence on the nutrient balances of Ca, Mg and K up to a depth of 90 cm. Accordingly, the liming effect was also not considered in the German-wide calculations of this study.

2.4. Uncertainty Estimations

Usually, calculating soil nutrient balances is associated with a high degree of uncertainty, mainly due sampling and measurement errors, errors of the predictive models for balance elements as well as regionalization errors and biases [44,94,95]. Although uncertainty estimation is an important part of model application, there are numerous challenges and pitfalls. A thorough, very detailed, statistical discussion of uncertainty may reduce acceptance by stakeholders [96] as would the concealment of prediction uncertainties [97]. Yanai et al. [98] demanded that element balances should be supplemented with uncertainty analyses as a standard tool, not least also to allow reliable statements about the significance of any presented results.

The Monte Carlo simulation method is a widely used technique for uncertainty analysis, which can be described as follows [99]: For a model Φ of arbitrary complexity, the calculation of the resulting variable Z is done according to:

$$Z_i = \Phi(X_i, Y_i) \tag{4}$$

where X and Y are normally distributed random variables and the index i refers to samples from these normal distributions. X and Y are assumed to be independent of each other and covariance terms are not taken into account directly [63]. However, Yanai et al. [98] recommended to consider each covariance structure in their joint probability distributions when randomizing the parameters. In our application, the variation of the parameters was generated with the function rmvnorm() from the R software package "splus2R" [100] directly incorporating the covariance between the variables. A compilation of statistical parameters used for the different parts of the element budget calculations and to account for regionalization errors can be found in Appendix C.

We performed repeated calculations of the nutrient balances according to the respective balance equation. An error value randomly selected from its known (or assumed) probability distribution was repeatedly added to the model prediction for the individual balance elements, or the sub-equations for their calculation (e.g., leaching, harvest removal). After 10,000 iterations, the magnitude of the total error was derived from the realized predictions and the corresponding result statistics (mean, standard deviation, quantiles, etc.) were calculated. To assess the level of significance for the occurrence of negative or positive element balances, we analysed the resulting probability densities of the nutrient balances. The terms significant and weakly significant were defined with error probabilities of $\alpha \leq 0.05$ and $0.05 < \alpha \leq 0.1$, respectively.

3. Results and Discussion

3.1. Nutrient Fluxes of Deposition, Weathering, Leaching and Harvest Removal

The influence of the various balance components on the total budget calculations was very element-specific (cf. Figure 3A–C). For Ca, a relatively even distribution among the balance components was observed. Loss by leaching and removal tended to be higher than gains from deposition and weathering, so that the overall balance is negative in many cases. Due to the high leaching losses, more than 25% of the NFI tracts had a negative Ca balance already when not considering the harvest removal. For Mg, harvest removal was less important, and the balance remained positive on average. In the case of K, leaching losses were very small and harvest removal was the dominating loss factor. At most of the NFI tracts, harvest export (scenario REAL) was compensated by weathering rate and negative balances were calculated for slightly less than 20% of the NFI tracts. The considered soil stocks (cf. Section 2.1.2) were generally of minor importance, especially for Mg and K. To give an idea of the magnitude and large-scale spatial differences of the balances and their input variables, the medians for different model regions (see Figure 2A) are compiled in Table 2.

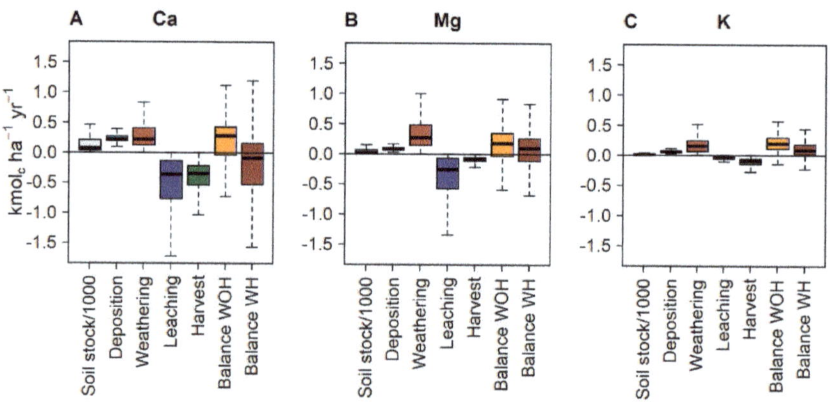

Figure 3. Median and range data for calcium (**A**), magnesium (**B**) and potassium (**C**) of soil stocks, deposition, weathering, leaching, harvest removal, nutrient balance without harvest (WOH), and nutrient balance with harvest (WH) for NFI tracts in Germany (for Ca and Mg excluding carbonate sites; harvest export for scenario REAL).

The weathering rates of the base cations (Ca, Mg, K) in the stratified model regions (see Figure 2A and Table 2) were, in ascending order, as follows: young moraines, north German lowland: 0.3, old moraines, north German lowland: 0.35, hills on clay- and silt schist bedrocks: 0.7, hills on crystalline bedrocks: 0.7, hills on sand, silt, clay bedrocks: 1.1, loess regions, fluvial valleys: 1.2, pre-alpine moraines and limestone Alps: 2.5, hills on limestone bedrocks: 3.9 kmol$_c$ ha^{-1} yr^{-1}. Calculations of the weathering rates, using an approximation from soil type and texture by Posch et al. [101], give weathering rates of 0.27–2.92 kmol$_c$ ha^{-1} yr^{-1} for the five main weathering rate classes (WRc) for non-calcareous soils in Germany, which agrees quite well with our results. Field weathering rates of base cations in Dutch sandy soils are reported to vary between 0.1 to 0.7 kmol$_c$ ha^{-1} [102–104] cited in van der Salm et al. [105] and 0.16 to 0.58 kmol$_c$ ha^{-1} yr^{-1} [106]. In the Netherlands and Germany, estimated weathering rates for loess ranged from 0.26–1.85 kmol$_c$ ha^{-1} yr^{-1} and 0.350–1.72 kmol$_c$ ha^{-1} yr^{-1}, respectively, and in river-clay soils from 0.76–5.3 kmol$_c$ ha^{-1} yr^{-1} [105]. These orders of magnitude are also quite comparable with the data for our model region "loess regions, fluvial valleys". De Vries et al. [7] classified the weathering rates for soil types from unconsolidated rocks in the Netherlands as follows: poor sand: 0.250, moderate poor sand: 0.385, rich sand: 0.520; loess: 0.600; clay: 1.300 kmol$_c$ ha^{-1} yr^{-1}.

It should be noted that the rates quoted from the above studies also include Na. They can be related to our values by multiplying by a factor of 0.7 for poor sandy soils and 0.85 for rich soils [107]. However, it is well known that data on weathering can vary widely, and calculations are associated with a high degree of uncertainty [108–112]. Kolka et al. [94] and Wesselink et al. [113] determined uncertainties of 25%. Somewhat larger uncertainties were found by Jönsson et al. [63] and Dultz [114] with 40% and 75%, respectively. According to Hodson and Langan [115], most methods for determining weathering rates have an accuracy of ±50%. Much higher uncertainties of 100% and 250% were reported by Hodson et al. [62,110]. Orders of magnitude above 100% are also given by the works of Klaminder et al. [108] with 98–110% and Futter et al. [109] with 33–300% when comparing different methods for estimating weathering rates. The high variation of weathering rates in the literature can partially be attributed to differences in the methodologies applied, for example different integration levels (soil profile to catchment) or the consideration of gravel content (cf. [108,116,117]). Therefore, when comparing weathering rates determined by different methods, the methodological approach must always be considered.

Table 2. Medians of the nutrient balances on the NFI tracts stratified by model regions (Ca and Mg only for carbonate-free soils) for the harvest scenario REAL. DEP: deposition; WEA: weathering; LEA: leaching; HAR: harvest removal; WOH: balance without harvesting; WH balance with harvesting.

Model Regions		STOCK	DEP	WEA	LEA	HAR	WOH	WH
		All Tracts					Carbonat-Free Tracts	
		[kmol$_c$ ha^{-1}]	[kmol$_c$ ha^{-1} yr^{-1}]				[kmol$_c$ ha^{-1} yr^{-1}]	
Pre-alpine moraines and limestone Alps	Ca	426.6	0.252	1.124	2.549	0.554	0.627	0.018
	Mg	134.9	0.063	1.174	1.249	0.108	0.406	0.283
	K	17.1	0.058	0.177	0.035	0.119	0.219	0.085
Hills on limestone bedrock	Ca	840.0	0.243	2.076	3.228	0.524	0.562	0.101
	Mg	75.8	0.072	1.581	0.923	0.104	0.768	0.657
	K	25.6	0.061	0.216	0.044	0.117	0.268	0.143
Hills on crystalline bedrock	Ca	46.8	0.292	0.223	0.529	0.406	0.102	−0.342
	Mg	23.9	0.087	0.295	0.471	0.091	−0.021	−0.128
	K	11.7	0.075	0.174	0.048	0.103	0.217	0.105
Hills on sand, silt, and clay bedrock	Ca	172.3	0.241	0.425	0.394	0.391	0.454	0.061
	Mg	92.7	0.076	0.402	0.322	0.088	0.215	0.122
	K	18.8	0.060	0.225	0.031	0.102	0.277	0.165
Old moraines, north German lowlands	Ca	57.2	0.201	0.136	0.099	0.240	0.298	0.045
	Mg	10.4	0.084	0.161	0.050	0.070	0.206	0.139
	K	7.2	0.071	0.050	0.026	0.065	0.110	0.043
Young moraines, north German lowlands	Ca	50.9	0.217	0.122	0.054	0.302	0.336	0.038
	Mg	6.4	0.103	0.139	0.037	0.082	0.210	0.134
	K	6.2	0.059	0.038	0.016	0.079	0.088	0.016
Loess region and fluvial valleys	Ca	275.3	0.201	0.476	0.702	0.396	0.343	−0.070
	Mg	52.9	0.073	0.522	0.340	0.077	0.232	0.147
	K	17.5	0.053	0.216	0.028	0.101	0.256	0.133
Hills on clay- and silt schist bedrock	Ca	57.2	0.228	0.189	0.764	0.396	−0.247	−0.705
	Mg	32.6	0.093	0.312	0.607	0.084	−0.160	−0.251
	K	12.6	0.068	0.192	0.055	0.107	0.215	0.105

The median Ca deposition in the stratified model regions ranges between 4 and 6 kg ha^{-1} yr^{-1}. Recent results from the Netherlands show similar magnitudes [7]. Except for sites strongly influenced by sea salt deposits and sites with very high precipitation, most sites are characterized by Mg depositions of about 1 kg ha^{-1} yr^{-1}. Median K deposition in the model regions varies between around 2 and 3 kg ha^{-1} yr^{-1} (Table 2) and is also quite similar to other studies [7]. The spatial distributions and regional patterns of the deposition input vary largely between the investigated elements (Figure A2). While Mg is strongly influenced by sea salt from the North Sea, this influence is weaker for K. Atmospheric deposits of K are more strongly influenced by local and regional sources [118], which can vary greatly from year to year [46]. In addition to the effect of precipitation and wind speed in the low mountain ranges, the importance of agriculture as a source of K emission is also evident here. Dämmgen et al. [118] noted that Na and Mg depositions now reached a magnitude that can be described as largely unaffected by anthropogenic factors. In the case of Ca, it is mainly the mountainous areas that show maximum inputs.

Compared to deposition, the leaching rates of Ca and Mg differ more strongly between the model regions (Table 2). In regions with higher fractions of carbonate soils the median leaching losses exceed 50 kg ha^{-1} yr^{-1} (Ca) and 10 kg ha^{-1} yr^{-1} (Mg). For all other upland and hilly areas, the leaching amounts to 8–15 kg ha^{-1} yr^{-1} (Ca) and 4–7 kg ha^{-1} yr^{-1} (Mg). In regions dominated by loess and fluvial valleys, the leaching losses amount to 14 and 4 kg ha^{-1} yr^{-1}, respectively. In the poor and moderately poor sandy soils of the regions with "old and young moraine deposits", the leaching losses are much lower and amount to 1.1–2.0 kg ha^{-1} yr^{-1} for Ca and 0.4–0.6 kg ha^{-1} yr^{-1} for Mg. In contrast, K leaching losses are much more homogeneous and differ only slightly between the individual model regions (between 1 and 2 kg ha^{-1} yr^{-1}). A similar magnitude for K with a leaching rate of <2 kg ha^{-1} yr^{-1} was also determined in the study of de Vries et al. [7]. Due to the high variability of German soils, a direct comparison of leaching losses with the study of de Vries et al. [7] is difficult, as a larger part of the Dutch forests is located on well-drained sandy soils with sometimes very different nutrient availability. The regions "old moraines" and "loess and fluvial valley" might be the most comparable to the Dutch conditions. The leaching rate for Ca and Mg in the Dutch forests was slightly higher than in our "old moraines" region, at about 4 and 1–4 kg ha^{-1} yr^{-1}, respectively, but significantly lower than in the "loess and fluvial valley" region.

The calculated harvest export rates of the sum of Ca, Mg and K ranged from 0.38 and 0.78 $kmol_c$ ha^{-1} yr^{-1} in the different model regions (Table 2). These values are at the upper end of those published in other studies [7,119–123]. It must be emphasized, however, that the current growth rates are higher due to increased N input by deposition [124] and that harvest intensity in many regions of Germany is above the level of many neighbouring countries [25].

3.2. *Nutrient Balances for Different Harvest Intensities*

The base cation content as well as biomass amount differ greatly between the wood, bark, and brushwood biomass compartments. This explains that nutrient export is much more influenced by the harvest intensity than the biomass export.

At the usual harvesting intensity (REAL), about 80% of the total aboveground biomass is removed from the stand, including 6.6% brushwood that accumulates in protective brushwood mats on the skid trails. About 20% of the biomass remains distributed within the stand area as harvest residues. The largest share of harvested biomass (including bark) is stem wood with 55.6%, followed by industrial wood with 13.7%, and the smallest share is fuel wood with 6.9%. The ratio between export with harvest and harvest residues is about 70:30% for the nutrient elements Ca, Mg and K (Figure 4A–C).

The "nutrient preservation potential" of the MIN scenario compared to REAL is mainly related to the assumption that debarking of stem wood and industrial wood is technically feasible during the harvesting process and that the bark remains distributed within the stand area. This assumption is an anticipation of recent developments in harvester technology [125]. In addition, the scenario MIN assumed no accumulation of brushwood on skid trails. When harvesting and timber logging is carried out with forestry machines, brush mats weighing 10–20 kg m^{-2} are required to protect the soil function on the skid trails [126]. Thus, brush accumulation on skid trails could be avoided either by increasing the use of motor-manual harvesting techniques and logging with cable cranes or by consecutive collecting, chipping, and re-distributing brush mats. Both alternatives are considered to be much more costly than conventional fully mechanized harvesting and logging.

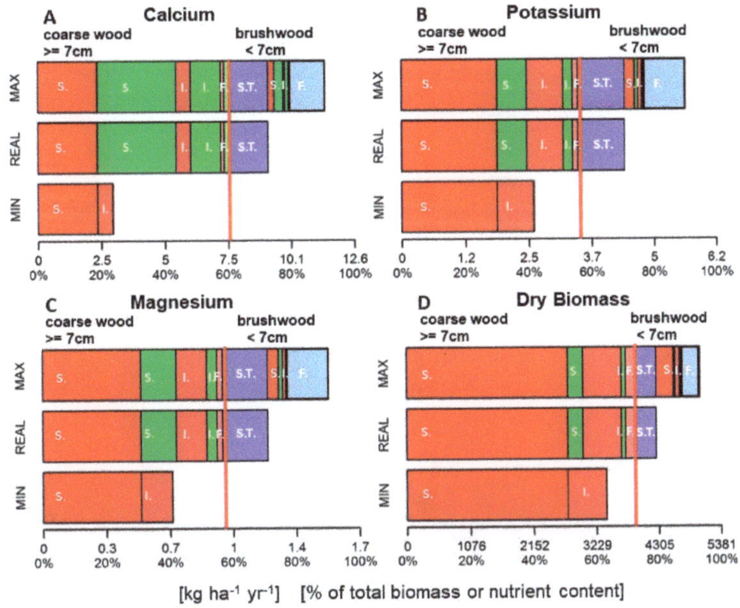

Figure 4. Average export of Ca (**A**), K (**B**), Mg (**C**) and biomass (**D**) at NFI tracts with harvested assortments for the three harvesting intensities MIN, REAL and MAX, differentiated into the biomass compartments wood (red), bark (green) and brushwood with bark (blue). Colour intensity decreasing from stem wood over industrial wood to fuel wood. Harvested wood assortments including bark (minimum diameter ≥ 7 cm) marked by a vertical red line. S.: stem wood, I.: industrial wood, F.: fuel wood, S.T.: brush mat on skid trails.

If it would be technically possible to leave all bark and brushwood biomass well distributed in the stand area (scenario MIN), Ca export could be reduced by about 70% compared to the REAL scenario, mainly due to the high Ca content of the bark. Additionally, 43% less Mg and 45% less K are exported in the MIN scenario compared to REAL. Compared to the nutrient loss, the harvest volume of the marketable assortments (stem, industrial and fuel wood) is reduced much less, by about 9% compared to the REAL scenario. Scenario MAX increased the biomass harvest rate by 13% compared to scenario REAL, because wood compartments with diameters < 7 cm (as industrial or fuel wood) are used and commonly occurring harvest losses of 10% are avoided. The average harvest of fuel wood, amounting to only 3.1% in the REAL scenario, could be increased up to 9.7% in the MAX scenario. In addition, stem and industrial wood from recovered harvest losses account for up to 7% of total biomass and could be used in the MAX scenario (Figure 4D). Thus, the fuel wood potential could be extended in this scenario up to 18% of the total aboveground biomass. However, the export for all three nutrients would be about 30% higher in the MAX scenario than in the REAL scenario.

The regional distributions of Ca, Mg, and K balances can reveal hot spots of balance deficits or positive balances and show effects of increasing harvesting intensity (Figure 5). Figure 5 also shows the uncertainty level of the balance calculation at each NFI tract (cf. Section 3.3). Striking regional patterns are found for balance deficits of Ca and Mg, which are most significant in the low mountain ranges with crystalline bedrock and in hill regions with clayey and/or silty shale. The variation caused by harvest intensity is of minor importance for Ca and Mg. The causes are base-poor bedrocks [127] combined with high seepage fluxes [67] and high acid deposition rates [46,128]. Particularly noteworthy here is the hilly 'Sauerland' region, which is characterized by very high conventional harvest intensities [25] and high atmospheric inputs of acidifying components (sulphur and nitrogen) in the past [45,46]. Akselsson et al. [58] also referred to high historical sulphur deposition and simultaneously high site productivity for their high risk classes. The northern lowlands are dominated by clearly positive Mg balances, and Ca balances also tend to be positive. The pine forests that predominate in this region are generally characterized by low uptake rates of base cations. For example the study of Akselsson et al. [20] indicated that uptake is clearly higher in spruce than in pine. However, the difference is mostly both a species and a site effect [36]. This is compounded by the close proximity of this region to the North Sea and the high deposition rates of base cations by sea spray (see Figure A2).

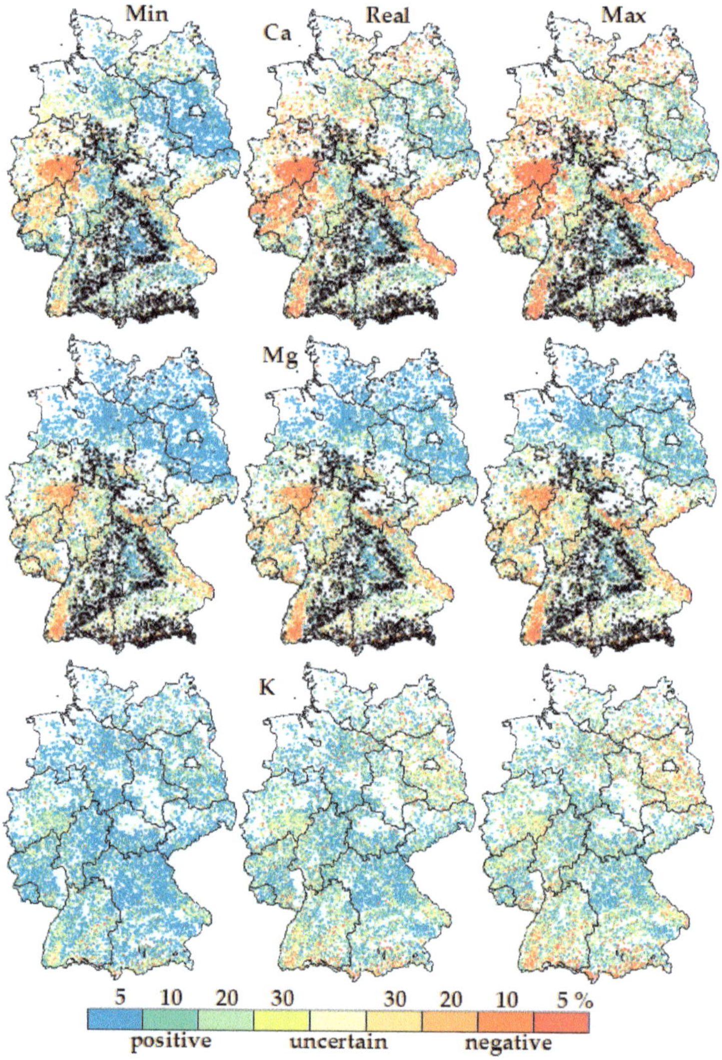

Figure 5. Effects of different intensities of biomass use on the nutrient balance of calcium (Ca), magnesium (Mg) and potassium (K) at the NFI tracts (scenario MIN: left column; scenario REAL: centre column; scenario MAX: right column). Balance deficits are shown in red, positive balances in blue. The colour intensity indicates the probability of error.

The K balances show much less pronounced regional differences than Ca and Mg. They are largely positive for the MIN scenario and not significantly different from zero for the REAL scenario. The higher export of biomass under the MAX scenario results in an increase in significant negative K balances for 24.8% of NFI tracts compared to the REAL scenario (18.6%). The K balances tend to be negative in the MAX scenario predominantly in the Alps, the Black Forest, the Swabian Alb, and the sandy sites of the northern lowlands. This is consistent with the results of other studies [7,88,129] and indicates that insufficient K supply may occur on shallow dolomite and limestone soils and on soils with high water permeability (e.g., poor sandy soils).

3.3. Uncertainties in the Calculated Balances and Methodological Limitations

When using model results to derive silvicultural management strategies, there is a great risk that policy or forest decision makers may view model results as "absolute" [130]. Therefore, it is important to also assess the uncertainties of modelling and communicate these to users [96,131]. On such a basis, improved decisions can be made, and the limits of model application can be more clearly demonstrated. There are numerous approaches to represent model uncertainties [99,132,133]. In our study, the Monte Carlo method was used because it is very easy to implement and generally applicable to various modelling approaches. The method has been applied to numerous forestry issues and a wide range of topics such as soil acidification [134], critical loads [135], silicate weathering [63], or soil organic carbon stocks [136]. The main disadvantages of the Monte Carlo method [99] are that the results are not available in analytical form and the error can only be determined from a large number of simulations (10,000 in our case) using appropriate statistical settings. Moreover, the multidimensional distributions for correlated variables are often unknown and/or difficult to derive. The latter was considered in the present study by deriving the covariance between the different parts of the nutrient balance on the NFSI plots. However, it should be noted that the derived correlations are affected by large-scale site differences between inventory points. Accordingly, e.g., acidity and regularity of chemical characteristics at the respective soil plots can only be represented to a limited extent. This is impressively illustrated by the relationship between the weathering rate and the element concentrations in the seepage flux/soil solution (see Figure A4). These and the other relationships shown in Figure A4 must be interpreted in such a way that more Ca, Mg, and K is available for leaching at sites with higher soil stocks of base cations and correspondingly higher weathering rates. In addition, the individual members of the nutrient balance were determined independently of each other and, therefore, the simulations could theoretically be carried out without taking the covariance into account. However, the results of the correlation analyses show that the elements of the nutrient balance are not statistically independent random variables and, accordingly, covariance should be considered. Therefore, the models are independent, but the data are not. Accordingly, the presented approach is a compromise that allows a first approximation to the real conditions and uncertainties. Further uncertainties result from the partly incomplete and inhomogeneous data basis for the NFSI points. Particularly problematic is that very important parameters like sulphate (SO_4^{2+}), nitrate (NO_3^-) and chloride (Cl^-) concentrations from 2:1 extracts are not available in some federal states of Germany and where the derived nutrient balances are therefore connected with larger uncertainties (e.g., Brandenburg).

The chosen balancing approach is very vulnerable to large uncertainties in the individual components of the nutrient balance. The simple model of nutrient balancing has so far ignored important interactions between the individual balance components. Such feedback effects particularly influence the already uncertain leaching losses. While the interactions between the cations held on the exchange sites in soils and the concentration of cation in seepage flux are considered in the present approach, the influence of different harvesting intensities on exchanger composition and saturation is not. For example, studies by Zetterberg et al. [137] found reduced exchangeable Ca stocks in the mineral soil and consequently, 40% lower Ca concentrations in seepage water 27–30 years after WTH compared to conventional harvesting. Paré and Thiffault [39] also discussed the large uncertainties (estimation of element fluxes, lack of feedback effects) in applying the nutrient balance approach as indicator for critical biomass harvesting. Löfgren et al. [138] have criticized the nutrient balance approach for being too uncertain for developing forest ecosystem management strategies and because this approach does not account for all relevant processes. It becomes apparent that the "Simple Mass Balance" approach does not consider key factors of the biogeochemical nutrient cycle and dynamics in forest management, such as changes in humus stocks. Accordingly, nutrient balances are likely to provide a more realistic assessment of the actual situation in short rotation forestry and agricultural land use, as nutrient imports and exports as well as nutrient cycling are easier to assess, with the high proportion of artificial nutrient additions—especially on agricultural land—meaning that interrelated factors internal to the system have a much lower influence on the nutrient balance [39]. The mid- to long-term process dynamics in forest soils would suggest to implement ecosystem process models based on ecosystem models like, e.g., ForSAFE [139], that are however extremely difficult and unsure to parameterize under the influence of fast changing environmental conditions. Therefore, it was decided in this study to confine the balance calculation to the time span of 10 years between two NFI campaigns. Further environmental change can be taken into account by re-calculating the nutrient balances on the basis of data from each new NFI campaign, using the algorithms developed in this study.

3.4. Effective Options for Nutrient Management

The balances of the nutrients Ca, Mg and K are influenced by different factors and boundary conditions. Projected changes in soil temperature largely influence the weathering rates of base cations [55]. On the other hand, increasing water limitations may restrict the positive effect of higher temperatures on weathering rates in forest soils as the water content could have a great influence on the weathering rate [62]. Temperature driven increased mineralization rates could lead to higher nitrogen leaching and thus, also to a higher leaching of cations [140]. At present, however, actually the main drivers of base cation leaching are the substantial nitrogen deposition, nitrification processes, and the remobilization of previously retained sulphur in the forest soils [141]. Knust et al. [36] also found for pine and spruce stands in Northeast Germany that Ca and Mg balances became negative due to high leaching rates caused by historically very high sulphur inputs, even with stem-only harvesting. These factors lead to unnaturally high leaching losses of Ca, Mg and K. On the other hand, current forest management strategies can increase or mitigate balance deficits of Ca, Mg and K (Figure 1).

Nutrient export with the harvested biomass can be regulated by forest management, either by adapting the harvest intensity to the vulnerability of the respective forest sites or by technically replacing the exported nutrients by soil protecting liming. Reducing harvesting intensity increases the amount of harvest residuals, which ideally should remain well distributed on the stand area and release nutrients to the soil nutrient pool during mineralization. Through soil protecting liming fine ground dolomite powder or a dolomite-wood ash mixture is applied to the soil surface, and Ca, Mg and K are released within 3–6 years, as the applied materials dissolve slowly [142,143]. Dolomite-wood ash mixture was developed as a new, standardized product for soil protecting liming and has been used in practical forest liming campaigns in the German federal state Baden-Wuerttemberg since 2008 [144]. The wood ash used for this purpose is subjected to strict quality control to ensure that no harmful substances are distributed with it [145]. Thus, both options, adjusting harvest intensity and compensating nutrient deficits through soil protecting liming, allow for ecosystem-compliant replacement of nutrient deficits. In the following, the option of adjusting harvest intensity is derived from the German-wide element balances, incorporating the three elements Ca, Mg, K via Liebig's Law of the Minimum at all approximately 23,000 NFI tracts included in this study (Figure 6).

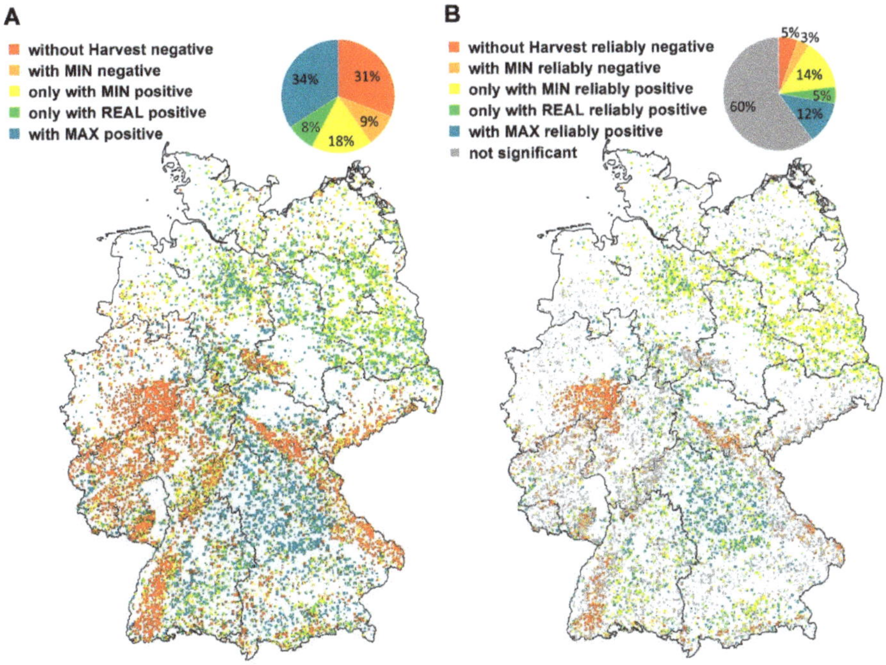

Figure 6. Spatial distribution of the base cations (Ca, Mg and K) that can be exported with the harvest scenarios MIN (yellow), REAL (green) and MAX (turquoise) without leading to negative nutrient balance of one or more elements (**A**). On the right map (**B**) the same data analysis is shown, incorporating the uncertainty of the models ($\alpha < 0.1$) to decide about possible harvest intensity. The sites that show negative (or significantly negative) balances without any harvest are marked in red and the sites where scenario MIN shows negative (or significantly negative) balances in orange.

Both maps in Figure 6 represent the same strategy approaches for active nutrient management through regional adjustment of harvest intensity by implementing only those harvesting scenarios that are supported by positive or balanced nutrient balances. If one of the three nutrient balances (Ca, Mg or K) becomes negative, the next more resource-preserving scenario is selected. However, at 31% of the NFI tracts (or at 5% if only the tracts with significantly negative nutrient balances are considered) nutrient balances are negative without any harvest export and therefore, cannot be closed even if harvesting was to completely abandoned. In addition, nutrient balances at 9% of NFI tracts are not large enough to support the lowest harvest intensity (MIN)—and 3% if only significant balances are considered. Satisfactory nutrient management by adjusting harvest intensity to the MIN scenario would be possible at 18% of NFI tracts or at 14% if only significant balances are considered. Conventional use (REAL) can be sustainably realized on 8% (significant on 5%) of the NFI tracts. Including the non-significant balances, the MAX scenario would be possible at 34% of the NFI tracts. If only the significant ones are considered, the MAX scenario is possible at 12% of the sites.

An alternative to reducing the harvest intensity would be to actively manage nutrients by recycling those amounts of nutrients that are required for closing the nutrient balances. The data basis for this approach is the same as for the harvest intensity adjustment. This approach has the advantage of also providing a realistic solution for sites where nutrient balances are negative without any harvesting or have such low positive balances that even the most resource-efficient harvesting scenario (MIN) does not ensure nutrient sustainability.

From our data, we can calculate the area where nutrient management by nutrient recycling is advisable if the present harvest intensity (REAL) is to be maintained. In Table 3, the first two rows show the area with a recycling requirement (based on expected significant nutrient balances only). The last two rows show the annual treatment areas depending on the amount of the yearly balance loss for the deficient nutrients at each tract of the NFI. The repetition time of liming campaigns was derived by cumulating the annual balance deficits until the mean nutrient content of a practical liming campaign with the dosage of 3 tons ha^{-1} dolomite rock powder or 4 tons ha^{-1} dolomite/wood ash mixture was reached, assuming a typical chemical composition of the applied materials. Possible counteracting measures for closing the nutrient balance depend on which nutrient is deficient. Liming with dolomite powder reduces Ca and/or Mg deficiency, whereas at sites with additional K deficiency, a K containing material such as wood ash should be applied in addition to the dolomite. At sites where only K is deficient, a formulation with a high K content should be considered. Summing the annual requirement of liming areas in our model (significantly negative balances for Ca + Mg and Ca + Mg + K), this is an annual requirement of 123,612 ha. This number agrees well with the approximately 100,000 ha yr^{-1} of limed forest area in Germany between 1980 and 2016 [146]. Our 23.6% higher estimate can be explained by the fact that our number refers to the required liming area, but not all forests with a liming demand have been actually limed in the past, and in some federal states no regular liming has been conducted since 1983.

Table 3. Areas with balance deficits for element combinations Ca and/or Mg, Ca and/or Mg and K and only K for the total forest area of Germany assuming harvest intensity REAL. Presented are all sites with balance deficits and those where balance deficits are at $\alpha < 0.1$ significant.

Consideration Level	Unit	Nutrient Element-Combination		
		CaMg	CaMgK	K
Area with balance deficits	ha	4,274,808 (39.4%)	1,318,645 (12.2%)	662,275 (6.1%)
Area with balance deficits at $\alpha < 0.1$	ha	1,625,254 (15%)	180,331 (1.7%)	208,948 (1.9%)
Required annual treatment area	ha yr^{-1}	113,740 (1.05%)	113,539 (1.05%)	46,346 (0.42%)
Required annual treatment area at $\alpha < 0.1$	ha yr^{-1}	84,961 (0.78%)	38,651 (0.36%)	41,452 (0.38%)

Although both discussed management options, adjusting harvest intensity and soil protective liming, can mitigate nutrient loss, it is evident that a combination of both options is required to close nutrient balances at many sites. This particularly concerns those sites that show negative nutrient balances already for the MIN scenario (Figure 5).

Potential consequences of a forest overuse include growth losses [26,27,147] which, among others, would turn short-term gains from fuel wood use into long-term losses. In this context, previous experience with extreme overuse of forests through forest grazing or litter harvesting may be a warning. Furthermore, it should be noted that on sites with low nutrient stocks due to historical overuse, positive balances serve to restore the natural site potential. When using dolomitic lime, complications from nutrient imbalances must also be considered. In this context, special reference should be made to a possible disturbance of the K supply due to calcium-potassium antagonism [148]. Summaries of numerous studies on this topic can be found in Hüttl and Zoettl [149]. Recent evaluations also show that spruce needles have lower P and K levels on limed NFSI sites than on unlimed sites [150]. Corresponding results were also observed on other experimental plots for forest liming [91,151]. Accordingly, the risks of liming to forest ecosystems and potential conflicts with other objectives, like e.g., nature preservation issues should also be included in the decision-making process on nutrient recycling measures [92,152–155].

The evaluations presented here allow a regional estimate of the magnitude of the recirculation requirement and from this also a rough calculation of possible costs. The uncertainty analyses provide the necessary confidence to justify the effort and cost required for nutrient management measures in practical planning. They also provide a reliable framework for prioritizing measures.

4. Conclusions

Sustainable management of forest soils implies that the nutrient export by wood and biomass harvesting, in the long term, does not exceed the nutrient replenishment from weathering and deposition. Based on our findings we recommend distinguishing between sites where negative nutrient balances are mainly due to harvest exports and those that are depleted in nutrients primarily due high atmospheric deposition of sulphur and nitrogen.

Nutrient balances and their uncertainties vary widely depending on geology, soils, climate, deposition history and forest stand characteristics and hence, reliable calculations must be based on local or regional information on those drivers of the nutrient balance. If necessary, the reduction of the harvest intensity can contribute to achieving a sustainable nutrient supply. Alternatively, or additionally, nutrient base cations can be recycled, e.g., by forest liming. However, possible negative side effects must also be considered here. The analysed harvest scenarios show that WTH has a high impact on base cation budgets in forest soils, as WTH removes additional biomass categories (twigs, branches, bark, needles) which have a much higher base cation content than stem wood. This implies a major risk of site degradation, which must be considered in any management. Nevertheless, the option to increase harvest intensity at sites where nutrient balances are significantly positive (12% of NFI sites) allows at least a partial compensation for reductions in harvest intensity at sensitive sites.

The results of this study provide valuable information for practitioners and environmental policy makers to enable spatiotemporal adaptive ecosystem management on the reliable and quality-assured basis of monitoring data. Nutrient balances can be readily adapted to changing environmental conditions by applying the evaluation algorithms developed in this study to data sets from future regular forest monitoring ampaigns.

Author Contributions: Conceptualization, B.A. and K.v.W.; methodology, B.A., W.W., C.V., D.Z. and H.P.; validation, B.A., W.W., C.V. and D.Z.; formal analysis, B.A., W.W., C.V. and D.Z.; investigation, B.A., W.W., C.V., H.P., G.K., C.S. and D.Z.; writing—original draft preparation, B.A., K.v.W. and C.V.; writing—review and editing, B.A., W.W., C.V., H.P., G.K., C.S. and D.Z.; visualization, B.A., W.W., D.Z., C.V. and H.P.; supervision, K.v.W., G.K. and C.S.; project administration, K.v.W., G.K. and H.P.; funding acquisition, K.v.W. and G.K. All authors have read and agreed to the published version of the manuscript.

Funding: This study was funded by BMEL/FNR (FKZ: 22006512, 22020212 and 22020312). Part of the data were co-financed by the European Commission. Partial funding of data collection and evaluation was provided by the European Union under Council Regulation (EEC) 3528/86 on the Protection of Forests against Atmospheric Pollution, the Regulation (EC) 2152/2003 concerning monitoring of forests and environmental interactions in the community (Forest Focus) and by the project LIFE 07 ENV/D/000218, Further Development and Implementation of an EU-level Forest monitoring Systeme (FutMon).

Institutional Review Board Statement: Not applicable.

Informed Consent Statement: Not applicable.

Data Availability Statement: The datasets related to this article are available from the corresponding author and co-authors on reasonable request. The original datasets for the NFI and NFSI plot and regionalized deposition are available on request from the institutions holding the data. Deposition data by UBA, Dessau, Germany. All NFI and NFSI data by Thünen Institute of Forest Ecosystems, Eberswalde, Germany. Data from German Intensive Forest Monitoring sites (mainly ICP Forest sites) fall under the publication policy of ICP Forests (Annex II of Seidling et al. [156]) and can be accessed from the ICP Forests database (http://icp-forests.net/page/data-requests (accessed on 15 February 2022)). upon request from the Programme Co-ordinating Center (PCC) in Eberswalde, Germany. Deposition calculated by dry deposition factor by Northwest German Forest Research Institute (NW-FVA).

Acknowledgments: For the far-reaching provision of data and the close cooperation in all necessary steps of data harmonization, we would like to express our sincere thanks to the representatives of the following state institutions: Research Institute for Forest Ecology and Forestry Rhineland-Palatinate; Hessen-Forest with the Service Center for Forest Management and Nature Conservation (FENA); State Office for Agriculture, Environment and Rural Areas Schleswig-Holstein; Hessian State Office for Nature Conservation, Environment and Geology; State Office for Mining, Energy and Geology Lower Saxony; State Office for Geology and Mining Saxony-Anhalt; Lower Saxony State Forests; Saxony-Anhalt State Forestry Office; State Institute for Ecology, Land Readjustment and Forests North Rhine-Westphalia; State Competence Center Forestry Eberswalde and Forestry Administration of Baden-Wuerttemberg. Maps of the soils and geology of Germany were made available to us by the Federal Institute for Geosciences and Natural Resources (BGR) for use. Data from the monitoring networks of the NFSI and the ICPForest-monitoring (Level II) were provided by the Thünen Institute for Forest Ecosystems in Eberswalde. Germany-wide wet deposition data were provided by the Federal Environment Agency. Further data of great value for the project work were provided by scientists from various research institutes. We would like to thank Hans Pretzsch and Ralf Moshammer (both Technical University of Munich), Joachim Block and Julius Schuck (both Rhineland-Palatinate Research Institute for Forest Ecology and Forestry), and Reinhard Stock (The German Federal Environmental Foundation), who provided extensive data on nutrient concentrations in tree compartments, which were collected in the framework of the project "Decision support Decision Support System for Nutrient Removal in Timber Harvesting" (Ref. 25966-33/0). The constructive comments by the four anonymous reviewers, which helped to improve the manuscript significantly, are gratefully acknowledged.

Conflicts of Interest: The authors declare no conflict of interest.

Appendix A

Table A1. Harvesting scenarios used for element balance calculations.

Scenario	Abbreviation	Description
Nutrient Saving	MIN	The harvest is limited to saw logs and industrial round wood. The utilization limit varies between top diameters of 12 to 14 cm for softwood and 12 to 17 cm for hardwood species depending on diameter at breast height. Harvest losses remaining on site are assumed to be 10% [33,157]. Brushwood and branch debris are not utilized and remain in the stand interior (not concentrated on the skid trails) through appropriate motorized or mechanized delimbing. In addition, as suggested by Heppelmann et al. [125], debarking of logs takes place inside the stand. This leaves the most nutrient-rich parts (crown material and bark) entirely on site, where they remain available as a source of nutrients.
Current harvest intensity	REAL	The main assortments are logs, industrial wood, and fuel wood. All solid volume including bark minus 10% harvest losses is harvested. The utilization limit is set to a top diameter of 7 cm. Harvesting is done by machine and the logging roads are reinforced by half of the arising brushwood volume for soil protection. It is assumed that 80% of the nutrients accumulated in the skid trails with biomass are not available to the system in the medium-term and are thus taken as losses.
Highest intensity	MAX	All solid volume is used as log, industrial or fuel wood assuming no technical harvesting losses in the coarse wood fraction. Furthermore, under the same assumptions as in the 'REAL' scenario, half of the produced brushwood is placed on the logging roads. In addition, the other half of the crown material, minus harvesting losses of 20% for conifers and 40% for broadleaved trees, is used for energy production or chemical conversion. This scenario thus largely corresponds to a whole-tree harvesting (WTH).

Appendix B

Estimation of Site Specific Yearly Dry Deposition Factors (DDF)

To explain the variation in dry deposition, we investigated the relationship between several stand, site and climate parameters and the DDF by generalized additive mixed effect models (GAMM). These models are used for model development taking the "pseudo-replicated" deposition and stand data at every single monitoring site into account. Standard software to parameterize this type of model is available in the form of the R package "mgvc" [158], which additionally calls for the libraries "MASS" [159] and "nlme" [160]. The model structure is as follows:

$$DDF_{y,p} = b_0 + f_1(x_{1,yp}) + f_2(x_{2,yp}) + \ldots + f_n(x_{n,yp}) + Z_p b_p + \varepsilon \quad \text{(A1)}$$

where DDF is the dry deposition factor in year y of plot p, b_0 is the intercept term, f_1, f_2, …, f_n: are spline smoothing functions, $x_{1,yp}$, $x_{2,yp}$, …, $x_{n,yp}$ are 1 … n predictor variables of year y of plot p, Z_p: is a linear model matrix including dummy variables for coding random effects for plots p with p = 1, …, 115. b_p is a vector of random effects and ε is an independent and identically normally distributed error term.

Table A2. Estimated coefficients and statistical characteristics of the dry deposition factor (DDF) model. Est.: estimated parameter value, SE: standard error of the parameter estimates, edf: effective degrees of freedom, BD_{Na}: bulk deposition of sodium; WLI: windward/leeward index; WS: wind speed, DN: distance to North Sea; H: stand height; signif. codes: <0.001 ***, <0.01 **.

	Est.	SE	Edf
Parametric coefficients			
Intercept	0.23369 ***	0.02431	
Oak	−0.15948 **	0.05046	
Spruce	0.25400 ***	0.03381	
Pine	0.14753 ***	0.04235	
Approximate significance of smooth terms			
$\log(BD_{Na})$			2.608 ***
WLI			1.000 **
WS			1.701 ***
DN			1.000 ***
H			1.000 ***
$R^2_{adj.}$	0.648		n = 928

To use the model independently from R, the following calculations can be performed:

$$DDF_{y,p} = b_0 + f_1(\ln(BD_{Na})) + f_2(WLI) + f_3(WS) + f_4(DN) + f_5(H) + b_1 \quad \text{(A2)}$$

where b_0 = 0.23369, and b_1 for beech, oak, spruce, and pine is: 0, −0.15948, 0.25400, 0.14753. The different linear and non-linear functions could be calculated as follows:

$$f_1(\ln(BD_{Na})) = 0.0592965032 \cdot (\ln(BD_{Na}))^2 - 0.56333208 \cdot \ln(BD_{Na}) + 0.65236948$$

$$f_2(WLI) = 0.3132540717 \cdot WLI - 0.3559040829$$

$$f_3(WS) = -0.0096723754 \cdot WS^2 + 0.174999468 \cdot WS - 0.5263822623$$

$$f_4(DN) = -0.0015403976 \cdot DN + 0.5616516592$$

$$f_5(H) = 0.0084502178 \cdot H - 0.2362130627$$

where BD_{Na}: bulk deposition of sodium [kg ha^{-1} yr^{-1}]; WLI: windward/leeward index [-]; WS: annual mean wind speed in 10 m [m s^{-1}]; DN: distance to North Sea [km]; H: stand height [m]. In case of negative values, the DDF should be set to zero.

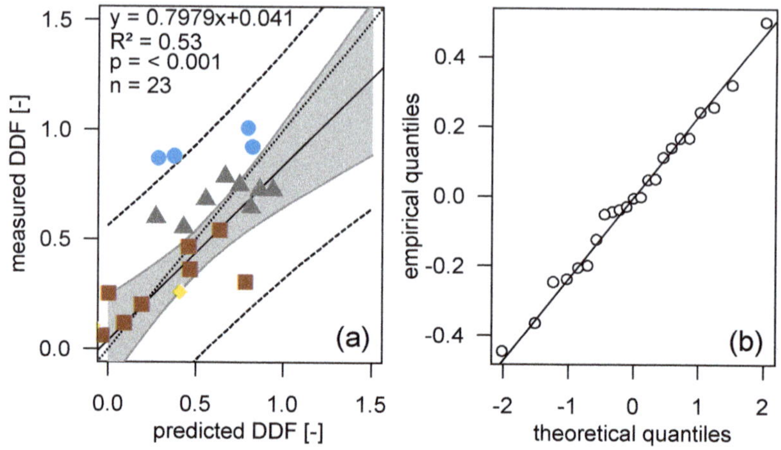

Figure A1. Relationship between predicted and measured dry deposition factors (DDF) for forest stands from two different age chronosequence studies in Germany [161,162] used for model evaluation. •: Norway spruce and Douglas fir, ■: European beech, ▲: Scots pine, ♦: Pedunculated/Sessile oak (**a**) Normal probability plot (Q-Q plot) for visual assessment of residuals (**b**).

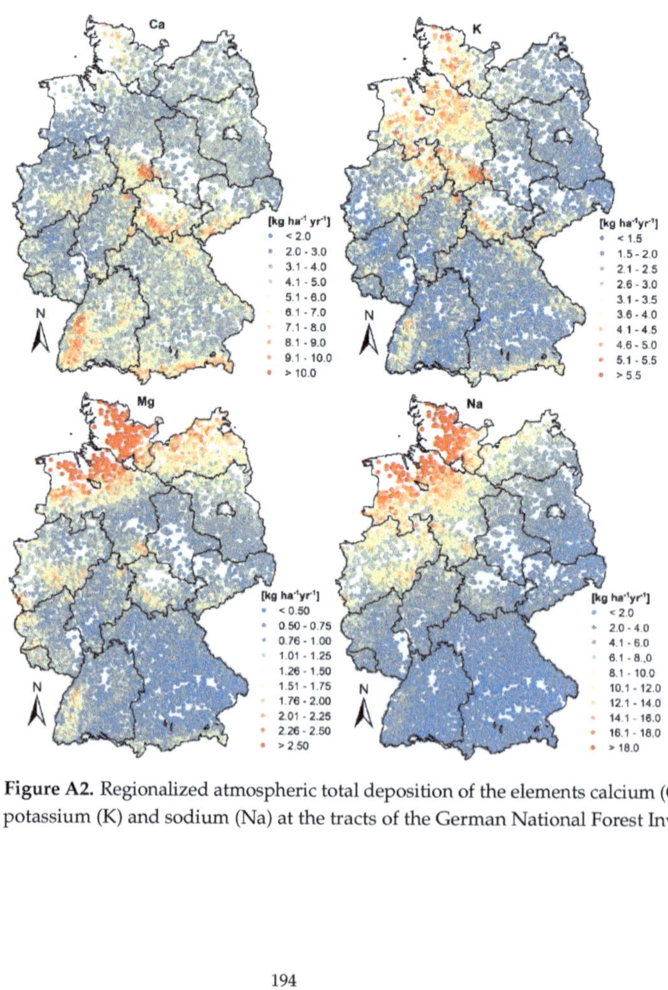

Figure A2. Regionalized atmospheric total deposition of the elements calcium (Ca), magnesium (Mg), potassium (K) and sodium (Na) at the tracts of the German National Forest Inventory (NFI).

Appendix C

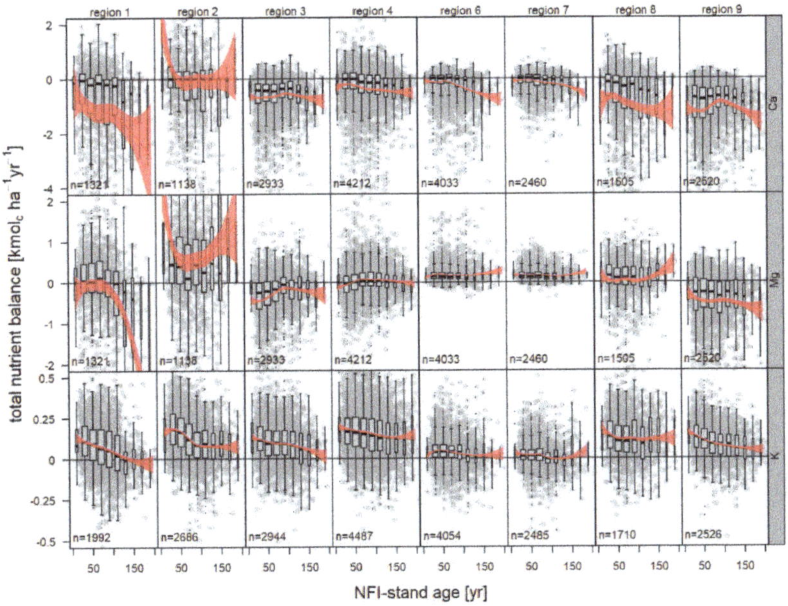

Figure A3. Nutrient balance (weathering + deposition—leaching—harvesting removal + soil stock/1000 [scenario REAL]) at NFI tracts over mean stand age at time NFI 2012, stratified by model regions (see Figure 2), for calcium (Ca) and magnesium (Mg) only on carbonate-free soils. Boxplots for age classes of 20 years; the width of the box is proportional to the number of tract corners in the age class; y-axis restricted to 90% of the data, but the data basis for the loess curve (red) is not. 1—pre-alpine moraines and limestone Alps; 2—hills on limestone bedrocks; 3—hills on crystalline bedrocks; 4—hills on sand, silt, clay bedrocks; 6—old moraines, north German lowland; 7—young moraines, north German lowland; 8—loess regions, fluvial valleys; 9—hills on clay- and silt schist bedrocks. Note: the number 5 was deliberately not assigned.

Appendix D

Experimental Design of the Monte Carlo Uncertainty Simulations

The uncertainty analysis was limited to the carbonate-free sites with respect to the elements Ca and Mg, as the balances for these elements are very uncertain on sites with carbonate soils, but sufficient Ca and Mg supply can be assumed. First, correlation analyses were performed for different variables (see Figure A4 as example from an NFSI plot) used for the derivation of the nutrient balance (deposition, weathering, soil stocks etc.) in the statistical program R using the function rcorr() from the package "Hmisc" [163] to identify possible covariance structures of the parameters in their probability distributions so that they can be taken into account, if necessary. Spearman's correlation coefficients (r_{Spear}) were calculated for the correlation analysis because the balance variables, in particular for Ca and Mg were not normally distributed [164]. We only considered significant correlations ($\alpha \leq 0.05$). When non-significant correlations were present, correlation coefficients were set to zero in the further analyses. In addition, mean values and standard deviations for the respective parameters were required. During the Monte Carlo (MC) simulation, random parameter draws were based on the model's expected values and, considering the detected covariance, were realized with the function rmvnorm() from the R package "splus2R" [100]. A compilation of these values (the mean values are site-specific and therefore cannot be presented in the Table A3) is provided in Table A3 for the NFSI grid points. Table A4 contains the parameters to characterize the uncertainties of the regionalization models (see Section 2.3) for the NFI tracts.

Table A3. Input data and uncertainty ranges of the balance variables as the basis for the Monte Carlo simulations of the soil balances components on the NFSI points. DEP: deposition; WEA: weathering rate; CSE: concentration in seepage water; SEF: seepage flux; STO: soil stock of plant available nutrients; mean μ: site mean; min: minimum value; SD: standard deviation; RMSE: root mean square error; R: data reference.

Parameter	Unit	Mean μ	Min	SD[%]	RMSE	R
DEP_{Ca}	kg ha^{-1} yr^{-1}	modelled results	0.001	-	2.68	[165]
DEP_{Mg}	kg ha^{-1} yr^{-1}	modelled results	0.001	-	0.43	[165]
DEP_{K}	kg ha^{-1} yr^{-1}	modelled results	0.001	-	1.54	[165]
WEA_{Ca}	kg ha^{-1} yr^{-1}	modelled with site specific data	0.001	23.3	-	[166]
WEA_{Mg}	kg ha^{-1} yr^{-1}	modelled with site specific data	0.001	23.3	-	[166]
WEA_{K}	kg ha^{-1} yr^{-1}	modelled with site specific data	0.001	23.3	-	[166]
CSE_{Ca}	$mmol_c$ l^{-1}	modelled with site specific data	0.001	53.9	-	[73]
CSE_{Mg}	$mmol_c$ l^{-1}	modelled with site specific data	0.001	57.5	-	[73]
CSE_{K}	$mmol_c$ l^{-1}	modelled with site specific data	0.001	97.1	-	[73]
SEF	mm	modelled with site specific data	0.001	8.3	-	[65]
STO_{Ca}	0.001 kg ha^{-1}	measured	0.001	50.0	-	[167]
STO_{Mg}	0.001 kg ha^{-1}	measured	0.001	50.0	-	[167]
STO_{K}	0.001 kg ha^{-1}	measured	0.001	50.0	-	[167]

Table A4. Coefficient of variation (based on RMSE of the validation) of the regionalized balance elements for the individual model regions and the global model as a basis for the Monte Carlo simulations of the soil balances on the NFI tracts. Model Region (see Figure 2A, main text): 1: pre-alpine moraines and limestone Alps; 2: hills on limestone bedrocks; 3: hills on crystalline bedrocks; 4: hills on sand, silt, clay bedrocks; 6: old moraines, north German lowland; 7: young moraines, north German lowland; 8: loess regions, fluvial valleys; 9: hills on clay and silt schist bedrocks. Note: the number 5 was deliberately not assigned. X: Global model, *: Global model used (cf. Section 2.3).

Model Region	Leaching			Weathering Rate			Soil Stocks		
cf. Figure 2A	Ca	Mg	K	Ca	Mg	K	Ca	Mg	K
1	87.9 *	125.7 *	80.5 *	321.8 *	89.2 *	55.8 *	72.7 *	80.1 *	51.9 *
2	57.3	106.9	76.8	109.8	153.8	50.8	57.7	127.3	48.6
3	77.5	83.3	79.1	90.7	87.2	55.8 *	139.9	126.8	55.1
4	103.8	105.8	69.7	321.8 *	132.8	54.5	110	107.1	60.2
6	80.8	95.8	107.6	107.5	73.4	78.6	117.9	127.6	56.6
7	102.9	96.8	105.1	321.8 *	75.9	65	71.6	103.2	51.7
8	80.6	76.1	81.3	47.8	47.9	51.1	89.1	86.5	50.1
9	119.2	125.7 *	95	321.8 *	261	65	112.2	65	136.8
X	87.9	125.7	80.5	321.8	89.2	55.8	72.7	80.1	51.9

Again, the relative root mean square error (RMSE) was used as an indicator for the (unobserved) standard deviation, as the models were fitted with log-transformed data. If only information on the maximum error range (e.g., ±25% for seepage flux) was available, the percent standard deviation σ was determined as follows (for the normal distribution, 99.7% of the values are in the ±3-σ range):

$$\sigma = (\text{error})/3 \qquad [\%] \tag{A3}$$

Figure A4 shows an example of the distribution of randomly selected values for the NFSI plot with the ID 30016 and the balance variables deposition, weathering, nutrient concentration in seepage water, seepage flux, and soil stock of Ca.

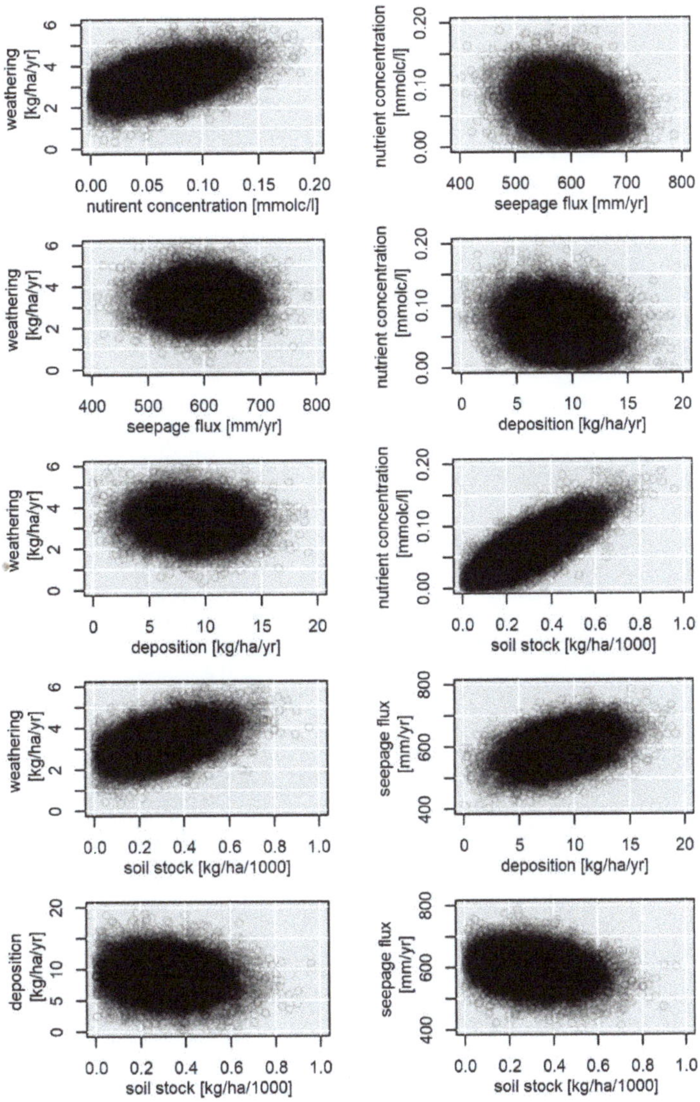

Figure A4. Example for randomly varied input data of the calcium (Ca) balance considering the covariance in the Monte Carlo procedure for the NFSI plot with the ID 30016.

As the function rmvnorm() does not offer the possibility to represent truncated parameter spans, but negative balance variables are implausible, 50,000 simulations were first performed for each NFI tract. From this data set, all simulations were used for which the balance variables reached values ≥ 0.001. Finally, a random sample (n = 10,000, [168,169]) was generated from this subset using the sample() function without replacement. Such a sample size allows the necessary reproducibility of the obtained results [63].

Except for deposition and harvesting removal, the calculated balance elements at the NFI sites exhibit—besides a methodological (model-related) error—an additional regionalization error. For deposition and harvest, the estimated model error at the NFI tracts was used. The example of weathering rates (WEA) is used to illustrate how the calculation of the i-th value within the Monte Carlo simulations $\left(WEA_i^{MC}\right)$ is based on the regionalized, site-specific value of WEA_{STA}, the regionalization error $\left(WEA_{i,REG}^{MC} - WEA_{STA}\right)$, and a methodological error $\left(WEA_{i,MET}^{MC} - WEA_{STA}\right)$. The terms $WEA_{i,REG}^{MC}$ and $WEA_{i,MET}^{MC}$ are the simulation terms drawn during the Monte Carlo simulation from the corresponding distribution of the regionalization and from the modelling derived weathering rates. Thus, the distribution of simulated weathering rates is generated as follows:

$$WEA_i^{MC} = WEA_{STA} + \left(WEA_{i,REG}^{MC} - WEA_{STA}\right) + \left(WEA_{i,MET}^{MC} - WEA_{STA}\right) \quad \text{(A4)}$$

The error in harvest removals was accounted for in a slightly different manner. The calculation of element removals is based on forest growth simulations, biomass functions and compartment specific nutrient levels and, therefore, very time consuming and not feasible to implement for 10,000 Monte Carlo iterations. Instead, for each NFI tract, the corresponding harvest removal was calculated, considering the error propagation of the prediction errors. Only the errors of the biomass functions and the models for nutrient levels in biomass were taken into account. Other elements, such as the error of the model WEHAM in predicting future forest development and of BDAT (program to calculate e.g., diameters, volume, assortments and double bark thickness for different tree species based on tree characteristics and sorting information [77]), as well as uncertainties in the input data for predicting nutrient levels were ignored. The time required for repeated evaluation of these models (updating of diameter at breast height, tree height and growth, possibly changing the dropping out collective and variety composition, re-evaluation of proximity element functions with varying predictors) is not proportionate to the expected gain in accuracy. In addition, the error rate in the estimation of biomass and approximate elements is much lower than the uncertainties of the other nutrient balance elements.

Many of the required error specifications could be derived directly from model validation of the various sub models. Assuming that the errors in the statistical models are normally distributed, the RMSE (root mean square error) was simplified to be equal to the standard deviation. For models with heteroscedastic errors (biomass models and log models of regionalization), the coefficient of variation based on RMSE, i.e., relative RMSE calculated as RMSE/E(Y) (E: expected value), was used.

For some parameters, data from the literature were used for a rough estimation. For soil nutrient stocks a coefficient of variation (CV) of 50% was assumed (cf. Table A3).

After the simulations, the significance level for the occurrence of negative or positive element balances was determined from the probability densities of the 10,000 Monte Carlo realizations. The terms significant and weakly significant are defined with error probabilities of $\alpha \leq 0.05$ and $0.05 < \alpha \leq 0.1$, respectively. Figure A5 shows this procedure as an example for three situations. In example B1 (red distribution curve), the distribution of the 10,000 simulations is mainly negative, but indeed encompasses also zero. More importantly, zero ranges between the 90% and 95% percentile (highlighted in red). Accordingly, the balance for this example is only weakly significantly negative ($0.05 < \alpha \leq 0.10$). Example B2 in Figure A5 shows slightly more positive than negative realizations, but this is not statistically significant: (the area between the 30% and 70% percentiles (highlighted in colour) includes zero). Example B3 in Figure A5 shows a situation of significantly positive ($0.05 < \alpha \leq 0.1$) balance simulations. Here, the value zero is excluded with over 95%

confidence (the 5% quantile of the distribution is highlighted in blue). Results (Figure 5, main text) were plotted using the R package "maptools" [170].

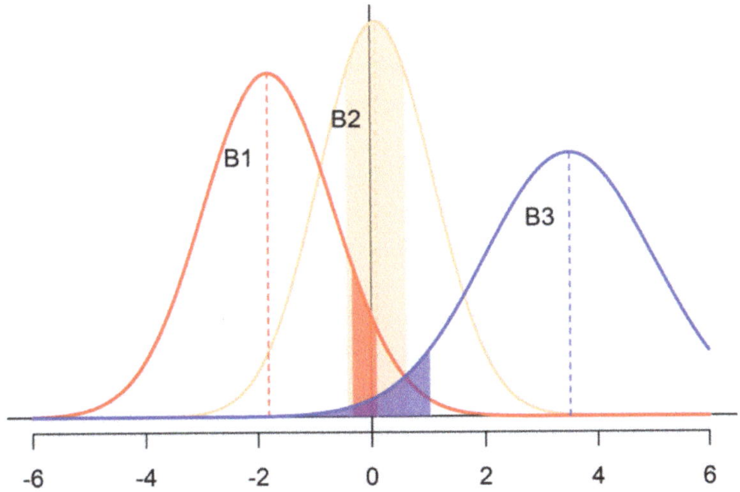

Figure A5. Examples (B1 to B3, see text) for deriving significance levels of negative or positive soil nutrient balances.

Appendix E

Validation of the Regionalized Nutrient Balances

The regression models developed at the test dataset of NFSI were applied with the predictors at the validation data set and compared with the measured NFSI data at each plot of the validation data individually. The assessment error of Ca and Mg balances were high (RMSE = 62.5 kg ha^{-1} yr^{-1} and RMSE = 23.2 kg ha^{-1} yr^{-1}, respectively), whereas it was low for K (RMSE = 4.3 kg ha^{-1} yr^{-1}). The relation between the balance values calculated from measured NFSI data and regionalized data at NFI sites follows closely the 1:1 line as it is presented in Figure A6 exemplarily for Ca and K.

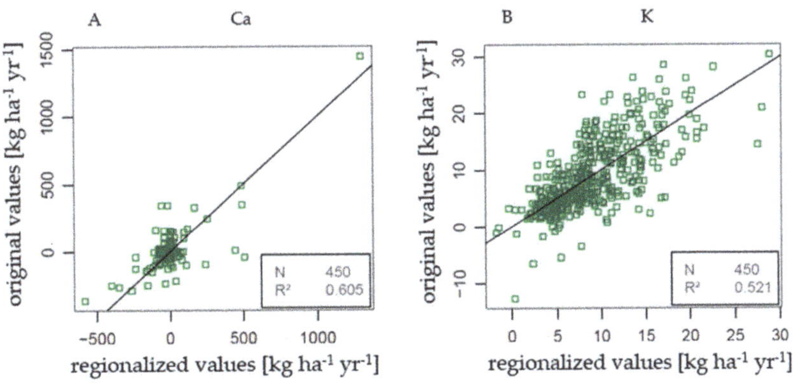

Figure A6. Scatter diagram of regionalized calcium (Ca, (**A**)) and potassium (K, (**B**)) balances versus balances calculated from measured NFSI data (y axis). Black line = 1:1 reference.

The spatial distribution of Mg balances calculated from measured NFSI data and regionalized Mg balances at NFI tracts is presented by the Figure A7. The comparison of the spatial distribution of Mg balances between those calculated with measured NFSI data and the regionalized balances in NFI sites demonstrates impressively that the regionalized data provide 26 times higher information density in space and much more consistent regional patterns of balance levels. It is also visible that regions with lacking NFSI data for balance calculation are supplemented by the regionalization process in a sensible way.

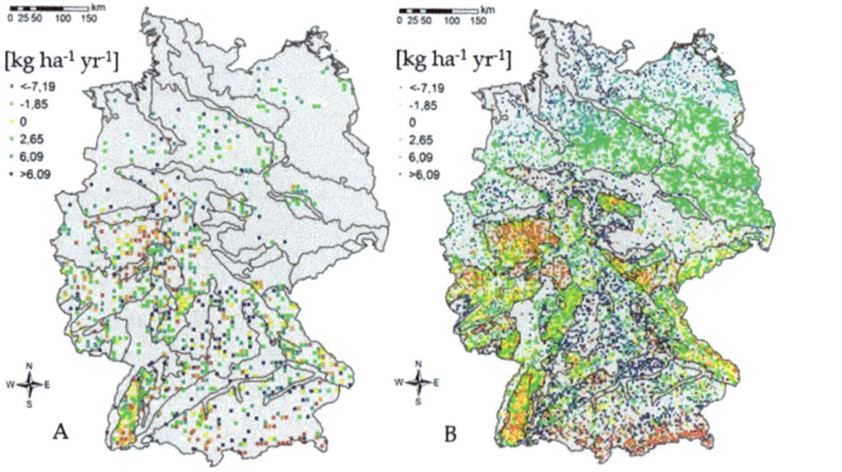

Figure A7. Spatial distribution of Mg balances calculated with measured NFSI data (**A**) and regionalized Mg balances at NFI sites (**B**). Black lines are boundaries of model regions (see Figure 2).

References

1. Lindberg, S.E.; Lovett, G.M.; Richter, D.D.; Johnson, D.W. Atmospheric deposition and canopy interactions of major ions in a forest. *Science* **1986**, *231*, 141–145. [CrossRef]
2. Hedin, L.O.; Granat, L.; Likens, G.E.; Buishand, T.A.; Galloway, J.N.; Butler, T.J.; Rodhe, H. Steep declines in atmospheric base cations in regions of Europe and North America. *Nature* **1994**, *367*, 351–354. [CrossRef]
3. Meesenburg, H.; Meiwes, K.-J.; Rademacher, P. Long term trends in atmospheric deposition and seepage output in northwest german forest ecosystems. *Water Air Soil Pollut.* **1995**, *85*, 611–616. [CrossRef]
4. Engardt, M.; Simpson, D.; Schwikowski, M.; Granat, L. Deposition of sulphur and nitrogen in Europe 1900–2050. Model calculations and comparison to historical observations. *Tellus Ser. B* **2017**, *69*, 1328945. [CrossRef]
5. Meesenburg, H.; Ahrends, B.; Fleck, S.; Wagner, M.; Fortmann, H.; Scheler, B.; Klinck, U.; Dammann, I.; Eichhorn, J.; Mindrup, M.; et al. Long-term changes of ecosystem services at Solling, Germany: Recovery from acidification, but increasing nitrogen saturation? *Ecol. Indic.* **2016**, *65*, 103–112. [CrossRef]
6. Vestreng, V.; Myhre, G.; Fagerli, H.; Reis, S.; Tarrasón, L. Twenty-five years of continuous sulphur dioxide emission reduction in Europe. *Atmos. Chem. Phys.* **2007**, *7*, 3663–3681. [CrossRef]
7. de Vries, W.; de Jong, A.; Kros, J.; Spijker, J. The use of soil nutrient balances in deriving forest biomass harvesting guidelines specific to region, tree species and soil type in the Netherlands. *For. Ecol. Manag.* **2021**, *479*, 118591. [CrossRef]
8. Rytter, L.; Ingerslev, M.; Kilpeläinen, A.; Torssonen, P.; Lazdina, D.; Löf, M.; Madsen, P.; Muiste, P.; Stener, L.-G. Increased forest biomass production in the Nordic and Baltic countries—A review on current and future opportunities. *Silva Fenn.* **2016**, *50*, 5. [CrossRef]
9. Busch, R.; Hirth, T.; Liese, A.; Nordhoff, S.; Puls, J.; Pulz, O.; Sell, D.; Syldatk, C.; Ulber, R. Nutzung nachwachsender Rohstoffe in der industriellen Stoffproduktion. *Chem. Ing. Tech.* **2006**, *78*, 219–228. [CrossRef]
10. Flückiger, W.; Braun, S. Nährstoffe im Wald lassen-oder recyclieren! *Wald. Und Holz* **2009**, *9*, 30–33.
11. Jacobsen, C.; Rademacher, P.; Meesenburg, H.; Meiwes, K.J. Gehalte chemischer Elemente in Baumkompartimenten—Literaturstudie und Datensammlung. In *Berichte des Forschungszentrums Waldökosysteme*; Reihe B 69; Selbstverlag des Forschungszentrums Waldökosysteme der Universität Göttingen: Göttingen, Germany, 2003; 81p.
12. Rumpf, S.; Schönfelder, E.; Ahrends, B. Biometrische Schätzmodelle für Nährelementgehalte in Baumkompartimenten. *Freibg. Forstl. Forsch.* **2018**, *101*, 33–73.
13. Aherne, J.; Posch, M.; Forsius, M.; Lehtonen, A.; Härkönen, K. Impacts of forest biomass removal on soil nutrient status under climate change: A catchment-based modelling study for Finland. *Biogeochemistry* **2012**, *107*, 471–488. [CrossRef]
14. Vanguelova, E.; Pitman, R.; Luiro, J.; Helmisaari, H.-S. Long term effects of whole tree harvesting on soil carbon and nutrient sustainability in the UK. *Biogeochemistry* **2010**, *101*, 43–59. [CrossRef]

15. Marčeta, D.; Petković, V.; Ljubojević, D.; Potočnik, I. Harvesting System Suitability as Decision Support in Selection Cutting Forest Management in Northwest Bosnia and Herzegovina. *Croat. J. For. Eng.* **2020**, *41*, 251–265. [CrossRef]
16. Block, J.; Meiwes, K.J. Erhaltung der Produktivität der Waldböden bei der Holz- und Biomassenutzung. In *Bodenschutz—Ergänzbares Handbuch der Maßnahmen und Empfehlungen für Schutz, Pflege und Sanierung von Böden, Landschaft und Grundwasser*; Bachmann, G., König, W., Utermann, J., Eds.; Erich Schmidt Verlag GmbH & Co. KG: Berlin, Germany, 2013; Volume 4200, pp. 1–50.
17. Titus, B.D.; Brown, K.; Helmisaari, H.-S.; Vanguelova, E.; Stupak, I.; Evans, A.; Clarke, N.; Guidi, C.; Bruckman, V.J.; Varnagiryte-Kabasinskiene, I.; et al. Sustainable forest biomass: A review of current residue harvesting guidelines. *Energy Sustain. Soc.* **2021**, *11*, 10. [CrossRef]
18. Stupak, I.; Raulund-Rasmussen, K. Historical, ecological, and governance aspects of intensive forest biomass harvesting in Denmark. *WIREs Energy Environ.* **2016**, *5*, 588–610. [CrossRef]
19. Fleck, S.; Eickenscheidt, N.; Ahrends, B.; Evers, J.; Grüneberg, E.; Ziche, D.; Höhle, J.; Schmitz, A.; Weis, W.; Schmidt-Walter, P.; et al. Nitrogen Status and Dynamics in German Forest Soils. In *Ecological Studies (Analysis and Synthesis)*; Wellbrock, N., Bolte, A., Eds.; Springer International Publishing: Cham, Switzerland, 2019; Volume 237, pp. 123–166.
20. Akselsson, C.; Westling, O.; Sverdrup, H.; Holmqvist, J.; Thelin, G.; Uggla, E.; Malm, E. Impact of Harvest Intensity on Long-Term Base Cation Budgets in Swedish Forest Soils. *Water Air Soil Pollut. Focus* **2007**, *7*, 201–210. [CrossRef]
21. Olsson, B.A.; Bengtsson, J.; Lundkvist, H. Effects of different forest harvest intensities on the pools of exchangeable cations in coniferous forest soils. *For. Ecol. Manag.* **1996**, *84*, 135–147. [CrossRef]
22. Joki-Heiskala, P.; Johansson, M.; Holmberg, M.; Mattsson, T.; Forsius, M.; Kortelainen, P.; Hallin, L. Long-term base cation balances of forest mineral soils in Finland. *Water Air Soil Pollut.* **2003**, *150*, 255–273. [CrossRef]
23. Forsius, M.; Akujärvi, A.; Mattsson, T.; Holmberg, M.; Punttila, P.; Posch, M.; Liski, J.; Repo, A.; Virkkala, R.; Vihervaara, P. Modelling impacts of forest bioenergy use on ecosystem sustainability: Lammi LTER region, southern Finland. *Ecol. Indic.* **2016**, *65*, 66–75. [CrossRef]
24. Likens, G.E.; Bormann, F.M.; Pierce, R.S.; Eaton, J.S.; Johnson, N.M. *Biochemistry of a Forested Ecosystem*; Springer: New York, NY, USA, 1977; p. 146.
25. Levers, C.; Verkerk, P.J.; Müller, D.; Verburg, P.H.; Butsic, V.; Leitão, P.J.; Lindner, M.; Kuemmerle, T. Drivers of forest harvesting intensity patterns in Europe. *For. Ecol. Manag.* **2014**, *315*, 160–172. [CrossRef]
26. Achat, D.L.; Deleuze, C.; Landmann, G.; Pousse, N.; Ranger, J.; Augusto, L. Quantifying consequences of removing harvesting residues on forest soils and tree growth—A meta-analysis. *For. Ecol. Manag.* **2015**, *348*, 124–141. [CrossRef]
27. Thiffault, E.; Hannam, K.D.; Pare, D.; Titus, B.D.; Hazlett, P.W.; Maynard, D.G.; Brais, S. Effects of forest biomass harvesting on soil productivity in boreal and temperate forests—A review. *Environ. Rev.* **2011**, *19*, 278–309. [CrossRef]
28. Zetterberg, T.; Köhler, S.J.; Löfgren, S. Sensitivity analyses of MAGIC modelled predictions of future impacts of whole-tree harvest on soil calcium supply and stream acid neutralizing capacity. *Sci. Total Environ.* **2014**, *494–495*, 187–201. [CrossRef] [PubMed]
29. Egnell, G.; Valinger, E. Survival, growth, and growth allocation of planted Scots pine trees after different levels of biomass removal in clear felling. *For. Ecol. Manag.* **2003**, *177*, 65–74. [CrossRef]
30. Evans, A.M.; Finkral, A.J. From renewable energy to fire risk reduction: A synthesis of biomass harvesting and utilization case studies in US forests. *GCB Bioenergy* **2009**, *1*, 211–219. [CrossRef]
31. Mason, W.L.; Mckay, H.M.; Weatherall, A.; Connolly, T.; Harrison, A.J. The effects of whole-tree harvesting on three sites in upland Britain on the growth of Sitka spruce over ten years. *For. Int. J. For. Res.* **2011**, *85*, 111–123. [CrossRef]
32. Kreutzweiser, D.P.K.P.; Hazlett, P.W.H.W.; Gunn, J.M.G.M. Logging impacts on the biogeochemistry of boreal forest soils and nutrient export to aquatic systems: A review. *Environ. Rev.* **2008**, *16*, 157–179. [CrossRef]
33. Wilpert, V.K.; Bösch, B.; Bastian, P.; Zirlewagen, D.; Hepperle, F.; Holzmann, S.; Puhlmann, H.; Schäffer, J.; Kändler, G.; Sauter, U.H. Biomasse-Aufkommensprognose und Kreislaufkonzepte für den Einsatz von Holzaschen in der Bodenschutzkalkung in Oberschwaben. *Ber. Freibg. Forstl. Forsch.* **2011**, *87*, 155.
34. Riek, W.; Russ, A.; Kühn, D. Waldbodenbericht Brandenburg. Zustand und Entwicklung der brandenburgischen Waldböden. Ergebnisse der landesweiten Bodenzustandserhebungen BZE-2 und BZE-2a. *Eberswalder Forstl. Schr.* **2015**, *60*, 172.
35. Ahrends, B.; Scheler, B.; Wagner, M.; Fleck, S.; Fortmann, H.; Meesenburg, H. Trends der Nährstoffbilanzen forstlich genutzter Intensiv-BDF in Niedersachsen. *Geoberichte* **2021**, *39*, 40–54. [CrossRef]
36. Knust, C.; Schua, K.; Feger, K.-H. Estimation of Nutrient Exports Resulting from Thinning and Intensive Biomass Extraction in Medium-Aged Spruce and Pine Stands in Saxony, Northeast Germany. *Forests* **2016**, *7*, 302. [CrossRef]
37. Borken, W.; Matzner, E. Nitrate leaching in forest soils: An analysis of long-term monitoring sites in Germany. *J. Plant Nutr. Soil Sci.* **2004**, *167*, 277–283. [CrossRef]
38. Brumme, R.; Ahrends, B.; Block, J.; Schulz, C.; Meesenburg, H.; Klinck, U.; Wagner, M.; Khanna, P. Cycling and retention of nitrogen in European beech (*Fagus sylvatica* L.) ecosystems under elevated fructification frequency. *Biogeosciences* **2021**, *18*, 3763–3779. [CrossRef]
39. Paré, D.; Thiffault, E. Nutrient Budgets in Forests Under Increased Biomass Harvesting Scenarios. *Curr. For. Rep.* **2016**, *2*, 81–91. [CrossRef]
40. Sutton, M.A.; Oenema, O.; Erisman, J.W.; Leip, A.; van Grinsven, H.; Winiwarter, W. Too much of a good thing. *Nature* **2011**, *472*, 159–161. [CrossRef]

41. Bol, R.; Julich, D.; Brödlin, D.; Siemens, J.; Kaiser, K.; Dippold, M.A.; Spielvogel, S.; Zilla, T.; Mewes, D.; von Blanckenburg, F.; et al. Dissolved and colloidal phosphorus fluxes in forest ecosystems—An almost blind spot in ecosystem research. *J. Plant Nutr. Soil Sci.* **2016**, *179*, 425–438. [CrossRef]
42. Missong, A.; Holzmann, S.; Bol, R.; Nischwitz, V.; Puhlmann, H.; Kaiser, K.; Siemens, J.; Klumpp, E. Leaching of natural colloids from forest topsoils and their relevance for phosphorus mobility. *Sci. Total Environ.* **2018**, *634*, 305–315. [CrossRef]
43. Ulrich, B. Natural and anthropogenic components of soil acidification. *J. Plant Nutr. Soil Sci.* **1986**, *149*, 702–717. [CrossRef]
44. Sverdrup, H.; Thelin, G.; Robles, M.; Stjernquist, I.; Sörensen, J. Assesing nutrient sustainability of forest production for different tree species considering Ca, Mg, K, N and P at Björnstorp Estate, Sweden. *Biogeochemistry* **2006**, *81*, 219–238. [CrossRef]
45. Builtjes, P.; Hendriks, E.; Koenen, M.; Schaap, M.; Banzhaf, S.; Kerschbaumer, A.; Gauger, T.; Nagel, H.-D.; Scheuschner, T.; Schlutow, A. *Erfassung, Prognose und Bewertung von Stoffeinträgen und ihren Wirkungen in Deutschland (Modelling of Air Pollutants and Ecosystem Impact—MAPESI)*; Federal Environmental Agency: Dessau-Rosslau, Germany, 2011; p. 97.
46. Schaap, M.; Wichink Kruit, R.J.; Hendriks, C.; Kranenburg, R.; Segers, A.; Builtjes, P.; Banzhaf, S.; Scheuschner, T. *Atmospheric Deposition to German Natural and Semi-Natural Ecosystems during 2009*; Umweltbundesamt: Dessau-Roßlau, Germany, 2015; p. 82.
47. Ulrich, B. Nutrient and Acid-Base Budget of Central European Forest Ecosystems. In *Effects of Acid Rain on Forest Processes*; Godbold, D., Hüttermann, A., Eds.; Wiley-Liss: New York, NY, USA, 1994; pp. 1–50.
48. Devlaeminck, R.; Schrijver, D.A.; Hermy, M. Variation in throughfall deposition across a deciduous beech (*Fagus sylvatica* L.) forest edge in Flanders. *Sci. Total Environ.* **2005**, *337*, 241–252. [CrossRef] [PubMed]
49. Staelens, J.; Houle, D.; De Schrijver, A.; Neirynck, J.; Verheyen, K. Calculating dry deposition and canopy exchange with the canopy budget model: Review of assumptions and application to two deciduous forests. *Water Air Soil Pollut.* **2008**, *191*, 149–169. [CrossRef]
50. Zuur, A.F.; Ieno, E.N.; Walker, N.J.; Saveliev, A.A.; Smith, G.M. *Mixed Effects Models and Extensions in Ecology with R*; Springer: New York, NY, USA, 2009; p. 574. [CrossRef]
51. Clarke, N.; Zlindra, D.; Ulrich, E.; Mosello, R.; Derome, J.; Derome, K.; König, N.; Lövblad, G.; Draaijers, G.; Hansen, K.; et al. Part XIV: Sampling and Analysis of Deposition. Available online: https://storage.ning.com/topology/rest/1.0/file/get/9995560266?profile=original (accessed on 17 February 2022).
52. Gauger, T.; Anshelm, F.; Schuster, H.; Draaijers, G.P.J.; Bleeker, A.; Erisman, J.W.; Vermeulen, A.T.; Nagel, H.-D. *Kartierung Ökosystembezogener Langzeittrends Atmosphärischer Stoffeinträge und Luftschadstoffkonzentrationen in Deutschland und deren Vergleich mit Critical Loads und Critical Levels*; Institut für Navigation, Universität: Stuttgart, Germany, 2002; p. 207.
53. RDevelopmentCoreTeam. *R: A Language and Environment for Statistical Computing*; R Foundation for Statistical Computing: Vienna, Austria, 2013; ISBN 3-900051-07-0. Available online: http://www.R-project.org (accessed on 7 June 2015).
54. Conrad, O.; Bechtel, B.; Bock, M.; Dietrich, H.; Fichter, E.; Gerlitz, L.; Wehberg, J.; Wichmann, V.; Böhner, J. System for Automated Geoscientific Analyses (SAGA) v. 2.1.4. *Geosci. Model Dev.* **2015**, *8*, 1991–2007. [CrossRef]
55. Sverdrup, H.; Warfvinge, P. Calculating field weathering rates using a mechanistic geochemical model PROFILE. *Appl. Geochem.* **1993**, *8*, 273–283. [CrossRef]
56. Stendahl, J.; Akselsson, C.; Melkerud, P.-A.; Belyazid, S. Pedon-scale silicate weathering: Comparison of the PROFILE model and the depletion method at 16 forest sites in Sweden. *Geoderma* **2013**, *211–212*, 65–74. [CrossRef]
57. Akselsson, C.; Belyazid, S.; Stendahl, J.; Finlay, R.; Olsson, B.A.; Erlandsson Lampa, M.; Wallander, H.; Gustafsson, J.P.; Bishop, K. Weathering rates in Swedish forest soils. *Biogeosciences* **2019**, *16*, 4429–4450. [CrossRef]
58. Akselsson, C.; Kronnäs, V.; Stadlinger, N.; Zanchi, G.; Belyazid, S.; Karlsson, P.E.; Hellsten, S.; Karlsson, G.P. A Combined Measurement and Modelling Approach to Assess the Sustainability of Whole-Tree Harvesting—A Swedish Case Study. *Sustainability* **2021**, *13*, 2395. [CrossRef]
59. Koseva, I.S.; Watmough, S.A.; Aherne, J. Estimating base cation weathering rates in Canadian forest soils using a simple texture-based model. *Biogeochemistry* **2010**, *2010*, 183–196. [CrossRef]
60. Phelan, J.; Belyazid, S.; Kurz, D.; Guthrie, S.; Cajka, J.; Sverdrup, H.; Waite, R. Estimation of Soil Base Cation Weathering Rates with the PROFILE Model to Determine Critical Loads of Acidity for Forested Ecosystems in Pennsylvania, USA: Pilot Application of a Potential National Methodology. *Water Air Soil Pollut.* **2014**, *225*, 2109. [CrossRef]
61. Whitfield, C.J.; Watmough, S.A.; Aherne, J.; Dillon, P.J. A comparison of weathering rates for acid-sensitive catchments in Nova Scotia, Canada and their impact on critical load calculations. *Geoderma* **2006**, *136*, 899–911. [CrossRef]
62. Hodson, M.E.; Langan, S.J.; Wilson, M.J. A sensitivity analysis of the Profile model in relation to the calculation of soil weathering rates. *Appl. Geochem.* **1996**, *11*, 835–844. [CrossRef]
63. Jönsson, C.; Warfvinge, P.; Sverdrup, H. Uncertainty in predicting weathering rate and environmental stress factors with the PROFILE model. *Water Air Soil Pollut.* **1995**, *81*, 1–23. [CrossRef]
64. Hammel, K.; Kennel, M. *Charakterisierung und Analyse der Wasserverfügbarkeit und des Wasserhaushalts von Waldstandorten in Bayern mit dem Simulationsmodell BROOK90*; Heinrich Frank: München, Germany, 2001; Volume 185, p. 148.
65. Ahrends, B.; Schmidt-Walter, P.; Fleck, S.; Köhler, M.; Weis, W. Wasserhaushaltssimulationen und Klimadaten. *Freibg. Forstl. Forsch.* **2018**, *101*, 74–94.
66. Schmidt-Walter, P.; Ahrends, B.; Mette, T.; Puhlmann, H.; Meesenburg, H. NFIWADS: The water budget, soil moisture, and drought stress indicators database for German National Forest Inventory (NFI). *Ann. Forest Sci.* **2019**, *76*, 39. [CrossRef]

67. Puhlmann, H.; Schmidt-Walter, P.; Hartmann, P.; Meesenburg, H.; von Wilpert, K. Soil Water Budget and Drought Stress. In *Status and Dynamics of Forests in Germany: Results of the National Forest Monitoring*; Wellbrock, N., Bolte, A., Eds.; Springer International Publishing: Cham, Switzerland, 2019; pp. 55–91.
68. Posch, M.; Kurz, D. A2M—A program to compute all possible mineral modes from geochemical analyses. *Comput. Geosci.* **2007**, *33*, 563–572. [CrossRef]
69. GAFA. *Handbuch Forstliche Analytik (HFA). Grundwerk des Gutachterausschuss Forstliche Analytik (GAFA)*; Federal Ministry of Food, Agriculture and Consumer Protection, Northwest German Forest Research Institute: Bonn, Germany, 2005.
70. GAFA. *Handbuch Forstliche Analytik (HFA). Grundwerk und 1.—4. Ergänzung des Gutachterausschuss Forstliche Analytik (GAFA)*; Federal Ministry of Food, Agriculture and Consumer Protection, Northwest German Forest Research Institute: Bonn, Germany, 2009.
71. GAFA. *Handbuch Forstliche Analytik (HFA). Grundwerk und 1.—5. Ergänzung des Gutachterausschuss Forstliche Analytik (GAFA)*; Federal Ministry of Food, Agriculture and Consumer Protection, Northwest German Forest Research Institute: Bonn, Germany, 2014.
72. Wellbrock, N.; Ahrends, B.; Bögelein, R.; Bolte, A.; Eickenscheidt, N.; Grüneberg, E.; König, N.; Schmitz, A.; Fleck, S.; Ziche, D. Concept and Methodology of the National Forest Soil Inventory. In *Status and Dynamics of Forests in Germany*; Wellbrock, N., Bolte, A., Eds.; Ecological Studies; Springer Nature Schwitzerland AG: Cham, Switzerland, 2019; Volume 237.
73. Weis, W.; Schäff, T.; Köhler, D.; Kölling, C. Stoffausträge mit dem Sickerwasser für die BZE II. *Freibg. Forstl. Forsch.* **2018**, *101*, 181–205.
74. Bösch, B.; Kändler, G. WEHAM II—Modelle und Algorithmen. Available online: https://www.bundeswaldinventur.de/fileadmin/SITE_MASTER/content/Downloads/Modelle_und_Algorithmen.pdf (accessed on 11 February 2022).
75. Rock, J.; Bösch, B.; Kändler, G. WEHAM 2012—Waldentwicklungs- und Holzaufkommensmodellierung für die dritte Bundeswaldinventur. Deutscher Verband Forstlicher Versuchsanstalten (DVFFA). In *Sektion Ertragskunde: Berichte zur Jahrestagung 2013*; Kohnle, U., Klädtke, J., Eds.; Forstliche Versuchsanstalt Baden-Württemberg: Freiburg, Germany, 2013; pp. 127–133.
76. Vonderach, C.; Kändler, G.; Dormann, C.F. Consistent set of additive biomass functions for eight tree species in Germany fit by nonlinear seemingly unrelated regression. *Ann. For. Sci.* **2018**, *75*, 49. [CrossRef]
77. Kublin, E. Einheitliche Beschreibung der Schaftform—Methoden und Programme—BDATPro. *Forstwiss. Cent. Ver. Mit Tharandter Forstl. Jahrb.* **2003**, *122*, 183–200. [CrossRef]
78. Block, J.; Schuck, J. Nährstoffentzüge durch die Holzernte und ihr Einfluss auf den Nährstoffhaushalt armer Standorte in Rheinland-Pfalz. *Forstl. Forsch. München* **2002**, *186*, 150–151.
79. NW-FVA. *Weichlaubholz—Ungenutztes Rohstoffpotenzial!? Abschlussbericht des Projekts "WEIPOL" im Rahmen des Förderungsprogramms "Nachwachsende Rohstoffe" des BMELV*; Technische Informationsbibliothek u. Universitätsbibliothek: Göttingen, Germany, 2013; p. 220.
80. Göttlein, A.; Dieler, D.; Baumgarten, M. Tree-internal Nutrient Distribution of Beech and Spruce at the Kranzberger Forst—Implications for Efficiency of Wood Production and for Nutrient Export with Different Harvest Intensities. *Nova Acta Leopold. NF* **2013**, *114*, 255–265.
81. Huber, C.; Borchert, H.; Kremer, J.; Weis, W.; Metthies, D.; Göttlein, A. *Biomasse-und Nährelementverschiebungen bzw.-verluste bei Harvesterdurchforstungen und deren Auswirkungen auf Bodeneigenschaften und Stoffhaushalt von Waldökosystemen. Abschlussbericht zum Projekt B65 des Kuratoriums für forstliche Forschung in Bayern*; LWF: Freising, Germany, 2011; p. 95.
82. Pretzsch, H.; Block, J.; Dieler, J.; Gauer, J.; Göttlein, A.; Moshammer, R.; Schuck, J.; Weis, W.; Wunn, U. Nährstoffentzüge durch die Holz- und Biomassenutzung in Wäldern. Teil 1: Schätzfunktionen für Biomasse und Nährelemente und ihre Anwendung in Szenariorechnungen. *Allg. Forst- Und Jagdztg.* **2014**, *185*, 261–285.
83. Rademacher, P.; Schönfelder, E.; Meiwes, K.J. Elementgehalte in Baumkompartimenten von Fichte (*Picea abies* [L.] Karst), Kiefer (*Pinus sylvestris* [L.] Karst), Buche (*Fagus sylvatica* [L.] Karst), (*Eiche Quercus robur und petraea* [L.] Karst) und Douglasie (*Pseudotsuga menziesii* [L.] Karst). *Forsch.: Möglichkeiten Und Grenzen Der Vollbaumnutzung (FKZ: 22015407)* **2012**, 125–154. Available online: https://www.fnr-server.de/ftp/pdf/berichte/22015407.pdf (accessed on 14 April 2022).
84. Rumpf, S.; Husmann, K.; Döbbeler, H.; Spellmann, H.; Nagel, J.; Meiwes, K.J. Ergebnisbericht. Verbundvorhaben: Bioenergie-Regionen Stärken (BEST)-Neue Systemlösungen im Spannungsfeld Ökologischer, Ökonomischer und Sozialer Anforderungen-Teilprojekt: Schwachholzpotenzial Wald (IO-H 4)-Entscheidungswerkzeuge zur Sicherung Einer Nachhaltigen Rohstoffversorgung für die Stoffliche und Energetische Verwertung; BMBF (FKZ: 003L033F). 2014. Available online: https://www.fisaonline.de/projekte-finden/details/?tx_fisaresearch_projects%5Bp_id%5D=5961&tx_fisaresearch_projects%5Baction%5D=projectDetails&tx_fisaresearch_projects%5Bcontroller%5D=Projects&cHash=c53771cb89d9664b6f6634165a180a60 (accessed on 20 February 2022).
85. Weis, W.; Gruber, A.; Huber, C.; Göttlein, A. Element concentrations and storage in the aboveground biomass of limed and unlimed Norway spruce trees at Höglwald. *Eur. J. For. Res.* **2009**, *128*, 437–445. [CrossRef]
86. Weis, W.; Göttlein, A. Appendix. In *Stoffliche Nachhaltigkeitskriterien für die Bayerische Forstwirtschaft. Abschlussbericht zum Projekt B67 des Kuratoriums für Forstliche Forschung in Bayern*; TUM: München, Germany, 2012; p. 140.
87. Weis, W.; Blumenthal, B.; Göttlein, A. Baumartenvergleich in intakten Altbeständen. *Forstl. Forsch. Ber. Münch.* **2014**, *212*, 13–38.
88. Weis, W.; Göttlein, A. Bewirtschaftungsmöglichkeiten zur Sicherung einer nachhaltigen Waldwirtschaft vor dem Hintergrund der Nährstoffversorgung kalkalpiner Standorte. *Forstl. Schr. Univ. Für Bodenkult. Wien* **2016**, *21*, 97–127.
89. Miller, D.M. Reducing Transformation Bias in Curve Fitting. *Am. Stat.* **1984**, *38*, 124–126. [CrossRef]

90. Forests, ICP. Manual on Methods and Criteria for Harmonized Sampling, Assessment, Monitoring and Analysis of the Effects of Air Pollution on Forests. Available online: http://icp-forests.net/page/icp-forests-manual (accessed on 12 February 2021).
91. Greve, M. Langfristige Auswirkungen der Waldkalkung auf den Stoffhaushalt. *Mitt. Aus Der Forsch. Für Wald. Und Forstwirtsch. Rheinl.-Pfalz.* **2015**, *73*, 333.
92. Huber, C.; Baier, R.; Göttlein, A.; Weis, W. Changes in soil, seepage water and needle chemistry between 1984 and 2004 after liming an N-saturated Norway sprcue stand at the Höglwald, Germany. *For. Ecol. Manage.* **2006**, *233*, 11–20. [CrossRef]
93. Zirlewagen, D. Regionalisierung der Bodenbilanz an BWI-Punkten. *Freibg. Forstl. Forsch.* **2018**, *101*, 227–241.
94. Kolka, R.K.; Grigal, D.F.; Nater, E.A. Forest soil mineral weathering rates: Use of multiple approaches. *Geoderma* **1996**, *73*, 1–21. [CrossRef]
95. Oenema, O.; Kros, H.; de Vries, W. Approaches and uncertainties in nutrient budgets: Implications for nutrient management and environmental policies. *Eur. J. Agron.* **2003**, *20*, 3–16. [CrossRef]
96. Cartwright, S.J.; Bowgen, K.M.; Collop, C.; Hyder, K.; Nabe-Nielsen, J.; Stafford, R.; Stillman, R.A.; Thorpe, R.B.; Sibly, R.M. Communicating complex ecological models to non-scientist end users. *Ecol. Model.* **2016**, *338*, 51–59. [CrossRef]
97. Arrouays, D.; McBratney, A.; Bouma, J.; Libohova, Z.; Richer-de-Forges, A.C.; Morgan, C.L.S.; Roudier, P.; Poggio, L.; Mulder, V.L. Impressions of digital soil maps: The good, the not so good, and making them ever better. *Geoderma Reg.* **2020**, *20*, e00255. [CrossRef]
98. Yanai, R.D.; Battles, J.J.; Richardson, A.D.; Blodgett, C.A.; Wood, D.M.; Rastetter, E.B. Estimating Uncertainty in Ecosystem Budget Calculations. *Ecosystems* **2010**, *13*, 239–248. [CrossRef]
99. Li, H.; Wu, J. Uncertainty analysis in ecological studies. In *Scaling and Uncertrainty Analysis in Ecology: Methods and Applications*; Wu, J., Jones, K.B., Li, H., Loucks, O.L., Eds.; Springer: Cham, Switzerland, 2006; pp. 45–66.
100. Constantine, W.; Hesterberg, T.; Wittkowski, K.; Song, T.; Kaluzny, S. splus2R: Supplemental S-PLUS Functionality in R. R Package Version 1.2-0. 2013. Available online: http://CRAN.R-project.org/package=splus2R (accessed on 14 April 2022).
101. Posch, M.; Vries de, W.; Sverdrup, H. Mass Balance models to derive critical loads of nitrogen and acidity for terrestrial and aquatic ecosystems. In *Critical Loads and Dynamic Risk Assessments*; de Vries, W., Hettelingh, J.-P., Posch, M., Eds.; Environmental Pollution; Springer: Dordrecht, The Netherlands, 2015; Volume 25, pp. 171–205.
102. de Vries, W.; Breeuwsma, A. Relative importance of natural and anthropogenic proton sources in soils in the netherlands. *Water Air Soil Pollut.* **1986**, *28*, 173–184. [CrossRef]
103. Breemen, N.V.; Visser, P.H.B.D.; Grinsven, J.J.M.V. Nutrient and proton budgets in four soil-vegetation systems underlain by Pleistocene alluvial deposits. *J. Geol. Soc.* **1986**, *143*, 659–666. [CrossRef]
104. Mulder, J.; Breemen, N.V.; Rasmussen, L.; Driscoll, C.T. Aluminum chemistry of acidic sandy soils with various inputs of acidic deposition in The Netherlands and in Denmark. In *Impact of Acid Atmospheric Deposition on Soils: Field Monitoring and Aluminum Chemistry*; Mulder, J., Ed.; United States Environmental Protection Agency: Washington, DC, USA, 1990; pp. 77–102.
105. van der Salm, C.; Köhlenberg, L.; de Vries, W. Assessment of weathering rates in Dutch loess and river-clay soils at pH 3.5, using laboratory experiments. *Geoderma* **1998**, *85*, 41–62. [CrossRef]
106. van der Salm, C. Assessment of the regional variation in weathering rates of loess and clay soils in the netherlands. *Water Air Soil Pollut.* **2001**, *131*, 217–243. [CrossRef]
107. Kauppi, P.; Kamari, J.; Posch, M.; Kauppi, L.; Matzner, E. Acidification of forest soils: Model development and application for analyzing impacts of acidic deposition in Europe. *Ecol. Model.* **1986**, *33*, 231–253. [CrossRef]
108. Klaminder, J.; Lucas, R.W.; Futter, M.N.; Bishop, K.H.; Köhler, S.J.; Egnell, G.; Laudon, H. Silicate mineral weathering rate estimates: Are they precise enough to be useful when predicting the recovery of nutrient pools after harvesting? *For. Ecol. Manag.* **2011**, *261*, 1–9. [CrossRef]
109. Futter, M.N.; Klaminder, J.; Lucas, R.W.; Laudon, H.; Köhler, S.J. Uncertainty in silicate mineral weathering rate estimates: Source partitioning and policy implications. *Environ. Res. Lett.* **2012**, *7*, 8. [CrossRef]
110. Hodson, M.E.; Langan, S.J.; Wilson, M.J. A critical evaluation of the use of the profile model in calculating mineral weathering rates. *Water Air Soil Pollut.* **1997**, *98*, 79–104. [CrossRef]
111. Langan, S.; Hodson, M.E.; Bain, D.; Hornung, M.; Reynolds, B.; Hall, J.; Johnston, L. The role of minerals weathering rate determinations in generating uncertainties in the calculation of critical loads of acidity and their exceedance. *Water Air Soil Pollut. Focus* **2001**, *1*, 299–312. [CrossRef]
112. Vadeboncoeur, M.A.; Hamburg, S.P.; Yanai, R.D.; Blum, J.D. Rates of sustainable forest harvesting depend on rotation length and weathering of soil minerals. *For. Ecol. Manag.* **2014**, *318*, 194–205. [CrossRef]
113. Wesselink, L.G.; Van Grinsven, J.J.M.; Grosskurth, G.F. Measuring and Modeling Mineral Weathering in an Acid Forest Soil, Solling, Germany. *Soil Sci. Soc. Am. Spec. Publi.* **1994**, *39*, 91–110. [CrossRef]
114. Dultz, S. Veränderung des mineralogischen Stoffbestandes in Böden aus pleistozänen Sedimenten Nordwestdeutschland und deren Bedeutung für die Elementfreisetzung. *Ber. Forsch.zent. Waldökosyst.* **2001**, *172*, 1–201.
115. Hodson, M.E.; Langan, S. Considerations of uncertainty in setting critical loads of acidity of soils: The role of weathering rate determination. *Environ. Pollut.* **1999**, *106*, 73–81. [CrossRef]
116. Starr, M.; Lindroos, A.-J.; Tarvainen, T.; Tanskanen, H. Weathering rates in the Hietajäri Integrated monitoring catchment. *Boreal Environ. Res.* **1998**, *3*, 275–285.

117. Starr, M.; Lindroos, A.-J.; Ukonmaanaho, L. Weathering release rates of base cations from soils within a boreal forested catchment: Variation and comparison to deposition, litterfall and leaching fluxes. *Environ. Earth. Sci.* **2014**, *75*, 5101–5111. [CrossRef]
118. Dämmgen, U.; Matschullat, J.; Zimmermann, F.; Strogies, M.; Grünhage, L.; Scheler, B.; Conrad, J. Emission reduction effects on bulk and wet-only deposition in Germany—Evidence from long-term observations. Part 2: Precipitation, potential sea salt, soil and fly ash constituents. *Gefahrst. Reinhalt. Der Luft Air Qual. Control* **2013**, *72*, 25–36.
119. McNulty, S.G.; Cohen, E.C.; Moore Myers, J.A.; Sullivan, T.J.; Li, H. Estimates of critical acid loads and exceedances for forest soils across the conterminous United States. *Environ. Pollut.* **2007**, *149*, 281–292. [CrossRef] [PubMed]
120. Mussche, S.; Bussche, B.; De Schrijver, A.; Neirynck, J.; Nachtergale, L.; Lust, N. Nutrient uptake of a mixed oak/beech forest in Flanders (Belgium). *Silva Gandav.* **1998**, *63*, 120–133. [CrossRef]
121. Zhao, Y.; Duan, L.; Larssen, T.; Hu, L.; Hao, J. Simultaneous Assessment of Deposition Effects of Base Cations, Sulfur, and Nitrogen Using an Extended Critical Load Function for Acidification. *Environ. Sci. Technol.* **2007**, *41*, 1815–1820. [CrossRef] [PubMed]
122. Duchesne, L.; Houle, D. Impact of nutrient removal through harvesting on the sustainability of the boreal forest. *Ecol. Appl.* **2008**, *18*, 1642–1651. [CrossRef] [PubMed]
123. Casetou-Gustafson, S.; Grip, H.; Hillier, S.; Linder, S.; Olsson, B.A.; Simonsson, M.; Stendahl, J. Current, steady-state and historical weathering rates of base cations at two forest sites in northern and southern Sweden: A comparison of three methods. *Biogeosciences* **2020**, *17*, 281–304. [CrossRef]
124. Etzold, S.; Ferretti, M.; Reinds, G.J.; Solberg, S.; Gessler, A.; Waldner, P.; Schaub, M.; Simpson, D.; Benham, S.; Hansen, K.; et al. Nitrogen deposition is the most important environment driver of growth of pure, even-aged and managed European forests. *For. Ecol. Manag.* **2020**, *458*, 117762. [CrossRef]
125. Heppelmann, J.B.; Labelle, E.R.; Wittkopf, S.; Seeling, U. In-stand debarking with the use of modified harvesting heads: A potential solution for key challenges in European forestry. *Eur. J. For. Res.* **2019**, *138*, 1067–1081. [CrossRef]
126. Labelle, E.R.; Poltorak, B.J.; Jaeger, D. The role of brush mats in mitigating machine-induced soil disturbances: An assessment using absolute and relative soil bulk density and penetration resistance. *Can. J. For. Res.* **2019**, *49*, 164–178. [CrossRef]
127. Richter, A.; Adler, G.H.; Fahrak, M.; Eckelmann, W. *Erläuterungen zur Nutzungsdifferenzierten Bodenübersichtskarte der Bundesrepublik Deutschland im Maßstab 1:1.000000 (BÜK 1000 N, Version 2.3)*; BGR: Hannover, Germany, 2007; p. 53.
128. Schaap, M.; Hendriks, C.; Kranenburg, R.; Kuenen, J.; Segers, A.; Schlutow, A.; Nagel, H.-D.; Ritter, A.; Banzhaf, S. PINETI-3: Modellierung atmosphärischer Stoffeinträge von 2000 bis 2015 zur Bewertung der ökosystem-spezifischen Gefährdung von Biodiversität durch Luftschadstoffe in Deutschland. *Texte Umweltbundesamt* **2018**, *79*, 149.
129. Mellert, K.H.; Ewald, J. Nutrient limitation and site-related growth potential of Norway spruce (Picea abies [L.] Karst) in the Bavarian Alps. *Eur. J. For. Res.* **2014**, *133*, 433–451. [CrossRef]
130. Beckers, J.; Smerdon, B.; Wilson, M. Review of hydrologic models for forest management and climate change applications in British Columbia and Alberta. *Forrex Ser.* **2009**, *25*, 1–179.
131. Ivanovic, R.F.; Freer, J.E. Science versus politics: Truth and uncertainty in predictive modelling. *Hydrol. Process.* **2009**, *23*, 2549–2554. [CrossRef]
132. Beven, K.; Freer, J. Equifinality, data assimilation, and uncertainty estimation in mechanistic modelling of complex environmental systems using the GLUE methodology. *J. Hydrol.* **2001**, *249*, 11–29. [CrossRef]
133. Zak, S.K.; Beven, K.; Reynolds, B. Uncertainty in the estimation of critical loads: A practical methodology. *Water Air Soil Pollut.* **1997**, *98*, 297–316. [CrossRef]
134. Kros, J.; Pebesma, E.J.; Reinds, G.J.; Finke, P.A. Uncertainty asseement in modeling soil acidification at the European scale: A case study. *J. Environ. Qual.* **1999**, *28*, 366–377. [CrossRef]
135. Reinds, G.J.; De Vries, W. Uncertainties in critical loads and target loads of sulphur and nitrogen for European forests: Analysis and quantification. *Sci. Total Environ.* **2010**, *408*, 1960–1970. [CrossRef]
136. Liski, J.; Palosuo, T.; Peltoniemi, M.; Sievänen, R. Carbon and decomposition model Yasso for forest soil. *Ecol. Model.* **2005**, *189*, 168–182. [CrossRef]
137. Zetterberg, T.; Olsson, B.A.; Löfgren, S.; Von Brömssen, C.; Brandtberg, P.-O. The effect of harvest intensity on long-term calcium dynamics in soil and soil solution at three coniferous sites in Sweden. *For. Ecol. Manag.* **2013**, *302*, 280–294. [CrossRef]
138. Löfgren, S.; Stendahl, J.; Karltun, E. Critical biomass harvesting indicator for whole-tree extraction does not reflect the sensitivity of Swedish forest soils. *Ecol. Indic.* **2021**, *132*, 108310. [CrossRef]
139. Wallman, P.; Svensson, M.G.E.; Sverdrup, H.; Belyazid, S. ForSAFE—An integrated process-oriented forest model for long-term sustainability assessments. *For. Ecol. Manag.* **2005**, *207*, 19–36. [CrossRef]
140. Fleck, S.; Ahrends, B.; Sutmöller, J.; Albert, M.; Evers, J.; Meesenburg, H. Is Biomass Accumulation in Forests an Option to Prevent Climate Change Induced Increases in Nitrate Concentrations in the North German Lowland? *Forests* **2017**, *8*, 219. [CrossRef]
141. Meesenburg, H.; Riek, W.; Ahrends, B.; Eickenscheidt, N.; Grüneberg, E.; Evers, J.; Fortmann, H.; Köng, N.; Lauer, A.; Meiwes, K.J.; et al. Soil Acidification in German Forest Soils. In *Status and Dynamics of Forests in Germany*; Wellbrock, N., Bolte, A., Eds.; Springer Nature Switzerland AG: Cham, Switzerland, 2019; Volume 237, pp. 93–120.
142. Jansone, L.; von Wilpert, K.; Hartmann, P. Natural Recovery and Liming Effects in Acidified Forest Soils in SW-Germany. *Soil Syst.* **2020**, *4*, 38. [CrossRef]

143. Sarenbo, S.L.; Claesson, T. Limestone and dolomite powder as binders for wood ash agglomeration. *Bull. Eng. Geol. Environ.* **2004**, *63*, 191–207. [CrossRef]
144. Von Wilpert, K.; Vonderach, C.; Zirlewagen, D. Enna-A project for sustainable harvesting wooden biomass. *VGB PowerTech* **2015**, *7*, 83–88.
145. von Wilpert, K.; Hartmann, P.; Schäffer, J. Quality control in a wood ash re-cycling concept for forests. *VGB Powertech* **2016**, *4*, 64–72.
146. Andreae, H.; Gemballa, R.; Jacob, F. Leitfaden zur Forstlichen Bodenschutzkalkung in Sachsen. *Sachs. Forst Schr.* **2020**, *31*, 1–102.
147. Helmisaari, H.-S.; Hanssen, K.H.; Jacobson, S.; Kukkola, M.; Luirod, J.; Saarsalmi, A.; Tamminen, P. Logging residue removal after thinning in Nordic boreal forests: Long-term impact on tree growth. *For. Ecol. Manag.* **2011**, *261*, 1919–1927. [CrossRef]
148. Baule, H.; Fricker, C. *Die Düngung von Waldbäumen*; BVL Verlagsgesellschaft: München, Germany, 1967; p. 259.
149. Hüttl, R.F.; Zoettl, H.W. Liming as a mitigation too in Germany's declining forests—Reviewing results from former and recent trials. *For. Ecol. Manag.* **1993**, *61*, 325–338. [CrossRef]
150. Talkner, U.; Riek, W.; Dammann, I.; Kohler, M.; Göttlein, A.; Mellert, K.H.; Meiwes, K.J. Nutritional Status of Major Forest Tree Species in Germany. In *Status and Dynamics of Forests in Germany: Results of the National Forest Monitoring*; Wellbrock, N., Bolte, A., Eds.; Springer International Publishing: Cham, Switzerland, 2019; pp. 261–293.
151. Huber, C.; Weis, W.; Göttlein, A. Tree nutrition of Norway spruce as modified by liming and experimental acidification at the Höglwald site, Germany, from 1982 to 2004. *Ann. For. Sci.* **2006**, *63*, 861–869. [CrossRef]
152. Lundström, U.S.; Brain, D.C.; Taylor, A.F.S.; Van Hees, A.W.; Geibe, C.E.; Holmström, S.J.M.; Melkerud, P.-A.; Finlay, R.; Jones, D.L.; Nyberg, L.; et al. Effects of acidification and its mitigation with lime and wood ash on forest soil processes in Southern Schweden. A Joint Multidisciplinary Study. *Water Air Soil Pollut. Focus* **2003**, *3*, 167–188. [CrossRef]
153. Reid, C.; Watmough, S.A. Evaluation the effects of liming and wood-ash treatment on forest ecosystems through systematic meta-analysis. *Can. J. For. Res.* **2014**, *44*, 867–885. [CrossRef]
154. Kreutzer, K. Effect of forest liming on soil processes. *Plant Soil* **1995**, *168–169*, 447–470. [CrossRef]
155. Hüttl, R.F. Liming and fertilization as mitigation tools in declining forest ecosystems. *Water Air Soil Pollut.* **1989**, *44*, 93–118. [CrossRef]
156. Seidling, W.; Hansen, K.; Strich, S.; Lorenz, M. Part I: Objectives, Strategy and Implementation of ICP Forests. In *Manual on Methods and Criteria for Harmonized Sampling, Assessment, Monitoring and Analysis of the Effects of Air Pollution on Forests, Manual Version 2016*; UNECE ICP Forests, Programme Coordinating Centre, Ed.; Thünen Institute of Forest Ecosystems: Eberswalde, Germany, 2016. Available online: https://www.icp-forests.org/pdf/manual/2016/ICP_Manual_2017_02_part01.pdf (accessed on 26 February 2022).
157. Pretzsch, H. *Forest Dynamics, Growth and Yield: From Measurement to Model*; Springer: Berlin, Germany, 2009; p. 664.
158. Wood, S.N. *Generalized Additive Models: An Introduction with R*; Chapman & Hall: Boca Raton, FL, USA, 2006; p. 410.
159. Venables, W.N.; Ripley, B.D. *Modern Applied Statistics with S*; Springer: New York, NY, USA, 2003; p. 495.
160. Pinheiro, J.; Bates, D.; Debroy, S.; Sarkar, D.; Team, R.C. nlme: Linear and Nonlinear Mixed Effects Models. R Package Version 3.1-103. 2008. Available online: https://cran.r-project.org/web/packages/nlme/index.html (accessed on 26 August 2016).
161. Mohr, K.; Meesenburg, H.; Horvath, B.; Meiwes, K.J.; Schaaf, S.; Dämmgen, U. *Bestimmung von Ammoniak-Einträgen aus der Luft und deren Wirkungen auf Waldökosysteme (ANSWER-Projekt)*; Bundesforschungsanstalt für Landwirtschaft (FAL): Braunschweig, Germany, 2005; Volume 279, p. 113.
162. Kues, J. Untersuchungen zum Stoffeintrag in den Göttinger Wald. *Ber. Forsch.zent. Waldökosyst./Waldsterben* **1984**, *6*, 206.
163. Harrell, F.E., Jr. Hmisc: Harrell Miscellaneous. R Package Version 3.13-0. 2013. Available online: http://CRAN.R-project.org/package=Hmisc (accessed on 21 May 2015).
164. Sachs, L. *Angewandte Statistik. 9 Auflage*; Springer: Berlin, Germany, 1999; p. 881.
165. Ahrends, B.; Wagner, M.; Klinck, U. Herleitung flächiger Depositionsdaten. *Ber. Freibg. Forstl. Forsch.* **2018**, *101*, 150–180.
166. Ahrends, B.; Klinck, U.; Klinck, C.; Weis, W. Herleitung flächiger Verwitterungsraten. *Freibg. Forstl. Forsch.* **2018**, *101*, 113–149.
167. Ahrends, B.; Vonderach, C.; Weis, W.; Wilpert, V.K. Unsicherheitsanalysen zur Nährstoffbilanzierung auf Umweltmessnetzen. *Freibg. Forstl. Forsch.* **2018**, *101*, 242–267.
168. De-Miguel, S.; Mehtätalo, L.; Durkaya, A. Developing generalized, calibratable, mixed-effects meta-models for large-scale biomass prediction. *Can. J. For. Res.* **2014**, *44*, 648–656. [CrossRef]
169. Metropolis, N.; Ulam, S. The Monte Carlo Method. *J. Am. Stat. Assoc.* **1949**, *44*, 335–341. [CrossRef]
170. Bivand, R.; Lewin-Koh, N. Maptools: Tools for Reading and Handling Spatial Objects. R Package Version 0.8-27. 2013. Available online: http://CRAN.R-project.org/package=maptools (accessed on 20 March 2017).

MDPI
St. Alban-Anlage 66
4052 Basel
Switzerland
Tel. +41 61 683 77 34
Fax +41 61 302 89 18
www.mdpi.com

Soil Systems Editorial Office
E-mail: soilsystems@mdpi.com
www.mdpi.com/journal/soilsystems

www.ingramcontent.com/pod-product-compliance
Lightning Source LLC
Chambersburg PA
CBHW061448210926
51850CB00008B/1645
* 9 7 8 3 0 3 6 5 5 0 6 3 3 *